Environmentally Sustainable Corrosion Inhibitors

Environmentally Sustainable Corrosion Inhibitors

Fundamentals and Industrial Applications

Edited by

Chaudhery Mustansar Hussain

Department of Chemistry and Environmental Science,
New Jersey Institute of Technology, Newark, NJ, United States

Chandrabhan Verma

Interdisciplinary Research Center for Advanced Materials,
King Fahd University of Petroleum and Minerals,
Dhahran, Saudi Arabia

Jeenat Aslam

Department of Chemistry, College of Science, Yanbu,
Taibah University, Al-Madina, Saudi Arabia

ELSEVIER

Elsevier
Radarweg 29, PO Box 211, 1000 AE Amsterdam, Netherlands
The Boulevard, Langford Lane, Kidlington, Oxford OX5 1GB, United Kingdom
50 Hampshire Street, 5th Floor, Cambridge, MA 02139, United States

Notices
Knowledge and best practice in this field are constantly changing. As new research and experience broaden our understanding, changes in research methods, professional practices, or medical treatment may become necessary.

Practitioners and researchers must always rely on their own experience and knowledge in evaluating and using any information, methods, compounds, or experiments described herein. In using such information or methods they should be mindful of their own safety and the safety of others, including parties for whom they have a professional responsibility.

To the fullest extent of the law, neither the Publisher nor the authors, contributors, or editors, assume any liability for any injury and/or damage to persons or property as a matter of products liability, negligence or otherwise, or from any use or operation of any methods, products, instructions, or ideas contained in the material herein.

British Library Cataloguing-in-Publication Data
A catalogue record for this book is available from the British Library

Library of Congress Cataloging-in-Publication Data
A catalog record for this book is available from the Library of Congress

ISBN: 978-0-323-85405-4

For Information on all Elsevier publications
visit our website at https://www.elsevier.com/books-and-journals

Publisher: Susan Dennis
Acquisitions Editor: Anita Koch
Editorial Project Manager: Allison Hill
Production Project Manager: Sruthi Satheesh
Cover Designer: Greg Harris

Typeset by MPS Limited, Chennai, India

Working together
to grow libraries in
developing countries

www.elsevier.com • www.bookaid.org

Contents

List of contributors

Ekemini D. Akpan   Department of Chemistry, School of Chemical and Physical Sciences and Material Science Innovation & Modelling (MaSIM) Research Focus Area, Faculty of Natural and Agricultural Sciences, North-West University, Mmabatho, South Africa

K.R. Ansari   Center of Research Excellence in Corrosion, Research Institute, King Fahd University of Petroleum and Minerals, Dhahran, Saudi Arabia

Afroz Aslam   Department of Chemistry, Faculty of Science, Aligarh Muslim University, Aligarh, India

Jeenat Aslam   Department of Chemistry, College of Science, Yanbu, Taibah University, Al-Madina, Saudi Arabia

Ruby Aslam   Corrosion Research Laboratory, Department of Applied Chemistry, Faculty of Engineering and Technology, Aligarh Muslim University, Aligarh, India

Megha Basik   Corrosion Research Laboratory, Department of Applied Chemistry, Faculty of Engineering and Technology, Aligarh Muslim University, Aligarh, India

Omar Dagdag   Laboratory of Industrial Technologies and Services, Department of Process Engineering, Height School of Technology, Sidi Mohammed Ben Abdallah University, Fez, Morocco

Amit Kumar Dewangan   Department of Chemistry, Government Digvijay Autonomous Post Graduate College, Rajnandgaon, India

Yeestdev Dewangan   Department of Chemistry, Government Digvijay Autonomous Post Graduate College, Rajnandgaon, India

R. Dhanalakshmi   PG Department of Chemistry, M.V. Muthiah Government Arts College for Women, Dindigul, India

Eno E. Ebenso   Institute for Nanotechnology and Water Sustainability, College of Science, Engineering and Technology, University of South Africa, Johannesburg, South Africa

Mustapha El Gouri   Laboratory of Industrial Technologies and Services, Department of Process Engineering, Height School of Technology, Sidi Mohammed Ben Abdallah University, Fez, Morocco

Younes El Kacimi   Laboratory of Materials Engineering and Environment: Modelling and Application, Faculty of Science, University Ibn Tofail, Kenitra, Morocco

Omolola E. Fayemi   Department of Chemistry, School of Chemical and Physical Sciences and Material Science Innovation & Modelling (MaSIM) Research Focus Area, Faculty of Natural and Agricultural Sciences, North-West University, Mmabatho, South Africa

Panneer Selvam Gayathri   PG Department of Chemistry, M.V. Muthiah Government Arts College for Women, Dindigul, India

Lei Guo   School of Material and Chemical Engineering, Tongren University, Tongren, P.R. China

Chaudhery Mustansar Hussain   Department of Chemistry and Environmental Science, New Jersey Institute of Technology, Newark, NJ, United States

R. Keerthana   PG Department of Chemistry, M.V. Muthiah Government Arts College for Women, Dindigul, India

Fahmida Khan   Department of Chemistry, National Institute of Technology Raipur, Raipur, India

Perla Akhil Kumar   Vishwavidyalaya Engineering College, Lakhanpur, Ambikapur, CC of CSVTU, Bhilai, India

Mine Kurtay Yildiz   Corrosion Research Laboratory, Department of Mechanical Engineering, Faculty of Engineering, Duzce University, Duzce, Turkey

Brahim Lakhrissi   Laboratory of Organic Chemistry, Catalysis and Environment, Faculty of Sciences, Ibn Tofail University, Kenitra, Morocco

Tengda Ma   School of Electronics and Information Engineering, Hebei University of Technology, Tianjin, P.R. China

Vivek Mishra   Vishwavidyalaya Engineering College, Lakhanpur, Ambikapur, CC of CSVTU, Bhilai, India

Mohammad Mobin   Corrosion Research Laboratory, Department of Applied Chemistry, Faculty of Engineering and Technology, Aligarh Muslim University, Aligarh, India

G. Nandhini   PG Department of Chemistry, M.V. Muthiah Government Arts College for Women, Dindigul, India

Lukman O. Olasunkanmi   Department of Chemistry, Faculty of Science, Obafemi Awolowo University, Ile Ife, Nigeria

Taiwo W. Quadri   Department of Chemistry, School of Chemical and Physical Sciences and Material Science Innovation & Modelling (MaSIM) Research Focus Area, Faculty of Natural and Agricultural Sciences, North-West University, Mmabatho, South Africa

M.A. Quraishi   Interdisciplinary Research Center for Advanced Materials, King Fahd University of Petroleum and Minerals, Dhahran, Saudi Arabia; Center of Research Excellence in Corrosion, Research Institute, King Fahd University of Petroleum and Minerals, Dhahran, Saudi Arabia

Susai Rajendran   Corrosion Research Centre, Department of Chemistry, St. Antony's College of Arts and Science for Women, Dindigul, India

Mohamed Rbaa   Laboratory of Organic Chemistry, Catalysis and Environment, Faculty of Sciences, Ibn Tofail University, Kenitra, Morocco

N. Renuga Devi   Department of Zoology, GTN Arts College, Dindigul, India

Marziya Rizvi   Corrosion Research Laboratory, Department of Mechanical Engineering, Faculty of Engineering, Duzce University, Duzce, Turkey

Tawfik A. Saleh   Department of Chemistry, King Fahd University of Petroleum and Minerals, Dhahran, Saudi Arabia

S. Senthil Kumaran   Department of Manufacturing Engineering, School of Mechanical Engineering, Vellore Institute of Technology, Vellore, India

Ambrish Singh   School of New Energy and Materials, Southwest Petroleum University, Chengdu, China

A. Suriya Prabha   Department of Chemistry, Mount Zion College of Engineering and Technology, Pudukkottai, India

Baimei Tan   School of Electronics and Information Engineering, Hebei University of Technology, Tianjin, P.R. China

Chandrabhan Verma   Interdisciplinary Research Center for Advanced Materials, King Fahd University of Petroleum and Minerals, Dhahran, Saudi Arabia

Dakeshwar Kumar Verma   Department of Chemistry, Government Digvijay Autonomous Post Graduate College, Rajnandgaon, India

Chenwei Wang   School of Electronics and Information Engineering, Hebei University of Technology, Tianjin, P.R. China

Mesut Yildiz   Corrosion Research Laboratory, Department of Mechanical Engineering, Faculty of Engineering, Duzce University, Duzce, Turkey

Da Yin   School of Electronics and Information Engineering, Hebei University of Technology, Tianjin, P.R. China

Abdelkader Zarrouk   Laboratory of Materials, Nanotechnology and Environment, Faculty of Sciences, Mohammed V University, Rabat, Morocco

Saman Zehra   Corrosion Research Laboratory, Department of Applied Chemistry, Faculty of Engineering and Technology, Aligarh Muslim University, Aligarh, India

Shihao Zhang   School of Electronics and Information Engineering, Hebei University of Technology, Tianjin, P.R. China

Preface

Generally, corrosion reduction has been controlled in a variety of ways, including process control, cathodic protection, metal impurity reduction, and application of surface treatment methods, as well as the incorporation of appropriate alloys. The use of environmentally sustainable corrosion inhibitors has been verified to be the simplest and cheapest method of corrosion protection and prevention in various media. Corrosion inhibitors slow down the rate of metallic dissolution and therefore avoid economic losses owing to the metallic losses on industrial liners, tools, or surfaces. Traditional organic and inorganic inhibitors are lethal and expensive and therefore the current focus has been turned to developing environmentally benign processes for corrosion retardation. Numerous scholars have lately focused on corrosion prevention methods using green inhibitors for metals or alloys. This book assembles the novel developments regarding corrosion inhibitors and their latest applications.

As can been seen throughout this book, the environmentally sustainable corrosion inhibitors are cost-effectively sufficient to ease the troubles caused via corrosion. Environmental regulations in industrialized countries are raising the pressure to remove, in the short term, many compounds broadly used in industry to protect against corrosion. Numerous alternatives of environmentally sustainable corrosion inhibitors are currently emerging, oriented toward minimizing the environmental impact and providing effective corrosion inhibition. Environmentally sustainable corrosion inhibitors contain natural products, plant extracts, and synthetic nontoxic materials. We wish that these products will be able to replace, in the near future, the toxic marketable products that are still being used in many industries worldwide.

A book to wrap up the developments in environmentally sustainable corrosion inhibitors is broadly overdue. It has been addressed by Dr. Hussain and coworkers, in a book which attends to the fundamental characteristics of environmentally sustainable corrosion inhibition, chronological growths, and the industrial applications of sustainable inhibitors. The book offers with the synthesis, characterization, inhibition mechanism and applications of environmentally sustainable corrosion inhibitors in industry. The corrosion inhibition applications are broad ranging. The examples provided in this book include areas such as food, the environment, electronics, oil, gas, and many more. The last chapters talk about the commercialization and economic thoughts which are currently of major significance. The book is divided into different sections, where each section contains various chapters. Section 1 "Sustainable corrosion inhibitors: design and developments" covers topics such as an overview of the corrosion chemistry, the fundamentals of corrosion chemistry, and a general introduction to corrosion inhibitors. Section 2 "Sustainable corrosion inhibitors: current approaches and experimental assessment" describes the synthetic environment-friendly corrosion inhibitors and the experimental methods of inhibitors assessment. Section 3

"Sustainable corrosion inhibitors: candidates and characterizations" entirely focuses on corrosion inhibitors for basic environments, corrosion inhibitors for neutral environments, and corrosion inhibitors for sweet (CO_2 corrosion) and sour (H_2S corrosion) oilfield environments. Section 4 "Sustainable corrosion inhibitors for environmental industry" explains in detail the various environmentally sustainable corrosion inhibitors. Section 5 "Sustainable corrosion inhibitors for electronics industry" gives a description of the environmentally sustainable corrosion inhibitors used in electronics industry. Finally, Section 6 "Sustainable corrosion inhibitors for oil and gas industry" is specific about corrosion inhibitors for refineries, environmentally sustainable corrosion inhibitors for the oil and gas industry, and provides a concise discussion of some high-temperature corrosion inhibitor used in the oil and gas industry.

Overall, this book is written for scholars in academia and industry, working corrosion engineers, and students of materials science, applied and engineering chemistry. The editors and contributors are well-known researchers, scientists, and true professionals from academia and industry. On behalf of Elsevier, we are very thankful to the authors of all chapters for their amazing and passionate efforts in the making of this book. Special thanks go to Dr. Anita Koch (acquisitions editor) and Dr. Allison Hill (editorial project manager) at Elsevier, for their dedicated support and help during this project. In the end, all thanks go to Elsevier for publishing the book.

Chaudhery Mustansar Hussain
Chandrabhan Verma
Jeenat Aslam

Sustainable corrosion inhibitors: design and developments

1

An overview of the corrosion chemistry

Saman Zehra[1], Mohammad Mobin[1], Jeenat Aslam[2]

[1]CORROSION RESEARCH LABORATORY, DEPARTMENT OF APPLIED CHEMISTRY, FACULTY OF ENGINEERING AND TECHNOLOGY, ALIGARH MUSLIM UNIVERSITY, ALIGARH, INDIA [2]DEPARTMENT OF CHEMISTRY, COLLEGE OF SCIENCE, YANBU, TAIBAH UNIVERSITY, AL-MADINA, SAUDI ARABIA

Chapter outline

Environmentally Sustainable Corrosion Inhibitors. DOI: https://doi.org/10.1016/B978-0-323-85405-4.00012-4

Abbreviations

GNP Gross National Product
NACE Corrosion Engineers National Association of Corrosion Engineers
GDP Gross Domestic Product
SCC Stress corrosion cracking
ICCP Impressed current cathodic protection

1.1 Introduction

Corrosion has been the subject of scientific research for ages, because of its devastating consequences and, in general, corrosion is a natural phenomenon explained as the decay of a material, usually a metal, or its properties caused due to the reaction to its environment [1]. The International Standard Organization's scientific definition of corrosion is the "physicochemical reaction between a metal and its environment, leading to modifications in the properties of the metal and often leading to degradation of the function of the metal, the environment or a scientific system of which it forms part" [2]. While common use usually correlates corrosion with metal oxidation, oxidation of all forms of natural and synthetic materials, including biomaterials and nanomaterials, is now included in the term corrosion. "Corrosion is an inevitable interfacial interaction of a material (metal, ceramic, and polymer) with its environment and may result in the consumption of the material or in degradation into the material of an environmental component" is a broader and commonly accepted alternative definition of corrosion [3]. The environment is composed of the whole surroundings which are in contact with the material. Physical state (either gas, liquid, or solid), chemical composition (constituents and concentrations), and temperature are the primary variables for defining the environment.

Metals tend to transform into much more thermodynamically stable substances (like oxides, hydroxides, salts, or carbonates) during corrosion. Metals have a tendency to shift to their lower energy, relatively natural ore state (e.g., iron ore) after production and shaping. The Law of Entropy controls this tendency. Naturally, it is very difficult to find metals in their pure state; they combine with other elements to form ores.

The corrosion subject has witnessed an inevitable transition from the state of isolation and obscurity to an established engineering discipline. There have already been significant advances over time in the field of corrosion and corrosion prevention. There are, however, still many problems that corrosion scientists/engineers need to resolve. Learned societies such as National Association of Corrosion Engineers (NACE), the European Corrosion Federation, the Japan Society of Corrosion Engineers, and others play a prominent role in the growth of education in corrosion engineering.

The primary aim of the chapter is to cover the fundamentals related to the corrosion chemistry in order to understand this catastrophic phenomenon more thoroughly. The chapter will cover the outline of the basics of corrosion precisely.

1.2 The cost of corrosion

In the 1960s the significance of corrosion was recognized when it was understood that damage was caused to the economies of developed nations, a useful life of manufactured goods were being reduced and resources are being wasted by antimetallurgical processes. In addition, corrosion also involved issues pertaining to human life and safety, huge environmental impact, and conservation of materials. Various impacts of corrosion are illustrated in Fig. 1−1.

Corrosion Awareness Day highlights the immense costs associated with corrosion worldwide on the April 24 each year. The economic losses and environmental impact have been the prime motives for much of the current research in the field of corrosion. In order to appreciate the economic effects of corrosion, corrosion cost studies have been undertaken by several countries. Various factors which contribute to the cost of corrosion are shown in Fig. 1−2.

Uhlig published the first important report on corrosion costs in 1949, where the annual corrosion cost in the United States was calculated to be 2.1% of the overall Gross National Product (GNP) [4]. Since then, a number of reports based on comprehensive studies have been published on the economic effects of corrosion. A landmark study on the economic effects of corrosion undertaken in the United States in the late 1970s showed that the total loss due to corrosion in the year 1975 was $70 billion, which was approximately 5% of GNP of that year [5]. In 2002 another breakthrough report was published by the US Federal Highway Administration, which measured the direct costs associated with metallic corrosion in the US industrial sector. The research was initiated by the NACE International, and as part of the Transportation Equity Act for the 21st century, has a mandate from Congress. The overall annual direct corrosion cost was found to be $276 billion, which is about 3.1% of

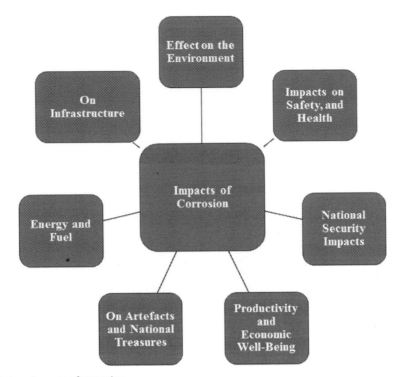

FIGURE 1–1 Various impacts of corrosion.

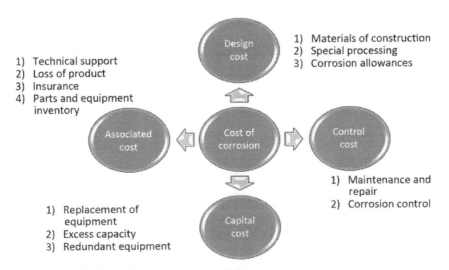

FIGURE 1–2 An outline of the factors impacting the cost of the corrosion.

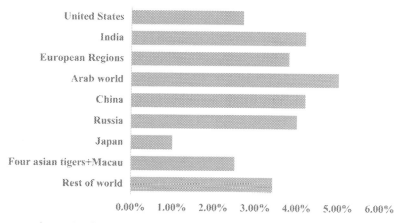

FIGURE 1–3 The cost of corrosion (3.4% of global GDP) published in the year 2013. *GDP, Gross Domestic Product.*

the GNP of the country [6]. This amount contained only the direct costs associated with the replacement of defective material and parts. Indirect costs, such as output losses, effects on the environment, disturbances to transport, accidents and fatalities, are calculated to be equal to direct costs. Other nations, such as the United Kingdom, Japan, Australia, Kuwait, Germany, Finland, Sweden, India, and China, have conducted similar corrosion cost studies. The research ranged from casual and modest attempts to formal and comprehensive attempts. The general conclusion of these studies was that the annual cost of corrosion ranged from approximately 1% to 5% of each nation's GNP. Global economic losses due to corrosion, as estimated by NACE International in 2016, have been recorded in some publications at $2.5 trillion [7,8]. In India, during the year 1984–85, the direct cost of corrosion was calculated to be Rs. 40.76 billion, of which Rs. 18.04 billion was considered avoidable [9].

Another report on the cost of corrosion in India published in 1997, estimated the annual losses due to corrosion to be Rs. 250 billion per year, which worked out to be 4% of GNP [10]. As per the latest global study by NACE International, the cost to India's economy on the account of corrosion is estimated to be 4.2% of Gross Domestic Product (GDP). An outline of the country wise contribution to the cost of corrosion, according to a report published in the year 2013, is illustrated in Fig. 1–3. According to which the combined cost of corrosion represents 3.4% of GDP. The various contribution are: United States contributes 2.7%, India 4.2%, European region 3.8%, Arab world 5.0%, China 4.2%, Russia 4.0%, Japan 1.0%, Four Asian Tigers + Macau 2.5%, and rest of the world approximately 3.4% of the total global cost to corrosion.

Apart from the economic consequences, corrosion results in structural failures that have serious consequences for human health and life and the surrounding environments. The safety and environmental concerns tend to be very difficult to define in terms of cost.

Therefore corrosion problems must be addressed for safety and to minimize environmental pollution.

1.3 Corrosion measurement units

The corrosion rate can be expressed as an increase per unit in the depth of corrosion time mils per year (mpy) (penetration rate, mils/yr) or weight loss per unit area per unit per unit area time, usually mdd (milligrams per square decimeter a day) or the current of corrosion $(mA\ cm^{-2}, cm^2)$. Although the preferred SI unit for expressing the corrosion rate is $mm\ yr^{-1}$ or $inch\ yr^{-1}$, The term mpy is the most (a mil is a thousand inches) commonly used and desirable corrosion rate expression in the United States, which can be estimated using Eq. (1.1) [11]:

$$\text{Corrosion rate (mpy)} = \frac{534W}{\rho At} \tag{1.1}$$

where W is weight loss in mg; ρ is the density of specimen in $g\ cm^{-3}$; A is the area of specimen in sq. in., and t is exposure time in hours.

1.4 Types of corrosion

Corrosion manifests in several forms and the basic understanding of the details of the type of corrosion is desirable, as it can help in the determination of the cause of corrosion and to select the most effective approach of preventing it. In general, classification is based on one of three factors: (1) the nature of the corrodent, (2) the corrosion mechanism, and (3) the appearance of the corroded metal [12]. On the basis of appearance the types of corrosion are identified as follows:

1.4.1 Uniform or general corrosion

The most commonly recognized form of corrosion is the uniform corrosion. It is usually known as the general corrosion. The attack, which usually occurs over the entire surface of the metal exposed to the corrosive medium, more or less uniformly, is called uniform corrosion. It usually takes place in the environments in which the rate of corrosion is comparatively low and well controlled (like chemically treated closed circulating system and some open water systems). By far the most prone metals to uniform corrosion are low alloy iron and magnesium alloys. The general attack results from local corrosion cell action, that is, multiple anodes and cathodes are active on the metal surface at any given time. Uniform corrosion over the bare metal surface progresses at nearly the same rate. When exposed to open spaces, soils, and natural waters, cast iron and steel corrode uniformly, contributing to the appearance of rust. The collection of suitable materials, chemical-resistant protective coatings, cathodic and anodic protection, and corrosion inhibitors typically regulate uniform corrosion.

1.4.2 Galvanic corrosion

In the electrolyte solution and an electron conductive path, galvanic corrosion, generally called two-metal or bimetallic corrosion, is an electrochemical behavior of two dissimilar metals. It arises if dissimilar metals or alloys with different electrode potentials are electrically connected in a standard electrolyte solution. The anode becomes more negative or an active metal, and the cathode of the galvanic corrosion cell becomes more positive or a noble metal. The less noble alloy normally shows an increase in corrosion and a decrease or suppression of corrosion in the more noble material. The potential difference between two metals, the nature of the environment, the polarization behavior of metals and the geometric relationship of the component metals affect the degree of galvanic corrosion. Because of the direct relation of brass valves to the carbon steel pipe, or between copper tubing and steel pipe where the steel serves as the anode, and brass or copper as the cathode, the most common example of such corrosion behavior, usually found in process plant operations, is due. Metal combinations where the constituents are as similar as possible in the respective galvanic series are chosen to prevent galvanic corrosion, using a seal, insulator, coating, etc. whenever possible to avoid direct contact between two different metals, preventing threaded junctions between materials widely separated in the galvanic series.

1.4.3 Pitting corrosion

Pitting seems to be the most devastating mode of corrosion attack which is often responsible for equipment failure in processing plants, accounting for roughly 90% of metal destruction caused by corrosion [13]. Corrosion pitting is a highly localized metal surface attack with no overall corrosion in the surrounding region. A pit may be defined as a cavity with a diameter equal to or less than the depth of a metal surface. It can cause the metal substrate to be penetrated rapidly. It is first identifiable as a white or gray powdery accumulation, comparable to dust, which blotches the surface. Tiny pits or holes could be seen on the surface as the coating is washed away. In certain cases, pitting is developed over the overall metal surface, creating an irregular or quite rough surface profile. Pitting corrosion, as it is much harder to trace, prevent, and manage, is considered to be more dangerous than uniform corrosion damage. Depassivation of a small region seems to be the driving force for pitting corrosion, which becomes anodic, whereas a huge area became cathodic, contributing to much more localized galvanic corrosion. Pitting corrosion can result in a variety of metal/environment combinations. Pitting is also susceptible to engineering alloys, such as stainless steel and aluminum alloys, which form protective passive films on the surface. Corrosion by pitting is most likely to occur in solutions containing ions of chloride, bromide, or hypochlorite. Also in the absence of oxygen, the existence of an oxidizing cation (Fe^{3+}, Cu^{2+}, Hg^{2+}, etc.) causes pits to develop. As an alloying factor, molybdenum is advantageous, and stainless steel containing molybdenum, such as types 316, 317, 904 and 254 SMO, is more resistant than stainless steel not containing molybdenum. In contrast, the metals susceptible to uniform corrosion may not tend to suffer from pitting. In seawater, standard carbon steel corrodes evenly, while stainless steel is pitted. Corrosion by pitting can be managed by the maintenance of clean surfaces, the application of protective coatings, and the use of immersion inhibitors or cathodic protection.

1.4.4 Crevice corrosion

Crevice corrosion, which also refers to corrosion in occluded areas, is one of the most damaging forms of localized material degradation [14]. This type of corrosion is comparable to the pitting corrosion occurring in the stagnant electrolytic conditions, and the crevice corrosion occurs due to the alteration of the local chemistry, that is, depletion of oxygen in crevice, rise in values of pH with increased hydrogen ion concentration, and increase in chloride ion concentration. Oxygen depletion implies that within the crevice region, cathodic reaction to oxygen reduction cannot be sustained and metal dissolution occurs as a consequence. On any metal and in any corrosive environment, crevice corrosion can take place. Metals such as aluminum and stainless steel that rely on their surface oxide film for corrosion resistance, however, are especially susceptible to crevice corrosion, especially in environments such as chloride ion-containing seawater. The substance responsible for the crevice corrosion formation does not have to be metallic. The causes of crevice corrosion have been identified in wood, plastic, rubber, glass, concrete, etc. It is possible to avoid crevice corrosion by properly designing installations to allow full drainage (no corners or stagnant zones), by using welds instead of bolted or riveted joints, by using only solid, nonporous seals, and by using solid, nonabsorbent gaskets such as Teflon.

1.4.5 Intergranular corrosion

Intergranular corrosion, also called intercrystalline corrosion, is a localized preferential attack on the grain boundary or the areas around it [15]. Little or no attack is noticed on the main body of the grain. The lack of strength and ductility results in this form of corrosion. Sometimes, the attack is quick, deeply penetrating into the metal and causing failure. The presence of impurities within the limits, or local enrichment or degradation of one or more alloying elements, is due to this form of corrosion. For instance, chromium corrosion in the grain boundary region arises when austenitic stainless steels are sensitized by heating in the range of temperature approximately $500°C-800°C$, leading to vulnerability to intergranular corrosion. Use of such low-carbon stainless steel grades or stabilized grades of stainless steel alloyed with strong carbide formers like titanium or niobium could avoid these types of corrosion. These components combine with the carbon to form the appropriate carbides, thereby preventing the depletion of chromium.

1.4.6 Stress corrosion cracking

Structural components that are subjected to a combination of sustained tensile stress and a corrosive environment can fail prematurely at a stress below the yield strength, known as stress corrosion cracking (SCC) [16]. Stress may occur as a result of loads being added, residual stresses from the production process (welding, heat treatment, machining, and grinding) or a combination of both. The group of commercial metals and alloys that are completely resistant to SCC is not identified. Standard austenitic stainless steels, such as AISI 304 and AISI 316, are typically vulnerable to SCC at temperatures above 60°C in Cl^- containing environments. This form of corrosion could be avoided or minimized by eliminating residual stresses through using heat treatments to alleviate stress, purifying the medium, selecting the most appropriate material, improving the surface quality, and applying external protection methods such as cathodic protection, inhibitors, and protective coatings.

1.4.7 Filiform corrosion

It is practically a particular form of crevice corrosion, often referred to as corrosion "under film." As moisture pervades the coating, such form of corrosion develops beneath painted or plated surfaces. Lacquers are most prone to the problem and "quick-dry" paints. Their use should be avoided until field experience has shown the absence of an adverse impact.

1.4.8 Erosion corrosion

The term "erosion" refers to decay due to mechanical forces. The assault is known as "erosion corrosion" when the factors leading to erosion increase the rate of corrosion of a metal. It is the product of a mixture of an active chemical environment and high surface velocities of fluid. This may be the consequence of a fast fluid flow past a stationary object, such as the example of the oilfield check valve, or it may be the product of the object's rapid motion in a stationary fluid, such as when the ocean is churned by a ship's propeller. Unlike surfaces from several other types of corrosion, surfaces that have suffered erosion corrosion are normally fairly clean. Many metals and alloys, including those which rely on the formation of a surface oxide film for corrosion resistance, are susceptible to erosion-corrosion damage. However, softer metals such as Cu, Al, Pb-alloys, and brass are naturally more vulnerable to corrosion by degradation than steel.

1.4.9 Selective leaching or de alloying

The selective displacement of an element from an alloy via corrosion leaving behind the elements that are much more resistant in that environment is called selective leaching. For instance, Zn is selectively leached out from the Cu-Zn alloy, leaving underneath a porous and brittle Cu-rich outer surface.

1.5 Electrochemical theory of corrosion

Corrosion is an electrochemical phenomenon that happens due to the existence of anodic and cathodic sites on the metals' surface, originating from the heterogeneous behavior of the bulk material and its surface. The presence of an electrolyte and electrical connection is also needed for electrochemical corrosion to occur. The bulk material and its surface offer the anode, cathode, and the electrical connection, while humid air, aqueous solution, etc., offer the electrolyte to complete the corrosion circuit. The arrangement of anode, cathode, and electrolyte is called a corrosion cell. The anode and the cathode can be of different metals or the same metal with heterogeneous surface structure. Corrosion occurs as a result of a difference in electrical potential between the two electrodes, which must be electrically connected. The following anodic and cathodic reactions occur in different media.

1.5.1 Anodic reaction

Considering only metals and alloys, the anode is the site at which the metal is corroded, that is, at which metal dissolution takes place. The anodic reaction involves the oxidation of a metal to its ion because of the electric charge difference at the solid–liquid interface. A metal's generic anodic process can be described by a metal's oxidizing reaction to its ions, which passes into solution:

$$M \rightarrow M^{n+} + ne^{-} \tag{1.2}$$

where "n" is the valence of metal, e^{-} indicates the electron, M the generic metallic material, and M^{n+} its ion which passes into the solution.

1.5.2 Cathodic reaction

The anodic dissolution reaction involves only the metallic phase, the cathodic reaction involves the environment. The generic cathodic reaction can be represented by the following reaction:

$$R^{+} + e^{-} \rightarrow R^{0} \tag{1.3}$$

where R^{+} is a positive ion in solution, e^{-} is an electron in the metal, and R^{0} is the reduced chemical species. Several different cathodic reactions (electron consuming) are possible and the one that occurs is determined by the environment. In an acidic aqueous environment and in the absence of dissolved oxygen, the main cathodic reaction is hydrogen evolution:

$$2H^{+} + 2e^{-} \rightarrow H_2 \tag{1.4}$$

While the same reaction in an alkaline aqueous solution occurs as:

$$2H_2O + 2e^{-} \rightarrow H_2 + 2OH^{-} \tag{1.5}$$

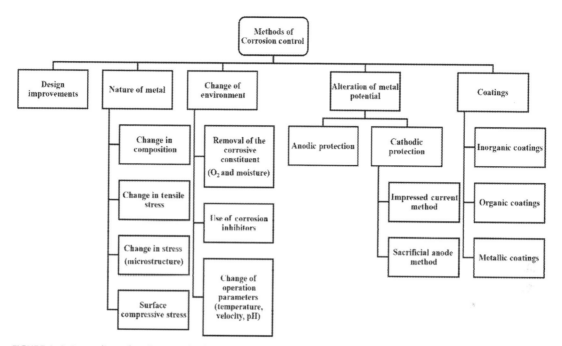

FIGURE 1–4 An outline of various method used in the corrosion control.

In the presence of dissolved oxygen, the cathodic reaction that is more thermodynamically favorable is oxygen reduction. In the acidic environment, it proceeds as:

$$2O_2 + 4e^- + 4H^+ \rightarrow 4H_2O \tag{1.6}$$

or in the alkaline environment, it proceeds as:

$$2O_2 + 4e^- + 2H_2O \rightarrow 4OH^- \tag{1.7}$$

The metal ion reduction and metal deposition are less common reactions and are most frequently found in chemical process streams.

$$M^{3+} + e^- \rightarrow M^{2+} \tag{1.8}$$

$$M^+ + e^- \rightarrow M \tag{1.9}$$

1.6 Methods of corrosion control

Corrosion management requires the implementation of engineering principles and techniques to, by the most economical method, mitigate corrosion to an appropriate degree. To mitigate or control corrosion, there are different practices that could be used (illustrated in Fig. 1–4). The use of the method relies upon multiple parameters, for example, the type and the location of the corrosion, the practical usage of the surface/structure to be protected, the local environment, etc. The objective of the corrosion control techniques is to minimize corrosion of materials to an acceptable limit, so that they are able to attain their normal desired lifetime. In a limited number of cases, corrosion control methods are designed to eliminate it completely. The subsequent text summarizes a few of most commonly used ways and methods to tackle corrosion [17].

1.6.1 Proper selection of materials

The methodology involves the selection and use of materials with high corrosion resistance to increase a structure's longevity in a specific setting. While there are no materials that are resistant to all corrosive conditions, it is important to choose suitable materials to avoid certain forms of corrosion failure. Titanium, for instance, is a highly corrosion-resistant material, but it is far more costly than steel. Furthermore, it is not as ductile as steel. Carbon steel is the material of choice in oil production systems, especially for equipment such as wells, pipelines, vessels, and tanks, due to its good mechanical properties and low cost. However, options such as stainless steel can be used for situations where more corrosion-resistant material is needed.

1.6.2 Environmental modification

A chemical reaction between the corrosive state of the metal and the local atmosphere causes corrosion. Metal corrosion can be regulated instantly by excluding the metal from, or modifying, the state of the atmosphere. This can be as simple as reducing contact with rain or seawater by indoor storage of metal materials or may be in the form of direct environmental manipulation affecting the metal. The rate of metal corrosion can be lowered by methods to minimize the sulfur, chloride, or oxygen content of the surrounding atmosphere. For example, in order to minimize corrosion in the interior of the unit, feed water for water boilers may be formulated with softeners or several other chemicals in order to alter the hardness, alkalinity, or oxygen content.

1.6.3 Cathodic protection

Cathodic protection involves polarization of the structure, to be protected, to potentials more negative than the corrosion potential, thus thermodynamically preventing the occurrence of anodic reaction. For this purpose, an external power source is used. Another common way is to use "sacrificial anodes." In this method the anode is made of a more active metal (e.g., magnesium) than the structure to be protected (e.g., iron or carbon steel). Thus the structure which needs to protected acts as the cathode of a new corrosion cell. This type of protection is widely used to protect some underground structures, for example, water storage tanks, buried pipelines, ship hulls, and marine facilities. There are two methods of cathodic protection:

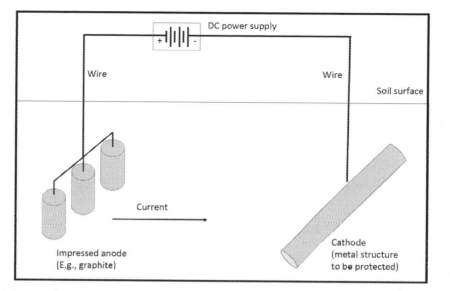

FIGURE 1–5 A general representation of impressed current cathodic protection method.

1.6.3.1 Impressed current method

The impressed current cathodic protection (ICCP) technique is widely used for the protection of buried pipelines and the hulls of ships immersed in seawater. A DC electrical circuit is used to apply an electric current to the metallic structure. The negative terminal of the current source is connected to the metal requiring protection. The positive terminal is connected to an auxiliary anode immersed in the same medium to complete the circuit. The electric current charges the structure with excess electrons and hence changes the electrode potential in the negative direction until the immunity region is reached. The layout for a typical ICCP system is shown in Fig. 1–5. ICCP is a specialized technology and can be very effective if correctly designed and operated. Typical materials used for anodes are graphite, silicon, titanium, and niobium plated with platinum. Coatings are often used in conjunction with ICCP systems to minimize the effect of corrosion on marine structures. One of the difficulties in designing a combined coating and ICCP system is that coatings deteriorate with time. Precious metals are used for impressed current anodes because they are highly efficient electrodes and can handle much higher currents. Precious metal anodes are platinized titanium or tantalum anodes; the platinum is either clad to or electroplated on the substrate. Impressed current systems are more complex than sacrificial anode systems and mostly used to protect pipelines.

1.6.3.2 Sacrificial anode method

The principle of this technique is to use a more reactive metal in contact with steel structure to drive the potential in the negative direction until it reaches the immunity region. Fig. 1–6 illustrates the principle, in which sacrificial metals used for cathodic protection consist of magnesium-base and aluminum-base alloys and, to a lesser extent, zinc.

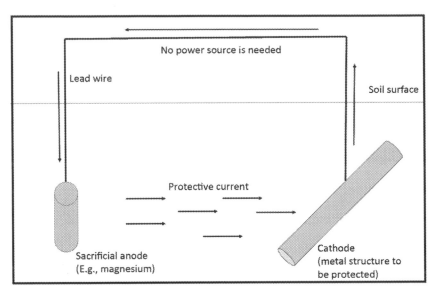

FIGURE 1–6 A general representation of sacrificial anode cathodic protection method.

No external power source is needed with this type of protection system and much less maintenance is required. These metals are alloyed to improve the long-term performance and dissolution characteristics. Sacrificial anodes serve essentially as sources of portable electrical energy. For cathodic protection of offshore platforms, aluminum anodes, made from aluminum−zinc alloys, are the preferred material. Most offshore petroleum-production platforms use sacrificial anodes because of their simplicity and reliability, even though the capital costs would be lower with impressed-current systems. Magnesium anodes have been used offshore in recent years to polarize the structures to a protected potential faster than zinc or aluminum alloy anodes. Magnesium tends to corrode quite readily in salt water, and most designers avoid the use of magnesium for permanent long-term marine cathodic protection applications. Zinc anodes are also used to protect ballast tanks, heat exchangers, and many mechanical components on ships, coastal power plants, and similar structures. In seawater, passivity can be avoided by alloying additions, such as tin, indium, antimony, or mercury. The three most common types of sacrificial anodes are activated aluminum, zinc, and magnesium. Aluminum is the most widely used material for anodes, as it has a higher current capacity in comparison to the other metals. Magnesium should be considered when the chloride content is less than 10,000 ppm.

1.6.4 Anodic protection

Anodic protection, which is an electrochemical method of controlling corrosion based on the phenomenon of passivity, is a comparatively new method suggested by Edeleanu [18,19]. To build a protective oxidized layer on the protected base material, also known as the substrate, an electrical current is used in the process. The passive potential is automatically maintained, usually electronically, by an instrument called the potentiostat. The method is applicable to metals having active passive transitions like nickel, titanium, iron, chromium, and their alloys. Its usefulness and its low current demands in highly corrosive environments are the significant benefits of anodic protection.

1.6.5 Coating, linings and nonmetallic piping

A coating is a thin material applied as a liquid or powder, which, on solidification, is firmly and continuously attached to the material to be protected. For an internal use, this method might be called lining. Coatings need to be flexible, to be resistant to chemical fluid attacks, to have strong adhesion, to have low or no porosity, and to be stable at working temperatures. The coatings are often applied in conjunction with cathodic protection systems to take care of any damage caused to the coating material. Application of nonmetallic piping is also good in many applications as they do not corrode, but the limitation is that they may deteriorate or be weakened by attack from the environment. Polyvinyl chloride (PVC) is one example of this type of piping; this material is usually repeatedly reheated, softened, and reshaped without destruction. However, where mechanical/structural integrity is important, such as in cases of high internal or external pressures and loads, PVC and other polymer-based materials cannot be used, and the use is then usually confined to metals. Metal coatings can also be incorporated in this category. For example, corrosion susceptible metals (e.g., carbon steels) could be coated with a thin chromium coating. Chromium is a highly corrosion-resistant material, and as long as the chromium coating is compact, it will protect the underlying structure from corrosion [20].

1.6.5.1 Organic coatings
Such coatings afford protection by providing a physical barrier between the metal and the environment. The most widely used protective coatings are used to protect aluminum, zinc, and carbon steel from corrosion. In these coatings, they may also contain corrosion inhibitors. Natural coatings are found in paints, resins, lacquers, and varnishes. Heavy organic coatings are often used, like mastics and coal tars, to coat aluminum structures embedded in soils and concrete.

1.6.5.2 Inorganic coatings
To provide a barrier between the atmosphere and the metal, these coatings are often used. Enamel, glass linings, and conversion coatings are all inorganic coatings. The treatment transforms the metal surface into a metallic oxide film or a compound that is more resistant to corrosion than the natural oxide film and provides an effective base or additional safety key, such as paints.

1.6.5.3 Metallic coatings
Another type of coating is metallic coatings that provide a barrier in between metal substrate and the atmosphere. Furthermore, when the coating is damaged, metallic coatings can often provide cathodic protection. Using a variety of techniques, including hot dipping,

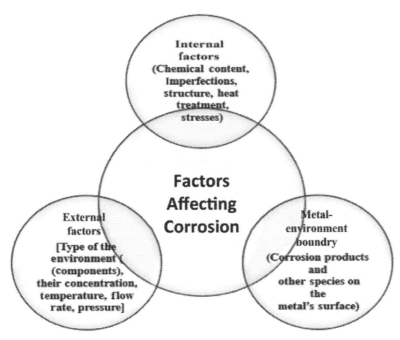

FIGURE 1–7 A general illustration of various factors affecting the corrosion.

electroplating, cladding, thermal spraying, and chemical vapor deposition, metallic coatings and other inorganic coatings are developed.

1.6.6 Use of corrosion inhibitors

In different industries, among the most efficient methods of effectively reducing corrosion is by using corrosion inhibitors. Corrosion inhibitors are chemical substances which stop or retard corrosion of the metallic surface when added to the corrosive media in small quantities. Most of the inhibitors actively employed are organic molecules [21,22]. Via adsorption on the metallic surface, these inhibitors work to protect the surface by forming a film that minimizes the contact between the metal surface and the aggressive environment. The physical blockage effect or the effects of the inhibitor on the mechanisms and kinetics of the corresponding corrosion reactions can be related to this minimization. Inhibitors are typically delivered from a solution or dispersion, but some are used in preparations for protective coating [23].

1.7 Factors affecting rate of corrosion

There are numerous internal and external factors which directly or indirectly affect the rate of corrosion. Some of the factors are shown in Fig. 1−7. We will focus on external influences, as other influences cannot be changed, and we will consider their impact on the corrosion of metals, namely the effect of electrolyte solutions on the chemical material.

In order to predict the corrosion behavior of metals under particular conditions, we should understand this effect. Some of the factors that contribute to the rate of corrosion are outlined below:

1.7.1 Effect of the electrical conductance of media

Due to the acceleration of the corrosion current if the electrical conductance is strong, the electrical conductance of media is a very significant factor in corrosion processes. The lower the medium's electric resistance R, the higher the corrosion current I_{corr}:

$$I_{corr} = V/R = (Ek - Ea)/R \tag{1.10}$$

All electrochemical (cathodic) safety parameters (shape selection, density of electrical current, etc.) depend on the electrical resistance of the media and the anode shape. The electrical resistance of the soil often depends on the distribution of stray electric currents. Of course, electrochemical measurements depend, for example, on the electrical resistance of the media to determine the electrical current as a function of the electrical potential. The ohmic potential drop (known as IR drop) occurs during an electrical current flow between the working electrode and a reference electrode in solution. This is a possible decrease due to the solution layer's electrical resistance. The higher a solution's electrical resistance, the greater the IR decrease. This error in potential measurements needs to be accounted for. An IR decrease may involve polarization measurements in pure water and nonaqueous solvents, as well as metal electrodes with coatings.

1.7.2 Influence of the pH on metallic corrosion

According to their relationship (corrosion resistance) to pH values, all metals can be split into five groups:

1. Corrosion resistance of gold (Au), silver (Ag), platinum (Pt), palladium (Pd), rhodium (Rh), ruthenium (Ru), mercury (Hg), tantalum (Ta), niobium (Nb), osmium (Os), and iridium (Ir) is not affected by the pH value. At pH = 0 to 14, they are resistant toward the corrosion.
2. Iron (Fe), chromium (Cr), and manganese (Mn) are strongly corroded at temperatures above 80°C in acidic solutions (pH < 4) and in very hot alkaline solutions (pH > 13.5). They often corrode, but at neutral pH (6−8) at a low rate, and are immune at pH 9−13.
3. Magnesium (Mg), titanium (Ti), hafnium (Hf), vanadium (V), and bismuth (Bi) are extremely resistant to neutral and alkaline solutions (strong pH) and corrode at a high rate at low pH (in acid solutions).
4. In acid and neutral solutions, molybdenum (Mo), tungsten (W), and rhenium (Re) are resistant to corrosion but corrode in alkaline solutions.

5. Beryllium (Be), aluminum (Al), copper (Cu), zinc (Zn), cadmium (Cd), tin (Sn), lead (Pb), cobalt (Co), nickel (Ni), zirconium (Zr), gallium (Ga), and indium (In). The last group is big. In both acidic and alkaline liquids, they corrode, and are thus called amphoteric metals. Only neutral solutions are resistant to these metals. It is evident that for various amphoteric metals, an area of pH where high resistance exists is distinct. Aluminum is pH 4.5−8.3 resistant, and zinc is pH 6.5−12 resistant. The pH region of the high resistance of amphoteric metals will significantly alter the temperature and presence of different ions in a solution. Strictly speaking, the amphoteric metals are often associated with iron and chromium as they dissolve at low and very high pH (above 13.5); the higher the temperature, the lower the pH when iron is dissolved in alkaline solutions.

1.7.3 Effect of substances forming chemical complexes with metals on corrosion

Substances of metals forming chemical complexes intensify their corrosion. Neutral molecules (ammonia NH_3) or ions (cyanides CN^-) can be these compounds. Ammonia is highly soluble in water, producing an ammonium hydroxide alkaline solution. In the presence of ammonia and dissolved oxygen in water, copper and copper alloys are not immune due to the formation of a complex:

$$4Cu(s) + 8NH_3(aq) + O_2(g) + 2H_2O(l) \leftrightarrow 4\left[Cu(NH_3)_2\right]^+(aq) + 4OH^-(aq). \tag{1.11}$$

Using copper condensers with water from bays near agricultural fields where ammonia fertilizers are used is therefore very risky. Due to their high corrosion resistance, we call gold and silver "noble" metals. Many people say they are immune to all media sources. But if you place them in an aqueous solution of sodium or potassium cyanide (NaCN or KCN), with these "noble" metals, anions of CN form chemical complexes:

$$4Au(s) + 8CN^-(aq) + 2H_2O(l) + O_2(g) \rightarrow 4\left[Au(CN)_2\right]^-(aq) + 4OH^-(aq), \tag{1.12}$$

$$4Ag(s) + 8CN^-(aq) + 2H_2O(l) + O_2(g) \rightarrow 4\left[Ag(CN)_2\right]^-(aq) + 4OH^-(aq). \tag{1.13}$$

Hence, in solutions containing cyanide anions, gold and silver are not resistant.

1.7.4 Effect of cations participating in cathodic reactions

In addition to the presence of dissolved oxygen (O_2) in neutral and alkaline solutions and H_3O^+ cations in acidic solutions, Fe^{3+} and Cu^{2+} cations may be involved in cathodic processes on the surface of the metal and accelerate corrosion. It is good to say that according to the reaction, Fe^{2+} cations formed in the cathodic reaction can be oxidized to Fe^{3+}:

$$4Fe^{2+}(aq) + O_2(g) + 4H_3O^+(aq) \leftrightarrow 4Fe^{3+}(aq) + 6H_2O(l) \tag{1.14}$$

So, Fe^{3+} cation regeneration continues all the time, and corrosion is intensified. It is therefore very important to recognize and track such cations in a medium that can take part in cathodic reactions and accelerate corrosion.

1.7.5 Effect of temperature

In chemistry, temperature is the big parameter for adjusting the rate of chemical reactions. Under large variations in environmental conditions, metals and alloys are used. The thermodynamics and kinetics of metallic corrosion may be influenced by temperature. Temperature rises typically accelerate anodic and cathodic reactions, decrease gaseous oxygen dissolution in the media, and accelerate the diffusion of cathodic agents (O_2, H_3O^+, Fe^{3+}, etc.). The dissolution and transformation of corrosion products forming on metal surfaces are also changed by temperature. The effect of temperature is complex, and it is not easy to predict the behavior of metals with changes in temperature. Temperature thus influences corrosion by two factors: it accelerates both anodic and cathodic reactions and decreases the concentration of dissolved oxygen (cathodic reaction as a result). These two variables are identical at a certain temperature, and corrosion reaches the maximum value. Then the decrease in oxygen concentration in water prevails over the rise in the anodic and cathodic reaction rates at higher temperatures. Dissolved oxygen disappears from the "scene" as a cathodic partner, and corrosion decreases. In the corrosion of metals, oxygen is so critical that, in specific cases, it can result in acceleration of corrosion and, in other cases, decrease and even stop it. In certain cases, the dissolved oxygen concentration in the media is nonuniform and induces the development of differential aeration cells.

1.7.6 Effect of the dissolved oxygen

In the corrosion of metals, dissolved oxygen plays a very crucial and complex role. In neutral, alkaline, and acidic media, oxygen takes part in cathodic procedures on the metal surface. Hence, for corrosion to occur, its existence is required. If dissolved oxygen is absent in water, corrosion in neutral and alkaline solutions decrease to almost zero. If the concentration of dissolved oxygen increases, as a consequence of oxygen involvement in the cathodic processes, corrosion accelerates. What would happen if we were to inject more and more water with oxygen? It was defined that oxygen under some particular conditions (in water of high purity) and high temperature may result in the formation of a passive protective dense film composed of metal oxides on the metal surface, and corrosion would decrease. One of the corrosion prevention strategies at power stations is the introduction of oxygen into water. If water contains ions (Cl^-, SO_2^-), passive films may be damaged and the passivation effect will not be obtained.

1.7.7 Effect of dissolved salts in water on corrosion

The chemical content of drinking water, cooling water, seas, oceans, rivers, lakes, and subterranean waters varies. For all these types of water, their almost neutral pH (usually between 5.5 and 8.3), the presence of inorganic and organic compounds, and dissolved gases are the common denominators. Slight changes in the chemical content of water species can result in drastic changes in metal corrosion rates. pH values from 5.5 to 8.3 have no effect on the corrosion of metals. Metal cations formed in anodic metal dissolution react with hydroxide anions OH^- formed in aqueous neutral solutions in a cathodic reduction of dissolved oxygen, and corrosion products are formed in the form of hydroxides. The rate of corrosion depends on the form of salt and its concentration.

1.8 Conclusive remark

Corrosion can be a horrific experience and can lead to some unpleasant effects when it applies to industry that relies on noncorroded metal to operate, which are not always as recognizable: lack of production, injuries, and major financial losses. Annually, industries are leaching billions of dollars out of their business, primarily due to corrosion. The overall cost of corrosion around the globe is huge; it sums up to approximately $2.5 trillion (USD), and that figure does not even involve human protection or environmental consequences. However, through obtaining and using highly qualified corrosion practitioners, harmonizing standards, along with continuing education, and training, all underpinned through fostering greater corrosion knowledge, the ability to reduce the cost by $875 billion (approximately) annually by effective implementation of current corrosion reduction technologies is easily attainable.

Useful links

https://www.wiley.com/en-us/Uhlig%27s + Corrosion + Handbook%2C + 3rd + Edition-p-9780470080320
https://onlinelibrary.wiley.com/doi/book/10.1002/9781118232163

References

[1] W.D. Callister, Materials Science and Engineering—An Introduction, 7th ed., John Wiley & Sons, Inc, 2007.

[2] Corrosion of Metals and Alloys, Basic Terms and definitions, European Committee for Standardization, EN ISO 8044, 1999.

[3] K.E. Heusler, D. Landolt, S. Trasatti, Electrochim. Acta 35 (1990) 295.

[4] H.H. Uhlig, The cost of corrosion to the U.S, Chem. Eng. N. Arch. 27 (39) (1949) 2764.

[5] L.H. Bennet, J. Kruger, R.I. Parker, E. Passiglia, C. Reimann, A.W. Ruff, et al., Economic Effects of Metallic Corrosion in the United States, 511, National Bureau of Standards Special Publication, Washington, DC, 1978.

[6] G.H. Koch, M.P.H. Brongers, N.G. Thompson, Y.P. Virmani, J.H. Payer, Corrosion costs and preventive strategies in the United States, Mater. Perform. 42 (2002).

[7] Available at http://insights.globalspec.com/article/2340/annual-global-cost-of-corrosion-2-5-trillion.

[8] J. Haque, V. Srivastava, C. Verma, M.A. Quraishi, Experimental and quantum chemical analysis of 2-amino-3-((4-((S)-2-amino-2-carboxyethyl)-1H-imidazol-2-yl)thio) propionic acid as new and green corrosion inhibitor for mild steel in 1 M hydrochloric acid solution, J. Mol. Liq. 225 (2016) 848–855. Available from: https://doi.org/10.1016/j.molliq.2016.11.011.

[9] K.S. Rajagopalan, in: 10th International Congress on Metallic Corrosion, Chennai, Oxford & IBH, New Delhi, 2 (1987) 1765.

[10] A.S. Khanna, News Letter, NACE, India, 4 (1997) 3.

[11] M. Parveen, M. Mobin, S. Zehra, RSC Adv. 6 (2016) 61235–61248.

[12] M.G. Fontana, N.D. Greene, The Eight Forms of Corrosion, Corrosion Engineering, McGraw-Hill, New York, 1978.

[13] A.I. Asphahani, W.L. Silence, Metal Handbook, 13, ASME, 1987, p. 113.

[14] J.R. Kearns, "Crevice", Corrosion Tests and Standards-Application and Interpretation, Editor, Robert Baboian, 19 (1995) 175.

[15] Intergranular Corrosion of Stainless Alloys, ASTM STP 656, American Society for Testing and Materials, West Conshohocken, PA, 1978.

[16] R.N. Parkins, Stress corrosion cracking, in: R.W. Revie (Ed.), Uhlig's Corrosion Handbook, 2nd Ed., Wiley, New York, 2000.

[17] H.H. Uhlig, Corrosion and Corrosion Control: An Introduction to Corrosion Science and Engineering, Wiley, 1965.

[18] C. Edeleanu, Method for the study of corrosion phenomena, Nature 173 (1954) 739.

[19] C. Edeleanu, Corrosion control by anodic protection, Metallurgia 50 (1954) 113.

[20] M. Aliofkhazraei, Developments in Corrosion Protection, InTech, 2014. ISBN 978-953-51-1223-5.

[21] M. Mobin, S. Zehra, R. Aslam, l-Phenylalanine methyl ester hydrochloride as a green corrosion inhibitor for mild steel in hydrochloric acid solution and the effect of surfactant additive, RSC Adv. 6 (2016) 5890–5902.

[22] M. Mobin, S. Zehra, M. Parveen, l-Cysteine as corrosion inhibitor for mild steel in 1 M HCl and synergistic effect of anionic, cationic and non-ionic surfactants, J. Mol. Liq. 216 (2016) 598–607.

[23] M. Mobin, R. Aslam, J. Aslam, Non toxic biodegradable cationic gemini surfactants as novel corrosion inhibitor for mild steel in hydrochloric acid medium and synergistic effect of sodium salicylate: experimental and theoretical approach, Mater. Chem. Phys. 191 (2017) 151–167.

2

Fundamentals of corrosion chemistry

Taiwo W. Quadri[1], Ekemini D. Akpan[1], Lukman O. Olasunkanmi[2],
Omolola E. Fayemi[1], Eno E. Ebenso[3]

[1]DEPARTMENT OF CHEMISTRY, SCHOOL OF CHEMICAL AND PHYSICAL SCIENCES AND MATERIAL SCIENCE INNOVATION & MODELLING (MaSIM) RESEARCH FOCUS AREA, FACULTY OF NATURAL AND AGRICULTURAL SCIENCES, NORTH-WEST UNIVERSITY, MMABATHO, SOUTH AFRICA [2]DEPARTMENT OF CHEMISTRY, FACULTY OF SCIENCE, OBAFEMI AWOLOWO UNIVERSITY, ILE IFE, NIGERIA [3]INSTITUTE FOR NANOTECHNOLOGY AND WATER SUSTAINABILITY, COLLEGE OF SCIENCE, ENGINEERING AND TECHNOLOGY, UNIVERSITY OF SOUTH AFRICA, JOHANNESBURG, SOUTH AFRICA

Chapter outline

Environmentally Sustainable Corrosion Inhibitors. DOI: https://doi.org/10.1016/B978-0-323-85405-4.00019-7

Abbreviations

ASTM	American Society for Testing and Materials
C_{dl}	Double layer capacitance
CR	Corrosion rate
E_{corr}	Corrosion potential
i_{corr}	Corrosion current density
ISO	International Standard Organization
IUPAC	International Union of Pure and Applied Chemistry
NACE	National Association of Corrosion Engineers
OCP	Open circuit potential
PDP	Potentiodynamic polarization
R_s	Solution resistance
R_{ct}	Charge transfer resistance
WL	Weight loss
Z_{IM}	Imaginary impedance
Z_{RE}	Real impedance

2.1 Introduction

Metals and alloys, such as steel, aluminum, copper, and zinc, have gained wide contemporary applications as structural materials in the industrial world [1,2]. The choice of interest in these metals is because of their ready availability in nature, high thermal conductivity, mechanical stability, and relative stability in commonly encountered environments [3]. However, most of these metals with the exception of silver, gold, and platinum are prone to corrosion because they are unstable in the atmosphere and always tend to return to their original states (ores). The term "corrosion" originates from the Latin word "corrodere" which means "gnawing to pieces" [4]. Literally, this definition suggests the eroding, wasting, and destruction of a material. According to reputable corrosion experts, corrosion is an electrochemical reaction occurring between a material or metal and its environment. The reaction is a phenomenon that is irreversible and leads to deterioration of the metal or material [5−14]. Environments such as moisture in air, acids, bases, salts, water, and aggressive polishes have been found to cause deterioration of material. In most cases, corrosion is often used in reference to metals and alloys, however, the phenomenon extends to all natural and synthetic materials such as polymers, concrete, ceramics, leather, plastics, wood, and paper [12,15]. Table 2−1 shows definitions of corrosion as proposed by some reputable scientific organizations.

This chapter deals with a brief historical development of corrosion and the impact of corrosion. A closer attention is paid to the classification and forms of corrosion, the chemistry behind electrochemical corrosion of common industrial metals, and the kinetics and thermodynamics of the corrosion phenomena. A brief account of popular corrosion protection methods and conventional monitoring techniques is also provided.

Table 2–1 Definitions of corrosion by scientific organizations.

Organization	Definition	References
National Association of Corrosion Engineers (NACE)/American Society for Testing and Materials (ASTM International)	The chemical or electrochemical reaction between a material, usually a metal, and its environment that produces a deterioration of the material and its properties	[4]
International Organization for Standardization (ISO)	Physicochemical interaction between a metal and its environment that results in changes in the properties of the metal, and which may lead to significant impairment of the function of the metal, the environment, or the technical system, of which these form a part	[16]
International Union of Pure and Applied Chemistry (IUPAC)	An irreversible interfacial reaction of a material (metal, ceramic, polymer) with its environment which results in consumption of the material or in dissolution into the material of a component of the environment	[17]

2.2 Historical developments of corrosion and its modern relevance

The development of the study of corrosion can be traced to notable works of several ancient philosophers, scientists, and writers. As far back as the 5th century BCE, Herodotus proposed the protection of iron with tin. Pliny the elder, a Roman philosopher who lived between CE 23 and 79 reported the deterioration of iron metal in his essay titled "*Ferrum Corrumpitur*." In ancient times, the Greeks were known to construct lead-protected ship decks using lead-coated copper nails while the Romans protected iron from rusting using materials such as tar and bitumen. The two published papers of Robert Boyle (1627–91) gave early insight into the causes and mechanism of corrosion. In 1819, Thenard discovered the electrochemical nature of the corrosion process. A decade later, Hall established that iron metal requires oxygen to rust and Davy discovered the principle of cathodic protection. Years later, Michael Faraday made more contributions that proved beneficial when he discovered a quantitative relation between electric current and chemical reaction. The laws of electrolysis he developed became the basis to calculate corrosion rate (CR) [10,11,18]. Over the years, several other scientists, research institutions and scientific bodies continued to make significant contributions to the field of corrosion science, engineering, and technology, which has helped to shape the understanding of corrosion and its control.

While corrosion has been a historical subject, it continues to gain much interest in the modern world because recent advancements in science and technology have made the reliance on metals indispensable [19]. These metals are known to undergo reverse extractive metallurgy, where, in order to attain stability or return to their oxidized form, they break into their constituent atoms and release energy. This process, otherwise known as corrosion, poses a huge environmental, engineering and economic problem in all nations of the world.

It is a destructive phenomenon that affects every facet of human life. The impact of corrosion on human life can be related to the adverse effect of natural disasters on human life and society. It is practically impossible to absolutely avoid the occurrence of corrosion just like natural disasters such as floods, earthquakes, and cyclones with their consequent economic and environmental implications on man. The deterioration of metal continues to result in damage to basic home appliances and utensils, contamination of drinking water systems in the neighborhood, and low industrial productivity due to failures in plant infrastructure. Metal-based industries such as petroleum production and refining, automobile, marine, medical, chemical processing, and construction industries continue to invest large sums annually to combat the negative consequences of corrosion. There have also been reported cases of the expulsion of gases from leaking containers, pipes and storage tanks, bridge collapses, and other forms of industrial accidents due to corrosion which has resulted in pollution, injuries, and death [8]. Fig. 2−1 shows the damage caused by corrosion on the gates of the dam on the Colorado River, Mexico and on the body work of a car. These direct and indirect costs incurred as a result of corrosion in industries and nations of the world are staggering [20,21]. A study conducted in 2013 by NACE on the economic impact of corrosion revealed that the cost of corrosion globally is about 3.4% of the global gross domestic product, which is equivalent to US$ 2.5 trillion [22].

2.3 Classification of corrosion

Corrosion can be classified based on these three factors [4,5]:

1. Nature of the corroding environment: based on this classification, corrosion can be said to **be** dry or wet. In dry corrosion, the deterioration of metals occurs under nonaqueous conditions, such as high temperatures in gaseous media, molten metals, and salts, while wet corrosion is a fast process involving an aqueous medium such as water/rain, moisture, or wet soil.
2. Mechanism of corrosion: based on the mechanism, metallic corrosion can either be classified as a chemical reaction where metals react directly with the corrosive environment, or an electrochemical reaction which occurs between a metal and an ion-conducting (electrolytic) medium.
3. Appearance of the corrodent: corrosion could either be broadly classified into uniform (general) or localized corrosion based on the nature of attack. In the case of general corrosion, deterioration of the entire exposed surface of the metallic material occurs, while in localized corrosion there is occurrence of uneven deterioration.

The basic forms of corrosion that have been identified by several experts include uniform/general, galvanic, crevice, filiform, pitting, intergranular, exfoliation, erosion, concentration cell corrosion, stress corrosion cracking, dealloying, fretting, corrosion fatigue, and microbial corrosion. Table 2−2 gives a summary of these common forms of corrosion found in metals. Fig. 2−2 depicts a pictorial representation of these forms of corrosion.

A

B

FIGURE 2–1 Corrosion effects on (A) the gates of the dam on the Colorado River, Mexico and (B) the body work of a car [15].

2.4 Mechanism of corrosion

When a piece of metal is exposed to an aqueous environment (electrolyte), electrochemical corrosion occurs, which involves two simultaneous half-cell reactions; an oxidation or metal dissolution at the anode and a reduction or hydrogen evolution at the

Table 2–2 Common forms of corrosion [4,5,10,18].

Forms	Mechanism
Uniform	The most common form of corrosion. It affects the exposed areas of the metallic material
Galvanic	A bimetallic corrosion occurring due to electron transfer between two different metals in an electrolyte. The more reactive metal often serves as the anode while the less active metal acts as the cathode
Crevice	Metallic deterioration resulting from deposits of corrosive fluid in between metals gaps and holes
Filiform	Corrosion caused by the breach of water through pores on coated/painted surfaces
Pitting	An extreme localized attack resulting in pits (holes) in metals, especially aluminum
Intergranular	Degradation of metals and alloys due to potential difference between the metal grain and its boundary
Exfoliation	A form of intergranular corrosion also called lamellar or layer corrosion. It is common in high strength aluminum alloys. It occurs when the grains are flattened by heavy distortion during the rolling process
Erosion	Wear effects due to relative motion between the metallic surface and the electrolytic medium
Concentration cell corrosion	Deterioration due to contact of two or more metals with different amounts of the same solution
Stress corrosion cracking	A severe form of attack that leads to fracture jointly caused by mechanical stress and a corrosive solution
Dealloying	Selective leaching or dissolution of one element from a solid alloy via corrosion
Fretting	A combination of corrosion and wear/friction
Hydrogen embrittlement	Corrosion caused by hydrogen atoms formed on the metallic surface in a corrosive solution
Corrosion fatigue	Deterioration caused by the joint effect of cyclic stress and corrosive solution
Microbial corrosion	Corrosion induced by microorganisms such as bacteria, fungi, and algae

cathode. The electrolyte, anode, and cathode form the three fundamental elements required for electrochemical corrosion to occur [24]. The electrolyte is an electrically conductive solution, the anode is the site where the metal experiences corrosion, while the cathode is either a part of the same metal surface or of another one placed in the electrolyte.

At the anodic half-cell reaction site, there is dissolution of the metal electrode as it passes into the conducting solution as positively charged ions, producing electrons which are used up at the cathodic site. Thus, the corrosion current between the anodic and cathodic sites comprises electrons' movement within the metal and ionic transfer within the electrolyte [25]. These half-cell reactions are believed to occur at microscopic anodes and cathodes covering a corroding surface which results in damage with time. The processes occurring at the anodic and cathodic site during electrochemical corrosion are illustrated using the reactions below [26,27]:

Anodic reaction:

$$M_{(s)} \rightarrow M^{n+}_{(aq)} + n\,e^{-} \tag{2.1}$$

Group I: Identifiable by visual inspection

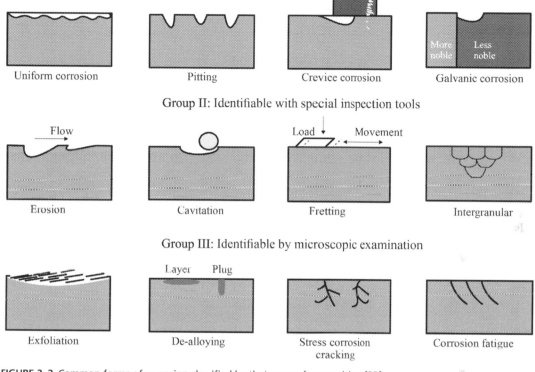

| Uniform corrosion | Pitting | Crevice corrosion | Galvanic corrosion |

Group II: Identifiable with special inspection tools

| Erosion | Cavitation | Fretting | Intergranular |

Group III: Identifiable by microscopic examination

| Exfoliation | De-alloying | Stress corrosion cracking | Corrosion fatigue |

FIGURE 2–2 Common forms of corrosion classified by their ease of recognition [23].

Cathodic reaction:
(A) Oxygen reduction (acid solution)

$$O_{2(g)} + 4H^+ + 4\,e^- \rightarrow 2H_2O \tag{2.2}$$

(B) Oxygen reduction (neutral or basic solution)

$$O_{2(g)} + H_2O_{(l)} + 2\,e^- \rightarrow 2(OH)^- \tag{2.3}$$

(C) Hydrogen evolution (acid solution)

$$2\,H^+_{(aq)} + 2\,e^- \rightarrow H_{2(g)} \tag{2.4}$$

(D) Metal ion reduction: metal ions present in the electrolytic solution may be reduced:

$$M^{n+} + e^- \rightarrow M^{(n-1)+} \tag{2.5}$$

Metal ion reduction occurs only if M^{n+} ions is in high concentration. Here, the M^{n+} ions decreases its valence state by accepting an electron.

(E) Metal deposition: metal may be reduced from its ionic form to a neutral metallic form:

$$M^{n+} + e^- \rightarrow M_{(s)} \qquad (2.6)$$

2.4.1 Corrosion of steel

Carbon steels are ferrous metals that contain a small percentage of carbon by weight (maximum 2.0%) that come in variants of low-carbon (mild) steel, plain (medium) carbon steel, and high-carbon steel. These classification of carbon steels are often based on the percentage by weight of carbon in the steel which ranges from 0.05% to 2.0% [28]. The major component of steel is iron with other alloying components such as phosphorus, chromium, nickel, molybdenum, vanadium, silicon, cobalt, manganese, and sulfur, all in trace quantities [4]. The required properties of the alloy needed in the carbon steel is a major determinant in the percentage composition of the carbon and alloy that will be included. Generally, mild steel is a type of carbon steel that contains low carbon content (0.05%−0.28% C). It is arguably the most commonly used and most versatile engineering material, which has found extensive application in numerous industries [29,30]. It enjoys such preference over other steel types for industrial application because it is inexpensive, has relatively low tensile strength, high mechanical strength, and good malleability and ductility [31,32]. In addition, mild steel has better corrosion-resistant properties in several media when compared to other forms of carbon steels.

Pickling of metals, which is aimed at the total removal of rust or oxide layers formed during various industrial production processes, cleaning of scales of thermal or oil refinery equipment in industrial sectors, and acidization of oil wells to enhance oil productivity, often requires the use of strong acid solutions [33]. Some of the mineral acids often employed for these industrial processes are HF, HCl, H_2SO_4, and HNO_3, which subject the metallic materials and products to severe corrosion attack to varying degrees. The use of HCl is highly favored because of its economical waste liquor recovery as compared to other mineral acids. The major advantage of HCl in these industrial processes is its ease of formation of residual metal chlorides which are readily soluble in aqueous media unlike nitrates, phosphates, and sulfates. These soluble metal chlorides result in the least polarizing effect without impeding the CR [30,34,35]. Mild steel, which essentially comprises iron, experiences deterioration when exposed to environmental conditions such as acid, air, water, and high temperatures and pressures. Generally, the reaction of iron with oxygen leads to the formation of iron oxide ($Fe_2O_3.xH_2O$), also known as rust, where x represents the amount of water present and defines the color of the rust. Fig. 2−3 illustrates the electrochemical process of rusting with an example in nature and the corrosion reactions of iron under different conditions.

2.4.2 Corrosion of aluminum

Aluminum, a silvery white material which represents about 8% of the total mineral component, is the third most abundant metal in the Earth's crust [38,39]. It is a metal of choice in many metal-based industries such as food packaging, pharmaceuticals, electronics, power, and transport industries because of its ready availability, light weight, and high electrical and thermal conductivity. Often, different elements are added to pure aluminum to form alloys in a bid to improve its

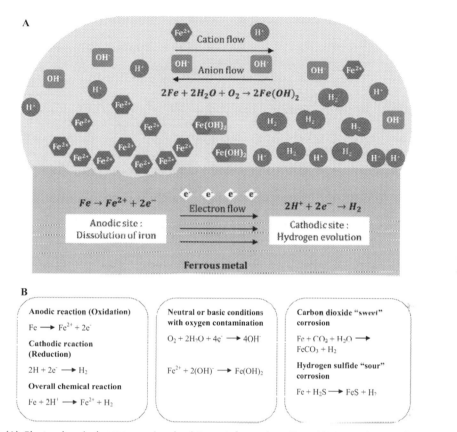

FIGURE 2–3 (A) Electrochemical processes involved in rust formation. From N.I.N. Haris, S. Sobri, Y.A. Yusof, N.K. Kassim, An overview of molecular dynamic simulation for corrosion inhibition of ferrous metals, Metals, 11 (2021) 46 and (B) corrosion of iron under different conditions. *From B.D.B. Tiu, R.C. Advincula, Polymeric corrosion inhibitors for the oil and gas industry: Design principles and mechanism, React. Funct. Polym. 95 (2015) 25–45.*

mechanical properties and corrosion resistance. The corrosion resistance of the aluminum alloy depends on the level of purity of the alloy, which makes it necessary to determine the quantity of the components present in the alloy. From literature, aluminum is a better corrosion-resistant metal than mild steel because it forms a protective oxide layer on its surface when exposed to the atmosphere which checks further oxidation. The thickness of this oxide layer depends on the temperature, corrosive environment and the chemical composition of the alloy. However, in the presence of strong acidic, alkaline and chloride-containing solutions (pH range of 4.5–8.5) [40], aluminum suffers corrosion. The most severe form of attack suffered by aluminum is pitting while other common ones are intergranular, exfoliation, galvanic, and stress-corrosion cracking. An increase in temperature also leads to rapid corrosion of aluminum. Fig. 2–4 is an illustration of pitting corrosion in aluminum.

The following chemical reactions depict the corrosion of aluminum:

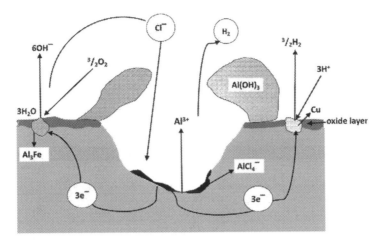

FIGURE 2-4 Pitting corrosion of an aluminum alloy [41].

According to the published reports of Pyun and Moon [42,43] in which the electrochemical behavior of aluminum in an alkaline solution was studied, it was found that at the anode, aluminum dissolves first to form hydroxide film, which is then chemically dissolved by the attack of hydroxide ions. The reaction leads to the formation of soluble aluminate ions.

Anodic reaction:

$$Al + 3OH^- \rightarrow Al(OH)_3 + 3e^- \tag{2.7}$$

$$Al(OH)_3 + OH^- \rightarrow Al(OH)_4 \tag{2.8}$$

The combination of the above equations gives the partial anodic dissolution of aluminum

$$Al + 4OH^- \rightarrow Al(OH)_4^- + 3e^- \tag{2.9}$$

Cathodic reaction:

$$\tfrac{3}{4}\,O_2 + 3/2\,H_2O + 3e^- \rightarrow 3OH^- \tag{2.10}$$

$$3H_2O + 3e^- \rightarrow 3/2\,H_2 + 3OH^- \tag{2.11}$$

The summary of the corrosion reaction is obtained by combining the partial anodic dissolution with Eqs. (2.10) and (2.11), giving

$$Al + OH^- + \tfrac{3}{4}\,O_2 + 3/2\,O_2 \rightarrow Al(OH)_4^- \tag{2.12}$$

$$Al + 3H_2O + OH^- \rightarrow 3/2\,H_2 + Al(OH)_4^- \tag{2.13}$$

In an acid chloride solution of pH of 3−4, aluminum oxide layer suffers attack from the chloride ions at its weakest part and the dissolution of aluminum occurs to form aluminum chloride [41].

Anodic reaction:

$$Al_{(s)} \rightarrow Al^{3+}_{(aq)} + 3e^-$$ (2.14)

$$Al^{3+}_{(aq)} + 3H_2O_{(l)} \rightarrow Al(OH)_{3(l)} + 3H^+_{(aq)}$$ (2.15)

Cathodic reaction:

$$AlCl_3 + 3H_2O \rightarrow Al(OH)_3 + 3HCl$$ (2.16)

$$3H^+ + 3e^- \rightarrow 3/2\ H_2$$ (2.17)

$$\frac{1}{2}\ O_2 + H_2O + 2e^- \rightarrow 2OH^-$$ (2.18)

2.4.3 Corrosion of copper

Copper is the fifth most abundant metal in the Earth's crust which is beneficial both in its pure and alloyed form. It is commonly used in industries because of its special features such as good corrosion resistance, thermal stability, high electrical conductivity, and malleability [5]. Copper and its several alloys are used in the electrical/electronic, marine, and chemical industries [4,44]. Just like aluminum, copper possesses some corrosion resistance in the atmosphere as it forms a passive or nonconductive layer on its surface [45]. However, in the presence of strong acidic and alkaline solutions, copper suffers pitting corrosion which leads to the loss of its essential properties.

The most common acid used in the pickling process in industry is HCl. Under this condition, copper is susceptible to anodic oxidation of the copper metal to cuprous ion which then reacts with Cl^- to precipitate insoluble CuCl and a soluble cuprous chloride complex as shown in the following equations [46,47]:

At the anode:

$$Cu_{(s)} \rightarrow Cu^+_{(aq)} + e^-$$ (2.19)

$$Cu^+_{(aq)} + Cl^- \rightarrow CuCl_{(aq)}$$ (2.20)

$$CuCl_{(aq)} + Cl^-_{(g)} \rightarrow CuCl^-_{2(aq)}$$ (2.21)

It is however possible for the dissolved complex to oxidize to cupric ions as shown below:

$$CuCl^-_{2(aq)} \rightarrow Cu^{2+}_{(aq)} + e^-$$ (2.21)

At the cathode:

$$4H^+_{(aq)} + O_{2(g)} + 4e^- \rightarrow 2H_2O_{(aq)}$$ (2.22)

2.4.4 Corrosion of zinc

Zinc, a nonferrous metal, is the fourth most abundant metal extracted from the Earth's crust. It has extensive applications in industries and is mostly used in zinc galvanizing to shield iron from deteriorating in atmospheric conditions [48]. Zinc is similar to aluminum because it possesses corrosion resistance properties in atmospheric environments, which is evidenced by the formation of an oxide layer on its surface. It is not prone to corrosion like steel because of its ability to form insoluble carbonate films that are firmly attached to its surfaces. As in other related metals, exposure to strong acidic, alkaline, and neutral solutions increases the CR of zinc and makes it less useful for industrial purposes.

Anodic reaction:

$$Zn_{(s)} \rightarrow Zn_{(aq)}^{+2} + 2e^- \tag{2.23}$$

Cathodic reaction:

$$2\,H_{(aq)}^+ + 2\,e^- \rightarrow H_{2(g)} \tag{2.24}$$

An overall reaction of zinc corrosion gives

$$Zn_{(s)} + 2HCl_{(aq)} \rightarrow ZnCl_{(aq)} + H_{2(g)} \tag{2.25}$$

2.5 Kinetics and thermodynamics of corrosion

The corrosion process generally leads to loss of essential properties in metals such as malleability, ductility, strength, and weight. The degree of corrosion occurrence can be measured as a function of weight loss (WL). The nature of the material and the environmental conditions are two key factors that determine CR. These main factors can be subdivided into the position of the metal in the galvanic series, nature of the metal, purity of the metal, relative area of the anode and cathode, nature of the surface film, atmospheric impurity, temperature, nature of the environment, concentration of the electrolyte, solution pH, flow velocity, etc. [6].

Insight into how metals behave in corrosive environment can be gained by investigating the influence of temperature on the rate of corrosion. The Arrhenius law provides an understanding on the relationship between the rate of corrosion constant and the temperature. Arrhenius postulates the following equation:

$$k = Ae^{-E_a/RT} \tag{2.26}$$

where k represents the rate constant, A is the preexponential factor or the prefactor, E_a is the activation energy, R is universal gas constant, and T denotes the absolute temperature. This equation suggests that the CR will increase as the temperature increases and E_a and A may vary with temperature.

The knowledge of thermodynamics plays an important role in understanding the corrosion behavior of metallic materials. While the rate of reaction cannot be determined using thermodynamics, it is possible to ascertain the spontaneity of a reaction which is advantageous in determining the theoretical feasibility of a corrosion process. Most industrial metals exist in an unstable thermodynamic state. According to Fontana [5], this instability will lead to a tendency for the metals to return to a natural state through the process of corrosion.

Studies have shown that metallic corrosion is a spontaneous reaction. A reaction is said to be spontaneous if its free energy change, ΔG, has a negative value, which indicates that energy is being released. The magnitude and sign of ΔG are both significant as they give insight into the feasibility of the corrosion taking place. For an electrochemical reaction such as corrosion, the overall free energy is expected to be negative. The standard free energy of reaction under standard conditions is represented by the following equation:

$$\Delta G^0 = -nF\Delta E^0 \tag{2.27}$$

where ΔG^0 represents the standard free energy change, n is the number of electrons transferred, F denotes the Faraday constant, and ΔE^0 is the standard energy change in the reaction.

Under nonstandard conditions, the electrode potential is related to the standard electrode potential using the Nernst equation:

$$E = E^0 + \left(\frac{RT}{nF}\right)\ln\left(\frac{a_{oxi}}{a_{red}}\right) \tag{2.28}$$

where a_{oxi} and a_{red} represent the activities of the oxidized and reduced species, respectively.

Thermodynamic consideration of electrochemical corrosion led to the development of the potential pH diagram also known as the Pourbaix diagram which characterizes the stability of a metallic material as a function of potential and pH [10]. These diagrams, which are derived from thermodynamic calculations, identify the conditions under which soluble and insoluble reaction products will be formed and a given metal is stable. A Pourbaix diagram for iron showing the region of corrosion, immunity, and passivation is shown in Fig. 2−5.

Other important thermodynamics parameters such as the adsorptive enthalpy, ΔH^0_{ads} can be generated. The Van't Hoff equation shown below can be employed:

$$\ln K = -\left(\frac{\Delta H^0_{ads}}{RT}\right) + \text{Constant} \tag{2.29}$$

where K is the adsorptive constant. The ΔH^0_{ads} obtained from this equation can also be confirmed using the Gibbs−Helmholtz equation.

$$\left[\frac{\partial(\Delta G^0_{ads}/T)}{\partial T}\right] = -\frac{\Delta H^0_{ads}}{T^2} \tag{2.30}$$

Standard entropy, ΔS^0_{ads} is another important thermodynamic parameter that can be obtained using the fundamental thermodynamic equation as follows:

$$\Delta G^0_{ads} = \Delta H^0_{ads} - T\Delta S^0_{ads} \tag{2.31}$$

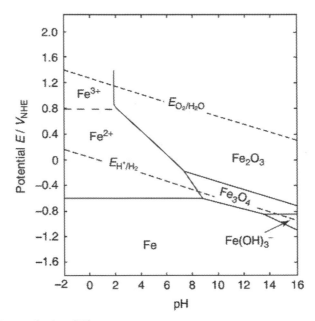

FIGURE 2–5 Pourbaix diagram for iron [49].

The equilibrium constant (K_{eq}) for the electrochemical reaction can be computed using the following equation:

$$RT \ln K_{eq} = -\Delta G^0 - nF\Delta E^0 \qquad (2.32)$$

2.6 Corrosion prevention methods: A brief view

Research efforts to mitigate the various forms of metallic corrosion have identified and adopted different corrosion control methods. The choice of control technique is dependent on different factors such as the form of corrosion, nature of metal, and nature of corrosive environment. The several possible methods of control involve either altering the material/metal, procedure, or the corrosive medium. It is noteworthy that there is no consensus among corrosion experts as to the best method of corrosion control for all forms of corrosion; therefore it is possible to employ one or more of these methods in a corrosion controlling system having put all necessary factors into consideration. The five popular techniques in controlling corrosion include [4,5]:

1. Proper design: this could also be referred to as excellent equipment design. This involves adopting a design of a material or system such that the metal escapes crevices and cracking. In this case, rational design principles are applied which can mitigate

metal deterioration and minimize the cost and time often related with corrosion management.

2. Material selection: this refers to the selective use of corrosion-resistant materials such as platinum, stainless steel, and special alloys to prolong the life span of the structure. The selected material is expected to possess high corrosion resistance, excellent mechanical strength, and be inexpensive. Efforts are being made to design new alloys and materials such as hastelloy nickel alloys and nirosta steels for corrosion prevention. However, several factors, such as operating conditions, environmental conditions, and economic consideration, are major setbacks in material selection.

3. Coatings: these are the most widely used method of controlling corrosion which entail the use of relatively inactive materials to protect a metal surface. Coatings are most commonly used in protecting metals from dissolution by providing a barrier between the material and the corrosive environment and often serve esthetic purposes as well. The different types of coating include greasing, electroplating, galvanizing, painting, and metallic coating. However, they are limited and are particularly useful when used with other corrosion control methodologies.

4. Cathodic and anodic protection: this method involves the alteration of the potential of the metal to be protected either by using direct current or galvanic current. In cathodic protection, the negative potential is shifted by impressing current to a region where the metal of interest is completely stable or by connection to a sacrificial anode from highly reactive metals. It is known to offer adequate protection to marine steel structures and underground oil and gas pipes or tanks. Anodic protection involves the formation of a passive film on a metallic surface by the application of external positive potential. As elegant as this form of protection is, it could be very expensive.

5. Corrosion inhibitors: these are chemical substances added in little amounts to the aggressive media to reduce the rate of corrosion thereby offering surface protection to the metal. They often impede the rate of corrosion by altering the environment, passivating, precipitating or adsorbing on the metal surface. It is a popular method of corrosion control in many industries. Over the years, much attention has been devoted to the study of corrosion inhibitors which is supported by a large repository of literature. The focus on corrosion inhibitors is because they offer practical, economic, and effective control of corrosion of metals in different corrosive environments and in a wide temperature range. Several classes of organic compounds, such as triazoles, imidazoles, quinoxalines, pyridines, and pyrimidines, and inorganic compounds, such as chromates, phosphates, and nitrites, have been explored as potential corrosion inhibitors for different industrial metal [50−53].

2.7 Corrosion monitoring methods: A brief account

Corrosion monitoring involves the adoption of quantitative methods to evaluate the effectiveness of corrosion control techniques and provide feedback needed to optimize corrosion control. Several methods exist to measure or monitor corrosion in the laboratory and industries which are broadly divided into direct (WL, hydrogen evolution and temperature variation methods) and indirect techniques [open circuit potential (OCP), linear polarization, potentiodynamic polarization (PDP), electrochemical impedance spectroscopy (EIS), cyclic voltammetry, and electrochemical frequency modulation]. The major corrosion monitoring techniques that are often reported in corrosion literature are discussed below:

1. WL: this is known to be the simplest, most reliable, and traditional technique of determining the performance efficiency of chemical inhibitors [53]. The procedure involves exposure of a polished metal sample to a corrosive medium for a specified duration, which is then removed for analysis. Parameters that are determined from this method include the WL expressed as CR, surface coverage, and percentage inhibition efficiency. This technique has the advantages of low cost, applicability to all corrosive media, visual inspection of metal specimen, and easy identification of localized corrosion. However, it gives less accurate results and is applicable only for low-conductivity solutions [54]. The CR can be expressed as the rate of WL measured in milligrams per square decimeter per day (mdd) or rate of penetration/metal thickness lost in metric units, millimeters per year (mmpy) or in American units, mils per year (mpy) [55].

$$CR(\text{mdd}) = \left(\frac{\Delta w}{At}\right) \tag{2.33}$$

where CR is the corrosion rate (g cm^{-2} h^{-1}), Δw denotes the change in WL (g), A is the total surface area of the metallic material (cm^2), and t is the immersion time (h).

$$CR\left(\text{mm year}^{-1}\right) = 8.76 \times 10^3 \left(\frac{\Delta w}{A\rho t}\right) \tag{2.34}$$

where CR is the corrosion rate (mmpy) and ρ is density of the metal (g cm^{-3}).

$$CR\ (\text{mpy}) = \frac{534\Delta w}{A\rho t} \tag{2.35}$$

where CR is the corrosion rate (mpy) and ρ is density of the metal (g cm^3).

2. PDP: PDP is a technique used to determine the CR and mechanism of corrosion inhibition using modern electrochemical-based instruments. Polarization measurements are often carried out in a conventional, three-electrode system comprising the reference electrode (saturated calomel electrode), working electrode (the metal sample), and the counter/auxiliary (platinum electrode). All three electrodes are usually immersed in the corrosive medium and the working electrode is varied over a potential range at a suitable scan rate relative to the OCP. After attaining the OCP, which is the difference in potential of the microcells, the external potential from a direct source is imposed on the system to monitor the current response at the anode and cathode simultaneously. An extrapolation of the linear portions of the anodic and cathodic branches to their point of

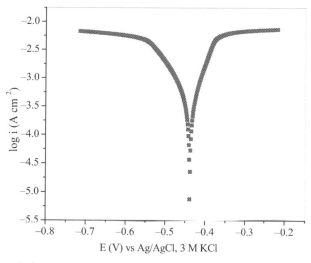

FIGURE 2–6 A typical potentiodynamic polarization plot.

intersection gives the corrosion potential, E_{corr}. The corrosion current density (i_{corr}), a critical electrochemical parameter is the current obtained at the point of the anodic and cathodic slope intersection. A typical PDP plot is shown is Fig. 2–6 [24]. The obtained i_{corr} can be used to determine the CR as shown in the following equation:

$$CR \text{ (mmpy)} = \frac{i_{corr} \times K \times EW}{\rho \times A} \qquad (2.36)$$

where CR is the corrosion rate (mmpy), K is the conversion factor of 3.27×10^{-3} mm g μA^{-1} cm^{-1} year^{-1} and EW is equivalent weight (g).

3. EIS: this is a sophisticated electrochemical method that provides insight into the physical processes occurring at the metal−electrolyte interface [56,57]. This technique which is also conducted in a conventional three-electrode system involves applying a small external potential (5−10 mV) to the system after attaining equilibrium to obtain response in the form of alternating current or voltage over a range of frequencies (100 kHz−10 mHz). Usually, a measured quantity, impedance is described by real, Z_{RE}, and imaginary impedance, Z_{IM}, which is presented as a semicircle on a plot of Z_{IM} versus Z_{RE}. This Nyquist plot shown provides information on the mechanism of adsorption of the organic compounds. The analyzed experimental data of the studied system is fitted using a simple equivalent electrical circuit (Randle circuit) as shown in Fig. 2−7. The circuit comprises a parallel connection of the charge transfer resistance (R_{ct}) and double-layer capacitance (C_{dl}) at the metal/electrolyte interface and is in series connection with the solution resistance (R_s). The double-layer capacitance is replaced with a constant phase element in practice.

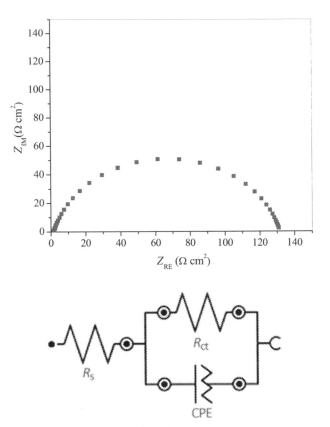

FIGURE 2-7 A typical Nyquist plot and Randle equivalent circuit.

2.8 Conclusions

From the above discussions, it can be concluded that corrosion is a natural and undesirable occurrence that has gained the attention of experts in academia, industry, and the governments of the nations of the world because of its adverse effect on man and the society. Corrosion is such a complex process that occurs in various forms and an appropriate understanding of corrosion chemistry of industrial metals is vital to the development of science for corrosion control. Studies on the electrochemical behavior of popular industrial metals, such as iron (steel), aluminum, copper, and zinc, in different corrosive environments to better understand their corrosion mechanism, kinetics and thermodynamics need to be further stimulated. The popular corrosion control techniques discussed need to be further explored and exploited with new technologies developed in a bid to control the rate of corrosion using several conventional corrosion monitoring techniques.

Acknowledgments

E.D.A. acknowledges the North-West University for providing financial support under the Postdoctoral Fellowship Scheme.

References

[1] M.A. Benvenuto, Metals and Alloys: Industrial Applications, Walter de Gruyter GmbH & Co KG, 2016.

[2] C.P. Sharma, Engineering Materials: Properties and Applications of Metals and Alloys, PHI Learning Pvt. Ltd., 2003.

[3] W.F. Smith, J. Hashemi, F. Presuel-Moreno, Foundations of Materials Science and Engineering, McGraw-Hill Publishing, 2006.

[4] J.R. Davis, Corrosion: Understanding the Basics, ASM International, 2000.

[5] M.G. Fontana, N.D. Greene, Corrosion Engineering, McGraw-Hill, 2018.

[6] R.W. Revie, Corrosion and Corrosion Control: An Introduction to Corrosion Science and Engineering, John Wiley & Sons, 2008.

[7] P.R. Roberge, Handbook of Corrosion Engineering, McGraw-Hill, 2000.

[8] E. McCafferty, Introduction to Corrosion Science, Springer Science & Business Media, 2010.

[9] L.L. Sheir, R.A. Jarman, G.T. Burstein, Corrosion: Metal/Environment Reactions, 8, Butterworths, London, 1994, pp. 3–8.

[10] Z. Ahmad, Principles of Corrosion Engineering and Corrosion Control, Elsevier, 2006.

[11] V.S. Sastri, Green Corrosion Inhibitors: Theory and Practice, John Wiley & Sons, 2012.

[12] E.E. Stansbury, R.A. Buchanan, Fundamentals of Electrochemical Corrosion, ASM International, 2000.

[13] S.A. Bradford, J.E. Bringas, Corrosion Control, Springer, 1993.

[14] D.E. Talbot, J.D. Talbot, Corrosion Science and Technology, CRC Press, 2018.

[15] B. Valdez, M. Schorr, R. Zlatev, M. Carrillo, M. Stoytcheva, L. Alvarez, et al., Corrosion control in industry, Environment and Industrial Corrosion, Practical and Theoretical Aspects, Books on Demand, 2012.

[16] ISO 8044, Corrosion of Metals and Alloys: Basic Terms and Definitions, European Committee for Standardization (CEN), Brussels, Belgium, 1999.

[17] K.E. Heusler, D. Landolt, S. Trasatti, Electrochemical corrosion nomenclature (Recommendations 1988), Electrochim. Acta 61 (1989) 19–22.

[18] E. Ghali, V.S. Sastri, M. Elboujdaini, Corrosion Prevention and Protection: Practical Solutions, John Wiley & Sons, 2007.

[19] W. Fürbeth, M. Schütze, Progress in corrosion protection as a requirement for technical progress, Mater. Corros. 60 (2009) 481–494.

[20] L.T. Popoola, A.S. Grema, G.K. Latinwo, B. Gutti, A.S. Balogun, Corrosion problems during oil and gas production and its mitigation, Int. J. Ind. Chem. 4 (2013) 35.

[21] M. Goyal, S. Kumar, I. Bahadur, C. Verma, E.E. Ebenso, Organic corrosion inhibitors for industrial cleaning of ferrous and non-ferrous metals in acidic solutions: a review, J. Mol. Liq. 256 (2018) 565–573.

[22] G. Koch, J. Varney, N. Thompson, O. Moghissi, M. Gould, J. Payer, International Measures of Prevention, Application, and Economics of Corrosion Technologies Study, NACE International, 2016, p. 216.

[23] R.A.M. Anaee, M.H Abdulmajeed, Tribocorrosion, Advances in Tribology, IntechOpen, 2016.

[24] A. Berradja, Electrochemical techniques for corrosion and tribocorrosion monitoring: methods for the assessment of corrosion rates, Corrosion Inhibitors, IntechOpen, 2019.

[25] A. Berradja, Electrochemical techniques for corrosion and tribocorrosion monitoring: fundamentals of electrolytic corrosion, Corrosion Inhibitors, IntechOpen, 2019.

[26] S. Abd El-Maksoud, The effect of organic compounds on the electrochemical behaviour of steel in acidic media: a review, Int. J. Electrochem. Sci. 3 (2008) 528–555.

[27] B.E. Brycki, I.H. Kowalczyk, A. Szulc, O. Kaczerewska, M. Pakiet, Organic corrosion inhibitors, Corrosion Inhibitors, Principles and Recent Applications, InTech Open, 2017.

[28] E. Osarolube, I.O. Owate, N.C. Oforka, Corrosion behaviour of mild and high carbon steels in various acidic media, Sci. Res. Essay 3 (2008) 224−228.

[29] S.R. Gupta, P. Mourya, M. Singh, V.P. Singh, Structural, theoretical and corrosion inhibition studies on some transition metal complexes derived from heterocyclic system, J. Mol. Struct. 1137 (2017) 240−252.

[30] M. Finšgar, J. Jackson, Application of corrosion inhibitors for steels in acidic media for the oil and gas industry: a review, Corros. Sci. 86 (2014) 17−41.

[31] D.K. Singh, S. Kumar, G. Udayabhanu, R.P. John, 4 (N,N-dimethylamino) benzaldehyde nicotinic hydrazone as corrosion inhibitor for mild steel in 1 M HCl solution: an experimental and theoretical study, J. Mol. Liq. 216 (2016) 738−746.

[32] I.A. Kartsonakis, P. Stamatogianni, E.K. Karaxi, C.A. Charitidis, Comparative study on the corrosion inhibitive effect of 2-mercaptobenzothiazole and Na_2HPO_4 on industrial conveying API 5L X42 pipeline steel, Appl. Sci. 10 (2020) 290.

[33] L.-F. Li, P. Caenen, J.-P. Celis, Effect of hydrochloric acid on pickling of hot-rolled 304 stainless steel in iron chloride-based electrolytes, Corros. Sci. 50 (2008) 804−810.

[34] P. Preethi Kumari, P. Shetty, S.A. Rao, Electrochemical measurements for the corrosion inhibition of mild steel in 1M hydrochloric acid by using an aromatic hydrazide derivative, Arab. J. Chem. 10 (2017) 653−663.

[35] C. Verma, E.E. Ebenso, I. Bahadur, M.A. Quraishi, An overview on plant extracts as environmental sustainable and green corrosion inhibitors for metals and alloys in aggressive corrosive media, J. Mol. Liq. 266 (2018) 577−590.

[36] N.I.N. Haris, S. Sobri, Y.A. Yusof, N.K. Kassim. An overview of molecular dynamic simulation for corrosion inhibition of ferrous metals, Metals, 11 (2021) 46.

[37] B.D.B. Tiu, R.C. Advincula, Polymeric corrosion inhibitors for the oil and gas industry: Design principles and mechanism, React. Funct. Polym. 95 (2015) 25−45.

[38] O.S.I. Fayomi, P.A.L. Anawe, A.J. Daniyan, The Impact of Drugs as Corrosion Inhibitors on Aluminum Alloy in Coastal-Acidified Medium, InTech Open, 2018, p. 79.

[39] K. Xhanari, M. Finšgar, M.K. Hrnčič, U. Maver, Ž. Knez, B. Seiti, Green corrosion inhibitors for aluminium and its alloys: a review, RSC Adv. 7 (2017) 27299−27330.

[40] A. Mak, Corrosion of steel, aluminum and copper in electrical applications, General Cable, 2015.

[41] K. Xhanari, M. Finšgar, Organic corrosion inhibitors for aluminum and its alloys in chloride and alkaline solutions: a review, Arab. J. Chem. 12 (2019) 4646−4663.

[42] S.-I. Pyun, S.-M. Moon, Corrosion mechanism of pure aluminium in aqueous alkaline solution, J. Solid State Electrochem. 4 (2000) 267−272.

[43] S.-M. Moon, S. Pyun, The formation and dissolution of anodic oxide films on pure aluminium in alkaline solution, Electrochim. Acta 44 (1999) 2445−2454.

[44] M.A. Amin, K.F. Khaled, Copper corrosion inhibition in O_2-saturated H_2SO_4 solutions, Corros. Sci. 52 (2010) 1194−1204.

[45] T. Suter, E. Moser, H. Böhni, The characterization of the tarnishing of Cu-15Ni-8Sn and Cu-5Al-5Sn alloys, Corros. Sci. 34 (1993) 1111−1122.

[46] E.-S.M. Sherif, Corrosion behavior of copper in 0.50 M hydrochloric acid pickling solutions and its inhibition by 3-amino-1,2,4-triazole and 3-amino-5-mercapto-1,2,4-triazole, Int. J. Electrochem. Sci. 7 (2012) 1884−1897.

[47] A. Fateh, M. Aliofkhazraei, A.R. Rezvanian, Review of corrosive environments for copper and its corrosion inhibitors, Arab. J. Chem. 13 (2017) 481−544.

[48] I. Odnevall Wallinder, C. Leygraf, A critical review on corrosion and runoff from zinc and zinc-based alloys in atmospheric environments, Corrosion 73 (2017) 1060−1077.

[49] N. Sato, Basics of Corrosion Chemistry, Green Corrosion Chemistry and Engineering: Opportunities and Challenges, Wiley-VCH, 2012.

[50] L.O. Olasunkanmi, E.E. Ebenso, Experimental and computational studies on propanone derivatives of quinoxalin-6-yl-4, 5-dihydropyrazole as inhibitors of mild steel corrosion in hydrochloric acid, J. Colloid Interface Sci. 561 (2020) 104–116.

[51] C. Verma, L.O. Olasunkanmi, E.E. Ebenso, M. Quraishi, Adsorption characteristics of green 5-arylaminomethylene pyrimidine-2,4,6-triones on mild steel surface in acidic medium: experimental and computational approach, Results Phys. 8 (2018) 657–670.

[52] C. Verma, L.O. Olasunkanmi, T.W. Quadri, E.-S.M. Sherif, E.E. Ebenso, Gravimetric, electrochemical, surface morphology, DFT, and Monte Carlo simulation studies on three N-substituted 2-aminopyridine derivatives as corrosion inhibitors of mild steel in acidic medium, J. Phys. Chem. C 122 (2018) 11870–11882.

[53] L.K.M.O. Goni, M.A. Mazumder, Green Corrosion Inhibitors, Corrosion Inhibitors, IntechOpen, 2019.

[54] M.A. Quraishi, D.S. Chauhan, V.S. Saji, Heterocyclic Organic Corrosion Inhibitors: Principles and Applications, Elsevier, 2020.

[55] G. Palanisamy, Corrosion inhibitors, Corrosion Inhibitors, InTech Open, 2019.

[56] T.W. Quadri, L.O. Olasunkanmi, O.E. Fayemi, M.M. Solomon, E.E. Ebenso, Zinc oxide nanocomposites of selected polymers: synthesis, characterization, and corrosion inhibition studies on mild steel in HCl solution, ACS Omega 2 (2017) 8421–8437.

[57] L.O. Olasunkanmi, I.B. Obot, M.M. Kabanda, E.E. Ebenso, Some quinoxalin-6-yl derivatives as corrosion inhibitors for mild steel in hydrochloric acid: experimental and theoretical studies, J. Phys. Chem. C 119 (2015) 16004–16019.

3

Corrosion inhibitors: an introduction

Saman Zehra, Mohammad Mobin, Ruby Aslam

CORROSION RESEARCH LABORATORY, DEPARTMENT OF APPLIED CHEMISTRY, FACULTY OF ENGINEERING AND TECHNOLOGY, ALIGARH MUSLIM UNIVERSITY, ALIGARH, INDIA

Chapter outline

Environmentally Sustainable Corrosion Inhibitors. DOI: https://doi.org/10.1016/B978-0-323-85405-4.00022-7

Abbreviation

VCI   Volatile corrosion inhibitors

3.1 Introduction

In order to control the corrosion of metals, several corrosion control methods are utilized. The methods of corrosion control involve the application of engineering principles and methods (represented by the Fig. 3−1), to impede the corrosion to an acceptable level by the aid of the most economical procedures. Each offers its own complexities and purposes. Generally, the corrosion control procedures involve understanding the involved corrosion mechanism. The objective of each of the corrosion control techniques is to minimize corrosion to an acceptable limit in order to allow the material to attain its normal or desired lifetime.

In a limited number of cases, the corrosion control method is designed to eliminate it completely. One of the most prevalent ways of mitigating internal corrosion is by employing corrosion inhibitors [1,2]. These are the chemicals that impede the rate of the corrosion of the metals. Corrosion inhibitors can be solids, liquids, and gases, and can be used in solid, liquid, and gaseous media. Solid media can be concrete, coal slurries, or organic coatings. Liquids may be water, aqueous solutions, or organic solvents. A gaseous medium is an atmosphere or water vapor. Depending on their dispersibility or solubility in the fluids that need to be inhibited, the chemical to be used as corrosion inhibitors is obtained [2,3].

The adsorption of the inhibitor's molecules onto the metallic surface is the mode of corrosion inhibition by the corrosion inhibitors. The inhibitor molecules form a protective layer or thin film on the surface of the metals that needs to be protected. The film thus formed minimizes the metal surface and the surrounding corrosive environment interaction [4,5]. Chemical substances are shown to be good inhibitors if added in small doses. The various factors which determine the performance of the applicability of various corrosion inhibitors are depicted by Fig. 3−2.

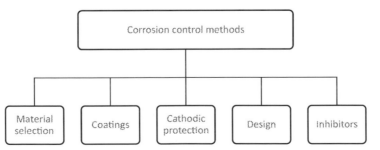

FIGURE 3–1 Various methods of corrosion control.

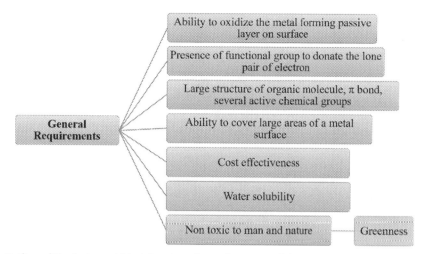

FIGURE 3–2 Outline of the factors which determines the performance of the corrosion inhibitors.

There is a range of other uses of the applicability of corrosion inhibitors like in the acid pickling, gasoline, gas and petroleum industries, water recirculation systems, chemical processing, water treatment, and additive materials industries etc [4,5]. In the oil production, refining, and chemical fields, the first line of protection is the application of corrosion inhibitors. The other application of the corrosion inhibitors is in the antifreeze of the automobiles [6].

A growing interest in area of the research of corrosion inhibitors can be seen by observing the increasing number of published articles in the corrosion literature. Therefore, in order to get the knowledge about the mechanistic aspect and application of the corrosion inhibitors, the main objective of this chapter is to establish the basics and fundamental understanding of corrosion inhibition.

3.2 Classification of the corrosion inhibitors

There are many ways of categorizing corrosion inhibitors. According to a well-known classification from one point of view, corrosion inhibitors adopt two forms [7]:

1. adsorption corrosion inhibitors; and
2. film-forming corrosion inhibitors.

3.2.1 Adsorption corrosion inhibitors

These corrosion inhibitors forms a chemisorptive bond with the metal surface and retards the ongoing electrochemical dissolution reactions. Chemisorption-type inhibitors are most organic inhibitors. Or we can say this is the class of the corrosion inhibitors which exhibit chemisorption and form chemisorptive bonds with the surface of the metals. This in turn retards or stops the ongoing electrochemical dissolution process. Most of the organic inhibitors act as adsorption corrosion inhibitors.

3.2.2 Film-forming corrosion inhibitors

1. Passivating inhibitors: passivating inhibitors act by the formation of a passive film on the metal surface. They may be oxidizing or nonoxidizing agents (oxidizing agents themselves got reduced in the process of oxidizing another molecule). For example, chromates are typical oxidizing inhibitors for iron or steels [8]. Chromate ion is reduced to Cr_2O_3 or Cr $(OH)_3$ on the metal surface, which produces a protective mixed oxide of chromium and iron oxides. Adsorption is also important with oxidizing inhibitors because they are usually adsorbed on the metal surface prior to their reduction and formation of the passive film.
2. Precipitation inhibitors: with precipitation inhibitors, a precipitation reaction deposits a 3-D barrier film on the metal surface between cations of the corroding metal and the inhibitor. Such a film is formed when the solubility product is exceeded for the salt formed between the cations of the metal and the anions of the inhibitor. Phosphates and silicates are examples of precipitation-type inhibitors.

From another point of view, inhibitors can be classified on the basis of their interaction into environmental conditioners or interface inhibitors, as illustrated in Fig. 3–3.

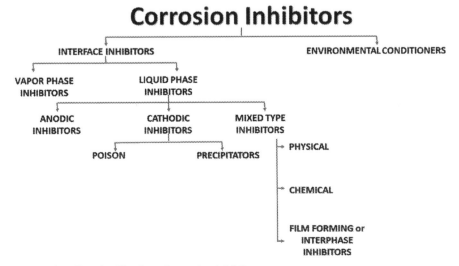

FIGURE 3–3 General outline classification of corrosion inhibitors.

1. Environmental conditioners

The corrosion of the near-neutral and alkaline solutions, where the oxygen reduction is a typical cathodic reaction, can be impeded by the elimination of the corrosive species from the medium. There are several inhibitors which are utilized for this purpose, these type of inhibitors, which reduce the corrosivity of the medium, are named as the environmental conditioners or the scavengers. These scavengers minimize the oxygen content of the medium. Hydrazine and sodium sulfate are examples of environmental conditioners. The application of sodium sulfate involves the following reactions:

$$2Na_2SO_3 + O_2 \rightarrow 2Na_2SO_4 \tag{3.1}$$

$$N_2H_4 + O_2 \rightarrow N_2 + 2H_2O \tag{3.2}$$

2. Interface inhibitors

Interface inhibitors hinder corrosion through the formation of film at the metal–solution interface. Interface inhibitors are classified into two classes:

a. Liquid-phase inhibitors: the classes of liquid-phase inhibitors are categorized, on the basis of the reaction to which they inhibit, as anodic, cathodic, and mixed type of corrosion inhibitors.

　i. Anodic inhibitors: anodic inhibitors are identified as those substances that reduce the anodic zone by acting on the anodic sites and polarizing the anodic reaction [9]. Typically, anodic inhibitors are utilized in the near-neutral solutions where corrosion products (like oxides, salts, or hydroxides) are sparingly soluble. Anodic inhibitors induce positive direction displacements in corrosion potential, reduce corrosion current, and hence decrease the rate of corrosion. If an anodic inhibitor is not accessible at a concentration level sufficient to obstruct all anodic sites, due to the oxidizing nature of the inhibitor that increases the metal potential and promotes the anodic reaction, localized attacks such as pitting corrosion can become a serious issue. As they can cause localized corrosion, anodic inhibitors are categorized as hazardous. They are also known as the passivating inhibitors. For example, chromate, nitrite, silicates, ferricyanide, orthophosphate, etc.

　ii. Cathodic inhibitors: cathodic inhibitors displace the corrosion potential in the negative direction and decrease the corrosion current, thereby restricting the cathodic reaction and limiting the rate of the corrosion process. They function by reducing the available area for the cathodic reaction. This is often achieved by precipitating an insoluble species onto the cathodic sites. They functions by reducing the available area for the cathodic reaction. They are of the following two types:

Cathodic poisons: these are the substances that interfere with the formation of hydrogen atoms or with the recombination of hydrogen atoms to form H_2 gas, thus delaying the reaction of hydrogen evolution. They are quite efficient in acid solutions, but are inefficient in the conditions where cathodic reactions are dominated by other reduction processes, such as oxygen reduction. Examples are sulfides and selenides (they both work by getting adsorbed onto the metal surface). The other examples are arsenic, bismuth, and antimony (they got reduced at the cathode to form metallic coatings).

Cathodic precipitators: the alkalinity of the cathodic sites is raised by the cathodic precipitators. They form insoluble compounds on the metallic surfaces. The carbonates of the Ca and Mg are the most widely applied cathodic precipitators.

$$Ca^{2+} + 2HCO^{3-} + OH^- \rightarrow CaCO_3\downarrow + HCO^{3-} + H_2O \qquad (3.3)$$

$$Zn^{2+} + 2OH^- \rightarrow Zn(OH)_2\downarrow \qquad (3.4)$$

iii. Mixed type inhibitors: approximately 80% of inhibitors are organic compounds that are known as mixed inhibitors and cannot be explicitly designated as anodic or cathodic. The efficacy of organic inhibitors is linked to the degree to which they adsorb and cover the surface of the metal. Adsorption depends on the inhibitor's structure, the metal's surface charge, and the electrolyte type. The metal is protected in three potential aspects by the mixed corrosion inhibitors: via physical adsorption, via chemisorption, and by the film formation.

Physical adsorption or physisorption: the positively charged surface facilitates the adsorption of the anions (negatively charged species), such as halide ions, which in turn provide aid to the adsorption of the positively charged molecules of the inhibitors. For example, through this synergistic effect, the corrosion of iron is impeded by the quaternary ammonium cations in the solution of sulfuric acid. The inhibitor molecules rapidly interact via the physical adsorption. But the inhibitor molecules adsorbed by physical adsorption get removed from the surface, very easily.

Chemical adsorption or chemisorption: chemisorption happens more slowly than physisorption. Adsorption as well as the inhibition increases, as the temperature increases. Chemisorption is unique and is not completely reversible.

Film forming: the polymeric films are formed as a result of the surface reaction that the adsorbed inhibitor molecules has undergone. As the film grows from 2D to 3D, the corrosion inhibition increases markedly. Inhibition is effective only if the film is adherent, is not soluble, and prevents the solution from approaching the metal. Protective films can be non-conducting (sometimes referred to as ohmic inhibitors because they increase the circuit's resistance, thereby inhibiting the process of corrosion) or conducting (self-healing).

b. Vapor phase inhibitors: these type of inhibitors are utilized for the temporary protection from the atmospheric corrosion (particularly in the closed enclosure). The compounds which have relatively low but significant vapor pressure, with inhibiting properties, are considered as effective. They are effective for the protection of ferrous and nonferrous metals.

3.3 Mechanistic aspect of the corrosion inhibition

For the majority of aqueous, or partially aqueous systems, inhibitor applications are concerned with four key categories of environment:

1. In acid solution
2. In near neutral water systems

3. In oil and gas industries
4. In atmospheric and gaseous corrosion

The corrosion inhibition mechanism can be explained in terms of the following environment:

3.3.1 Corrosion inhibition in acid solution

The inhibition of metal corrosion in acid solution can be inhibited by various substances for instance halide ions, CO and organic compounds containing heteroatoms of elements of Groups V and VI of the Periodic Table (i.e., N, P, As, O, S, and Se), and organic compounds having multiple bonds, especially triple bonds, in their structures, act as the most effective inhibitors for metal corrosion in the acid solution. The first step of the inhibition in the acid solution is generally considered to be adsorption onto the surface of the metals (generally free of any oxides in the acidic solutions), which impedes the cathodic and/or anodic electrochemical corrosion reactions. Acid solution's corrosion inhibitors may interfere with metals and cause the corrosion reaction in a range of ways, some of which may occur concurrently. And it is not always necessary to apply an inhibitor to a single general mode of action because, with test conditions, the mechanism may change. In the acid solution the corrosion inhibitors may interact in various ways and these interactions will purely depend on various factors such as the pH of the solution, concentration, nature on anions of acid used, and the other species which are present in the solution, nature of the metal to be protected from the corrosion, and the extent of the reaction to form the secondary corrosion inhibitors. For the inhibitors having same functional groups, the mechanism will additionally depends on the factors like the molecular structure on the electron density of the functional group and the size of the hydrocarbon portion of the molecule [10].

3.3.1.1 Mechanism of inhibition action of organic inhibitors

The very first encounter of the metal with inhibitors normally results in its physical adsorption onto the surface of the metal that might or might not be accompanied by chemisorption. Whether it is mere physical adsorption or if chemisorption has also occurred can be determined by characteristics such as adsorption kinetics and heat of adsorption. Physisorption is the consequence of electrostatic attraction amongst organic ions or dipoles that have been inhibited and metal surfaces that are electrically charged, while chemisorption requires the exchange of the charge or the movement of the charge from the organic molecule to the metal surface in order to create a bond of the coordinate type. Three entities are involved in the inhibition process, the metal being protected, the inhibitor, and the aggressive medium or electrolyte. Whether inhibition occurs by physical adsorption or chemisorption, the nature and properties of all the above three influence the inhibition process. It is presumed that water molecules of the electrolyte are adsorbed on the metal surface before the addition of the inhibitor because water displays a weak bipolarity. Adsorption of certain aggressive ions, for example, chlorides of the electrolyte cannot be ruled out. However, after the addition of inhibitor, water molecules or molecules of some other adsorbent are displaced by organic molecules according to the equation [11]:

$$Org_{(sol)} + nH_2O_{(ads)} \rightarrow Org_{(ads)} + nH_2O_{(sol)} \tag{3.5}$$

where n is the quantity of water molecules that have been substituted by every molecule of organic inhibitor adsorbed. "n" will depend on the relative cross-sectional area of organic molecule and the water molecules. The preferential adsorption of organic molecules takes place because the interaction energy between organic molecule and metal surface is higher than the interaction energy between the water molecule and the metal surface.

3.3.1.2 Adsorption of the inhibitor on metal surface

The general assumption is that, the first step in the mechanism of corrosion inhibition of metals by organic molecules is their adsorption onto the metal (substrate) surface via their functional group(s). Therefore, in order to understand the mechanisms of corrosion inhibition, the adsorption behavior of organic molecules (inhibitors) on the metal surface must be known [12,13]. The adsorption process depends upon (1) the topography, uniformity, and physicochemical properties (charge, conformation, chemistry) of the metal surface; (2) the propagation of the charge in the molecule; (3) the aggressive medium (electrolyte); and (4) the chemical composition of the organic molecule. The adsorption of ions or neutral molecules on bare metal surfaces immersed in solution is determined by the mutual interactions of all species present at the phase boundary. These include electrostatic and chemical interactions of the adsorbate with the surface, adsorbate—adsorbent and adsorbate—solvent interactions. Two modes of adsorption are present depending on the existence of the forces involved [13]: (1) chemisorption—in this type of adsorption, a single layer of molecules, atoms, or ions is attached to the surface by chemical bonds and is essentially irreversible; or (2) physical adsorption—in this mode of adsorption, the bonding is by the weaker Van der Waals forces, whose energy levels are comparable to those of condensation. Chemisorption has a higher heat of adsorption than adsorption involving electrostatic/van der Waals forces. In the case of physisorption, coulombic interactions between ions or dipoles of the inhibitor and charged metal surface are present, and other interactions also exist (Van der Waals forces, hydrogen bonding). The adsorption process in this case is rapid but the adsorbed species can easily be removed from the surface. Chemisorption commonly takes place through the donation of electrons from species with loosely bonded electrons, such as π-electrons in aromatic rings, or multiple bonds, or unpaired electrons in functional groups that contain atoms such as oxygen, nitrogen, sulfur, phosphorus, to the vacant d-orbital of transition metals (metal substrate which need to be protected). The performance of corrosion inhibition of the sequence homology set of organic substances differing only in the heteroatom is usually as follows: phosphorus > sulfur > nitrogen > oxygen. Factors influencing the absorption of ion inhibitors on metal surfaces are [14,15]:

1. Metal surface charge: adsorption may arise electrostatic force of attraction between the ions or dipole loads on the adsorbed inhibitor and the electrical charge on metal at the interface of the metal-solution.

2. The functional groups and structure of inhibitor: by electron transfer, inhibitors can bind to the metal surface and form a coordinated type of bond that contributes to good binding and successful inhibition. Species having comparatively loosely bound electrons in anions, neutral compounds, lone pairs of electrons, triple bond or organic ring systems-related π-electron systems, and functional groups having periodic table elements of group V or VI prefer fast transfer of electrons and better bond forming and thus successful inhibition. The propensity to form stronger coordinate bond increases with declining electronegativity and follows the order: oxygen < nitrogen < sulfur < phosphorus.

3. The interaction between adsorbed inhibitor species (synergism and antagonism): adsorbed species may enter into various interactions on the surface of an electrode that may significantly influence their inhibitive properties and the mechanism of their action.

4. Reaction of adsorbed inhibitors: the adsorbed inhibitor species may react usually by electrochemical reaction to form a product which is also inhibitive. Inhibition due to the added substance is termed as primary inhibition and that due to reaction products as secondary inhibition.

5. The adsorption to metal surfaces of organic inhibitors is widely known as isothermal adsorption. Adsorption isotherms are widely agreed to offer invaluable insights into the mechanism of corrosion inhibition and provide essential details about the nature of interactions amongst the metals and inhibitors. It is possible to deduce the strength of the adsorption bond from the adsorption isotherm, which shows the contrast between the inhibitor surface concentration and bulk solution. To illustrate the mode of adsorption of inhibitors onto the metal surface, various isotherms of adsorption (Langmuir, Temkin, Frumkin, Flory Huggins, Dhar-Flory-Huggins, Bockris-Swinkels, Freundlili-Swinkels, and Freundlich) have been reported.

3.3.2 Corrosion inhibition in near-neutral solutions

The corrosion of metals in nearly neutral solutions (and open to the air) differs from that in acidic solutions for two reasons. First, in acid solutions, the metal surface is oxide free, but in neutral solutions, the surface is oxide covered. Second, in acid solutions, the main cathodic reaction is hydrogen evolution, but in air-saturated neutral solutions, the cathodic reaction is oxygen reduction.

In an acid solution, corroding metal surfaces are oxide-free, while metal surfaces are coated in acid solutions with films of oxides, hydroxides, or salts, resulting in reduced solubilization of these species. Because of these differences, substances that inhibit corrosion in acid solution by adsorption on oxide-free surfaces do not generally inhibit corrosion in neutral solution. Typical inhibitors for near-neutral solutions are the anions of weak acids, some of the most important in practice being chromate, nitrite, benzoate, silicate, phosphate, and borate. Passivating oxide films on metals offer high resistance to the diffusion of metal ions, and the anodic reaction of metal dissolution is inhibited. These inhibitive anions are often referred to as anodic inhibitors, and they are more generally used than cathodic inhibitors to inhibit the corrosion of iron, zinc, aluminum, copper, and their alloys in near-neutral solutions. Inhibition in neutral solutions can also be due to the precipitation of compounds, on a metallic surface, that can form or stabilize protective films. The inhibitor may form a surface film of an insoluble salt by precipitation or reaction.

3.3.3 Corrosion inhibition in oil and petroleum industries

In the early days, many chemical compounds were added by oil companies, often effectively, to mitigate corrosion damage to oil wells and surface handling machinery. Corrosion prevention in oil fields became significantly more effective after amines and imidazolines came into use. Current inhibitors are introduced in the field at concentrations of 15−50 ppm, constantly or in intermittent batches, on the basis of overall liquid output. To tackle oil-field corrosion, a much broader range of inhibitor chemistry is available today than existed just a decade ago. In recent years, organic molecules containing sulfur, phosphorus, and nitrogen in various combinations have been developed. These types of inhibitors have increased the efficiency of oil field inhibitors, especially in the direction of oxygen contamination tolerance and control corrosion associated with high CO_2, low H_2S conditions [16].

Organic nitrogenous compounds are the majority of the inhibitors currently used in the development of wells. As a part of the structure, the basic forms have long-chain hydrocarbons (usually C18). Many of the inhibitors that are being used effectively today are either focused on long-chain aliphatic diamine or imidazolines from the long carbon chain. Several modifications of these structures are being introduced to alter the material's physical properties, such as the typical reaction of ethylene oxide with these compounds in different molecular percentages to provide polyoxy-ethylene derivatives with varying degrees of miscibility of the salt. In order to produce salts from these amines or imidazolines, many carboxylic acids are used. To tackle oilfield corrosion, a much broader range of inhibitor chemistry is available today than existed just a decade ago. The so-called sandwich theory is one of the classical ideas in which the bottom component of the sandwich is the bond between the molecule's polar end and the surface of the metal. The effectiveness of the defensive action hinges on this bond.

The center part of the sandwich is the molecule's nonpolar end, and the degree to which this portion of the molecule can cover or wet the surface is its contribution to defense. The hydrophobic layer of oil attached to the long carbon tail of the inhibitor is the top portion of the defensive sandwich. This oil layer acts as the external protective surface, shielding the surface of the inhibitor and providing a barrier to both ferrous ion outward diffusion and corrosive species inward diffusion. Water or water salt solutions alone will not cause harmful corrosion. Unless unique corrodents are contained, such as CO_2, H_2S, and their dissolution products. The wells for oil and gas are either sweet or sour [14]:

1. Sweet corrosion: corrosion can be grouped into three ambient temperatures in CO_2 gas wells. The corrosion product is nonprotective under $60°C$ and there will be high corrosion rates. Magnetite is formed above about $150°C$, and the wells are not corrosive even in the presence of high brine levels. The iron carbonate corrosion product layer is protective in the middle temperature regime, in which most gas well conditions lie, but is adversely affected by chlorides and fluid velocity. The dispersibility feature of the oil and the brine being produced is among the significant physical properties of oilfield inhibitors. A properly selected inhibitor based on the corrosion mechanism will not be successful unless it has access to the corrosive metal. There are also some major variations when it comes to handling oil and gas wells. There is no clear cut in the difference between an oil well and a gas well. The fact that a large volume of gas is generated by many oil wells and a large volume of liquid is produced by many gas wells, plus the fact that wells frequently undergo a shift in production over their lifetime, makes it difficult to create a technical distinction. There are more substantial distinctions, however. Typical gas wells are much hotter than oil wells, and they are much lighter in terms of hydrocarbon liquids. Normally, gas wells are much deeper and generally produce brines of lower total dissolved solids. In gas well corrosion, oxygen is not a consideration, but may cause major issues in artificial lift oil wells. Corrosion mechanisms may change due to the wide temperature gradient in many gas wells, resulting in various forms of corrosion in the same well, whereas oil wells do not exhibit this behavior. Oil wells usually contain more liquid than gas wells, resulting in a shorter batch treatment existence. Since corrosion in oil wells is electrochemical in nature, an electrolyte must be present for corrosion to occur. The source of water in oil wells is almost always the formation source, and at concentrations ranging from trace to saturation, the water may contain dissolved salts. Corrosion-related water can be in a thin layer, in droplets, or even the major process. Results of the corrosion control analysis by inhibitors in the development of oil wells in fields flooded with carbon dioxide showed that imidazolines are effective in protecting CO_2 brines. It was found that the inhibitor was integrated into the layer of the carbonate corrosion product, but if the surface film contained sulfide, it was even more successful. Inhibitors, such as nitrogen−phosphorus compounds or sulfur compounds in organic molecules, have also had better performance.

2. Sour corrosion: aldehydes, cyanamide thiourea, and urea derivatives are corrosion inhibitors used in the past to tackle the problem of corrosion faced by sour wells. Organic amines are by far the most frequently employed inhibitors in the sour wells. While organic amines are assumed to be somewhat less effective inhibitors of acid solution, amine inhibition is significantly accelerated by the presence of hydrogen sulfide [16]. Hydrogen sulfide is the main corrosive agent in sour wells, and carbon dioxide is also often present. The effect of multiple iron sulfides at varying hydrogen

sulfide concentrations in the corrosion products has also been established. Oilfield inhibitors act by integrating corrosion products on the metal surface into a thin layer. This surface film can be anaerobic or partly oxidized and may be a sulfide or a carbonate.

3. Acidizing: acidizing is an effective approach for promoting the production of oil and gas wells. These are not able to flow readily into the well due to the very poor permeability of such hydrocarbon-containing formations. If the rock is sandstone, these limestone/dolomite formations can be treated with HCl or a mixture containing HF. In the acidifying procedure, the acid is pumped down the tubing into the well where it reaches the perforations and hits the structure (e.g., HCl, at a concentration of 7%−28%); the acid etches channels that provide a path for oil and gas to enter the pipe. Several inhibitors like high-molecular-weight nitrogenous compounds alkyl or alkylaryl nitrogen compounds and acetylenic alcohols, such as 1-octyn-3-ol, are utilized for well acidizing operations. But since they are very hazardous, these materials pose severe handling problems; this will determine which product is ultimately used by an operator. In addition, both in productivity and time, their effectiveness is limited. Acid soaks usually last between 12 and 24 hours, after which time the efficiency of the inhibitor can start to drop alarmingly. Cinnamaldehyde and alkynols containing unsaturated groups in conjunction with the oxygen role defined as alphaalkenylphenones are oxygen-containing inhibitors which are efficient in concentrated HCl [17]. They provide protection similar to that obtained from acetylenic alcohols, especially when combined with small amounts of surfactants.

3.3.4 Corrosion inhibition in atmospheric and gaseous corrosion

Volatile corrosion inhibitors (VCIs), a subset of corrosion inhibitors, are also a very cost-effective and potent method for reducing the damage to metals caused by atmospheric or gaseous corrosion. Volatile corrosion inhibition basically depends on the environmental condition with a very small amount of inhibitors. Apart from volatility, many other electrochemical effects, such as changes in the potential of the diffusion part of the double layer that regulates the migration of the electrode reaction components, are also needed to act as a VCI in order to achieve the protective effect. The first condition for good efficiency of a vapor phase inhibitor is its capability to reach the metallic surface to be protected. The second is that the rate of transfer of the molecule should not be too slow to prevent an initial attack of the metal surface by the aggressive environment before the inhibitor can act. These two conditions are related partly to the vapor pressure of the inhibitor, partly to the distance between the source(s) of the inhibitor and the metal surfaces, and partly to the accessibility of the surfaces. The vapor pressure of a chemical compound will depend upon the structure of the crystal lattice and the character of the atomic bonds in the molecule. In this respect, organic components of the molecule will generally ensure its volatility. A convenient volatile inhibitor should not have too high a vapor pressure, because it will be lost as a result of the fact that enclosures are generally not airtight; protection will then drop. A convenient partial vapor pressure for efficient compounds will lie between 10^{-5} and 10^{-1} mm Hg (i.e., $10^{-3}-10$ Pa). Under certain atmospheric conditions the compounds which exhibit appreciable vapor pressure behave as electrolyte layer inhibitors. They act so by altering the electrode reaction kinetics and are basically classified into the category of VCIs. Neutralizing amines possess an appreciable vapor pressure but they are not classified as VCIs. They are commonly effective for the protection of ferrous metals. The mechanism of the inhibition depends on the adjustment of the electrolyte's pH values. This causes several conditions that are inhospitable for the formation of rust. The volatile compound usually reaches the protective vapor concentration rapidly, but the protective effectiveness can be of a very short span of time and the consumption is very excessive if the enclosure in which they are utilized is not airtight. In the case of the volatile compounds with very low vapor pressure they exhibit longer protection, however the problem with such compounds is that more time is required in order to acquire the protective vapor pressure. The other problem is the mere possibility of the corrosion which may occur during the initial period of saturation, and even the effective inhibitor concentration can never be achieved if the enclosure is not hermetically sealed. Hence, the volatile compounds utilized as corrosion inhibitors must neither possess too high nor too low vapor pressure. They must possess optimum vapor pressure.

3.4 Industrial applications of corrosion inhibitors

There are several applications of corrosion inhibitors known so far (as given in the Fig. 3–4). Some of them are discussed below:

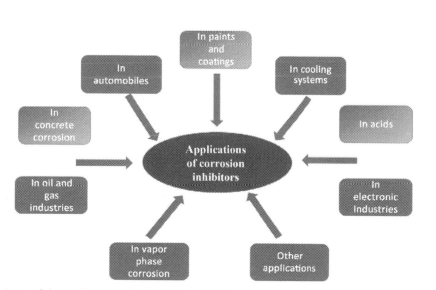

FIGURE 3–4 Some of the applications of the corrosion inhibitors.

3.4.1 In acid

Some of the major application of acids, in which metals or their alloys are in direct or indirect contact, are acid picking, in the acid well acidizing processes, the cleansing of oil gas industrial equipment, and heat exchangers. Mostly in order to protect them from corrosion mixed type of corrosion inhibitors are utilized. Some of the most utilized inhibitors in an acidic environment are summarized as follows:

1. Ethylene: utilized for the protection of Fe in HCl medium (0.5% Concentration).
2. Mercaptobenzotriazole (MBT): utilized for the protection of Fe in HCl medium (1% concentration).
3. Pyridine + Phenylhydrazine: utilized for the protection of Fe in HCl medium (0.5% + 0.5% concentration).
4. Rosin amine + ethylene oxide: utilized for the protection of Fe in HCl medium (0.2% concentration).
5. Phenylacridine: utilized for the protection of Fe in H_2SO_4 medium (0.5% concentration).
6. NaI: utilized for the protection of Fe in H_3PO_4 medium (200 ppm concentration)

3.4.2 In water

1. Potable water: the potable water is corrosive unless a protective film or deposit is formed on the surface of metal or alloys. Potable water is generally saturated with the dissolved oxygen. Some of the commonly utilized corrosion inhibitors for the potable water are:
   a. Ca $(HCO_3)_2$: utilized for the protection of cast iron and steel (used concentration 10 ppm).
   b. Polyphosphate: utilized for the protection of Fe, Zn, Cu, and Al (used concentration 5–10 ppm).
   c. Ca $(OH)_2$: utilized for the protection of Fe, Zn, and Cu (used concentration 10 ppm).
   d. Na_2SiO_3: utilized for the protection of Fe, Zn, and Cu (used concentration 10–20 ppm).

3.4.3 In cooling water system

In a recirculating system, evaporation is the chief source of cooling. As evaporation proceeds, the dissolved mineral salt content increases. Cooling systems may consist of several dissimilar metals and nonmetals. Metals picked up from one part of the system can be deposited elsewhere, producing galvanic corrosion. Corrosion is controlled by anodic (passivating) inhibitors, including nitrate and chromate, as well as by cathodic (e.g., zinc salt) inhibitors. Organic corrosion inhibitors are also utilized. Some of the most utilized corrosion inhibitors in the cooling water system are listed below:

1. $Ca(HCO_3)_2$: utilized for the protection of cast iron and steel (used concentration 10 ppm).
2. Na_2CrO_4: utilized for the protection of Fe, Zn, and Cu (used concentration 0.1%).

3. $NaNO_2$: utilized for the protection of Fe (used concentration 0.05%).
4. NaH_2PO_4: utilized for the protection of Fe (used concentration 1%).
5. Morpholine: utilized for the protection of Fe (used concentration 0.2%).

3.4.4 In boiler water

Inhibitors are used to control corrosion in boiler waters, some of the commonly used inhibitors are:

1. NaH_2PO: utilized for the protection of Fe, Zn, and Cu (used concentration 10 ppm).
2. Polyphosphate: utilized for the protection of Fe, Zn, and Cu (used concentration 10 ppm).
3. Morpholine: utilized for the protection of Fe (used at variable concentration).
4. Hydrazine: utilized for the protection of Fe (used as O_2 scavenger).
5. Ammonia: utilized for the protection of Fe (used as neutralizer).
6. Octadecylamine: utilized for the protection of Fe (used at variable concentration).

3.4.5 In automobiles

Inhibitors are used to minimize internal corrosion (fluid system corrosivity), usually caused by coolant aeration temperature, pressure, etc. Nitrites, nitrates, phosphates, silicates, arsenates, chromates (anodic inhibitors), amines, benzoates, mercaptans, and organic phosphates (mixed inhibitors); and polar or emulsifiable oils (film formers) are some of the widely used antifreeze corrosion inhibitors. Another use of corrosion inhibitors in automobiles is to protect from external corrosion (corrosion of metal surfaces which are exposed to the environment).The atmosphere to which automobiles are exposed contains moist air, wet SO_2 gas (forming sulfuric acid in the presence of moist air), and deicing salt (NaCl and $CaCl_2$). The corrosion-proofing formulations that are used include grease, wax resin, and resin emulsion along with metalloorganic and asphaltic compounds to control external corrosion. Typical inhibitors used in rust proofing applications are fatty acids, phosphonates, sulfonates, and carboxylates.

3.4.6 In organic coatings and paints

In the primers finely powdered inhibiting pigments (polar compounds) are incorporated. They usually displace the water and orient themselves in a such a way that their hydrophobic tails are towards the environment. Red lead (Pb_3O_4) is commonly used in paints on iron. It deters the formation of local cells and helps to preserve the physical properties of the paints. Lead azelate, calcium plumbate, and lead suboxide are other inhibitors which are utilized.

3.4.7 Internal corrosion of steel pipelines

The assembly pipelines working between the oil and gas wells and the processing plants also face the same problem as the refineries and petrochemical plants. The corrosion rate of the pipelines are affected by the flow of the multiphase fluids. At very high rates, flow-induced corrosion and erosion corrosion occurs, the high flow tends to cause the sweep of the sediments out of the pipelines. However, at low flow rates, the common form is pitting, as low velocity is related to the settlement of the sediments at the bottom. The corrosion of the pipelines or pigging is controlled by cleaning them by adding continuous or batches of corrosion inhibitors.

3.4.8 Petroleum production

To manage wet corrosion, refineries and petrochemical industries employ a number of film-forming inhibitors. Many of the inhibitors, including amines and amides, are long-chain nitrogenous organic materials. Water-soluble and water-soluble dispersible oil inhibitors and oil-soluble and oil-soluble water dispersible inhibitors (batch inhibitors) are continuously incorporated to mitigate the problem of corrosion. They are active at liquid—liquid and/or liquid—gas interfaces since inhibitors are interfacial in nature, and can lead to emulsification. As a consequence, in the presence of inhibitors, foaming is rarely observed. Film-forming inhibitors, via their polar group, anchor to the metal and the nonpolar tail protrudes out vertically. The physical adsorption of hydrocarbons (oils) on such nonpolar tails improves film thickness and the efficacy of hydrophobic barriers for corrosion inhibition.

3.4.9 Others

Inhibitors have been used for corrosion mitigation in fuel oil reservoirs, hot chloride dye baths, cooling brines, and steel reinforcement in concrete. Several corrosion inhibitors are also used for the protection of artifacts from the corrosion.

3.5 Toxicity of corrosion inhibitors

The use of corrosion inhibitors in different industries for corrosion control is a well-known practical technique, especially in the oil industry. However, some of the corrosion inhibitors that are still in use are considered toxic and harmful to the environment. Some of the examples are nitrate and chromate-based corrosion inhibitors. They were utilized in the treatment of industrial water, that is, in the recirculation of cooling water. Unfortunately both are highly toxic. Thiourea derivatives form an essential class of corrosion inhibitors used in the past; however, they enhance hydrogen uptake by metals and they also isomerize into toxic compounds, thus they cannot be safely used. Another example is corrosion protection by employing n-alkanthiols on iron and copper. In spite of the good performance exhibited by the n-alkanthiols, their widescale applicability is not acceptable due to their toxic nature. Similarly another class is the acetylene alcohols, they are applied as the acidizing inhibitors. They are commercially viable and effective. But they are applicable only at very high concentrations. They are highly toxic, and cause problems in the handling and disposal of the waste.

Several chemical compounds that are still used for the inhibition of corrosion are known to be extremely toxic. Fighting this problem and identifying the inhibitors that must play an effective role in controlling corrosion and are nonhazardous in nature remains an important subject for the researchers. The environment-friendliness of the organic compounds has paved many ways for the replacement of the existing corrosion inhibitors which are highly toxic.

3.6 Environment-friendly or green corrosion inhibitors

It is widely known that there is no proper definition associated to the term "environment-friendly" or "green" corrosion inhibitors [18]. Recently, studies related to the corrosion inhibitors have been directed toward the safety and environmental considerations. In order to select the compounds to perform as the corrosion inhibitor, several other factors are also accountable, such as their availability and economic viability. In view of these factors the current research in the development of the corrosion inhibitor is directed toward the evaluation of the nontoxic, environmentally sound, economical chemical compounds. Some of them are shown with the help of Fig. 3–5.

The increasing number of publications on corrosion inhibition of various metals during the past decade clearly indicates the interest in exploring the new inhibitors for a variety of corrosive environments. However, the number of publications appearing on green corrosion

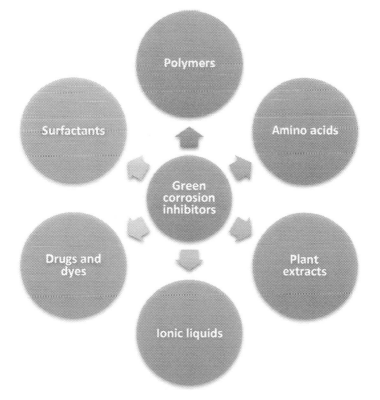

FIGURE 3–5 Green corrosion inhibitors.

inhibitors is quite limited [19]. So far several organic and inorganic compounds have been evaluated as green and eco-friendly inhibitors and a lot of work has been devoted to the development of efficient green corrosion inhibitors extracted from natural plants [20−22]. Plant extracts are environmentally benign and ecologically acceptable compounds. And in addition they are economically viable, easily available, and biodegradable. Their inhibition efficiency is attributable to the complex organic substances involved in their formulation (tannins, alkaloids and nitrogen bases, carbohydrates, and proteins, and also some hydrolysis products). They typically include heteroatoms in their molecular structures, such as nitrogen, sulfur, or oxygen atoms, as well as triple or conjugated double bonds or aromatic rings.

Polymers, both natural and synthetic, also have been examined in various corrosive conditions as corrosion inhibitors for various materials and therefore have gained significant widespread popularity. Like plant extracts they have also been found to be attractive as corrosion inhibitors as they are cost-effective, inherently stable, and nontoxic. Similarly both the natural and synthetic polymers have received wide acceptance and have been evaluated as corrosion inhibitors in the different corrosion environment for various metals. The applications of polymer are also cost-effective, stable, and nontoxic.

In comparison to monomer analogs, polymers exhibit a superior inhibition effect because of the fact that they possess long-chain carbon linkage and multiple adsorption sites and thus block large surface area of the corroding metal, thereby blanketing the surface and protecting the metal from corrosive agents present in the solution. They cover the surface, ultimately protecting the metal from the aggressive solution's corrosivity. Polymers have two significant advantages:

1. A single chain of the polymer displaces several water molecules from the metallic surface thus making the process entropically significant.
2. The presence of multiple bonding sites makes desorption of polymer molecules a slower process.

The protection offered by the polymers was observed to be related to their chemical composition, molecular weight, and distinct molecular and electronic structures. The polymeric inhibitors, which have been investigated as corrosion inhibitors include polyethylene glycol, polyvinyl alcohol, polyvinyl pyrrolidone, polyethylenimine, polyacrylic acid, poly acrylamide, starch, gum arabic, guar gum, xanthan gum, gellan gum, hydroxypropyl cellulose, hydroxyethyl cellulose, psyllium polysaccharide, carboxymethyl cellulose, pectin, and chitosan, etc.

3.7 Selection of the efficient corrosion inhibitors and challenges related

In order to select the corrosion inhibitors, their method of application and the characteristic of the system in which they are utilized must be considered. In order to make the proper choice of the inhibitors several factors should be considered, among which the primary decision is to match the appropriate inhibitor chemistry with the corrosion conditions. The selection of the inhibitors starts with the choice of the physical properties, that is, whether the inhibitor should be in the solid or liquid state? The importance of the melting and freezing point must also be considered. Another point which must also be considered is if their degradation with time and temperature is critical or not? It should be compatible with other additives of the system. Are specific solubility characteristics required? This list can be extensive but it is important because it paves the ways for the possible corrosion inhibitors. For the evaluation of any new system, it is regarded as the first step. The physical quantifications are those customarily done as part of the least quality affirmation testing. The simplest corrosion test should be executed in order to ascertain the performance of the inhibitors and to rule out the unsuitable candidates among them. The aim of the initial screening is that the inhibitors which do not perform well must not be carried out further. The inhibitor user must employ test procedures that rigorously exclude inferior inhibitors even though some good inhibitors may also be excluded. The most serious challenge in the evaluation of the performance of the inhibitor system is the designing of the conditions (real world) of the experiments which are needed to estimate their performance. The variables that must be considered include temperature, pressure, and velocity as well as metal properties and corrosive environment chemistry.

The practice of corrosion inhibition requires that the inhibitive species should have easy access to the metal surface. Ideally, surfaces should therefore be clean and not contaminated by oil, grease, corrosion products, water hardness scales, and so forth. And further measures should be taken to avoid the deposition of solid particles. This conditioning is often difficult to achieve, and there are many cases where less than adequate consideration has been given to the preparation of systems to receive inhibitive treatment. It is also necessary to ensure that the inhibitor reaches all parts of the metal surfaces. Care should be taken, particularly when first filling a system, that all dead ends, pockets, and crevice regions are contacted by the inhibited fluid. This will be encouraged in many systems by movement of the fluid in service, but in nominally static systems it will be desirable to establish a flow regime at intervals to provide a renewed supply of inhibitor.

Inhibitors must be chosen after taking into account the nature and combinations of metals present, the nature of the corrosive environment, and the operating conditions in terms of flow, temperature, and heat transfer. Inhibitor concentrations should be checked on a regular basis and losses restored either by appropriate additions of inhibitor or by complete replacement of the whole fluid as recommended, for example, with engine coolants. Where possible, some form of continuous monitoring should be employed, although it must be remembered that the results from monitoring devices, probes, coupons, and so forth, refer to the behavior of that particular component at that particular part of the system. Nevertheless, despite this caution, it must be recognized that corrosion monitoring in an inhibited system is well established and widely used.

3.8 Conclusion

As it is already discussed above that utilization of corrosion inhibitors is one of the most prevalent ways of combating the internal corrosion in various industries. Mostly they work by getting adsorbed onto the surface of the metals, thereby forming a thin barrier onto the metal surface which need to be protected. One of the challenges associated with corrosion inhibitor is their selection. Proper choices of inhibitors should be made by matching the appropriate inhibitor chemistry with the corrosion conditions and by selection of appropriate physical properties for the application conditions. Method of application and system characteristics must be considered when selecting physical properties of an inhibitor. Inhibitors must be chosen after taking into account the nature and combinations of metals present, the nature of the corrosive environment, and the operating conditions in terms of flow, temperature, and heat transfer. Inhibitor concentrations should be checked on a regular basis and losses restored either by appropriate additions of inhibitor or by complete replacement of the whole fluid as recommended.

Acknowledgments

Authors acknowledge the financial support from Council of Scientific & Industrial Research (CSIR), New Delhi, India, through the major research project [file number: 22(0832)/20/EMR-II].

Useful links

https://www.sciencedirect.com/topics/materials-science/corrosion-inhibitor
https://www.sciencedirect.com/topics/engineering/corrosion-inhibitor
https://link.springer.com/article/10.1007/s40735-019-0240-x#:~:text = Inhibitors%20decrease%20the%20dissolution%20rate,species%20and%20the%20metal%20surface.

References

[1] M.A. Bedair, S.A. Soliman, M.S. Metwally, J. Ind. Eng. Chem. 41 (2016) 10.

[2] C.C. Nathan, J.I. Bregman, National Association of Corrosion Engineers, Corrosion Inhibitors, National Association of Corrosion Engineers, 1973.

[3] M.A. Bedair, S.A. Soliman, M.S. Metwally, J. Ind. Eng. Chem. 41 (2016) 10.

[4] M.S. Morad, Corros. Sci. 50 (2) (2008) 436.

[5] R.W. Revie, H.H. Uhlig, Uhlig's Corrosion Handbook, Wiley, 2011.

[6] E. McCafferty, M.K. Bernett, J.S. Murday, Corros. Sci. 28 (1988) 559.

[7] E. Vuorinen, E. Kálmán, W. Focke, Surf. Eng. 20 (4) (2004) 281.

[8] P.B. Raja, G.S. Mathur, Natural products as corrosion inhibitor for metals in corrosive media—a review, Mater. Lett. 62 (2008) 113—116.

[9] M. Mobin, S. Zehra, R. Aslam, RSC Adv. 6 (2016) 5890—5902.

[10] M. Mobin, S. Zehra, M. Parveen, J. Mol. Liq. 216 (2016) (2016) 598—607.

[11] M. Mobin, R. Aslam, J. Aslam, Mater. Chem. Phys. 191 (2017) 151—167.

[12] A.S. Khanna, N.NACE Letter, India 4 (1997) 3.

[13] S.K.D. Karakaş, Corros. Sci. 77 (2013) 37.

[14] M. Benabdellah, R. Souane, N. Cheriaa, R. Abidi, B. Hammouti, J. Vicens, Pigment Resin Technol 36 (6) (2007) 373.

[15] M.A. Hegazy, H.M. Ahmed, A.S. El-Tabei, Corros. Sci. 53 (2) (2011) 671.

[16] L.W. Jones, Corrosion and Water Technology for Petroleum Producers, Oil and Gas Consultants International, Tulsa, Okla, 1988.

[17] O. Lahodny-Sarc, Corrosion inhibition in oil and gas drilling and production operations, A Working Party Report on Corrosion Inhibitors, The Institute of Materials, London, UK, 1994, pp. 104−120.

[18] R. Aslam, M. Mobin, S. Zehra, I.B. Obot, E.E. Ebenso, ACS Omega 2 (2017) 5691−5707.

[19] I. Obot, D. Macdonald, Z. Gasem, Corros. Sci. 99 (2015) 1−30.

[20] E.E. Oguzie, Mater. Chem. Phys. 99 (2−3) (2006) 441.

[21] A.M. Abdel-Gaber, B.A. Abd-EL-Nabey, I.M. Sidahmed, A.M. El-Zayaday, M. Saadawy, Corros. Sci. 48 (9) (2006) 2765.

[22] M. Parveen, M. Mobin, S. Zehra, RSC Adv. 6 (2016) 61235−61248.

Sustainable corrosion inhibitors: current approaches and experimental assessment

4

Synthetic environment-friendly corrosion inhibitors

Amit Kumar Dewangan[1], Yeestdev Dewangan[1], Dakeshwar Kumar Verma[1], Chandrabhan Verma[2]

[1]DEPARTMENT OF CHEMISTRY, GOVERNMENT DIGVIJAY AUTONOMOUS POST GRADUATE COLLEGE, RAJNANDGAON, INDIA [2]INTERDISCIPLINARY RESEARCH CENTER FOR ADVANCED MATERIALS, KING FAHD UNIVERSITY OF PETROLEUM AND MINERALS, DHAHRAN, SAUDI ARABIA

Chapter outline

4.1 Introduction

Metal corrosion is an electrochemical phenomenon, which arises due to the potential difference in the metal surface. Generally, metals such as mild steel, copper, and aluminum are used for industrial applications, as they are relatively hard and stable and are used as a base material for any construction or equipment. Before these metals are used, rust and scale are removed from them using various processes such as acid pickling and descaling; mainly HCl, H_2SO_4, and HNO_3 are mainly used for aggressive media [1–5]. Apart from excess rust during these processes, there is also a loss of a bare metal surface, which is very important to prevent. Generally, corrosion inhibitors are molecules that protect their metal surface from corrosion and maintain their mechanical, electrical, and physical properties. Synthetic organic molecules predominantly used as excellent potential corrosion because they contain electron-donating moiety such as unsaturated pi electrons and heteroatoms (O, S, P, N) due to which they behaved a like Lewis base and able to donate electron to the vacant d orbital of metal surface that behaved like a Lewis acid [6–9]. Generally, synthetic green inhibitors are those which can be applied in the environment and are without any adverse effects on organism. These are cost-effective and mainly a one-pot multicomponent reaction method is used to make them [10,11]. Usually hydroxamic acids [12,13], benzothiazole [14,15], Quinoxaline [16], imidazole [17], benzothiazine [18], chitosan [19], 8-Hydroquinoline [20], etc. containing molecules are used as synthetic green corrosion inhibitors due to their potential nature and less toxic properties. Nowadays theoretical calculation along with the experimental technique are used to investigate the potential properties of inhibitor molecules [21]. Traditional corrosive monitoring techniques include gravimetric analysis and electrochemical investigation, which reveal the adsorption properties of any inhibitors and the electrochemical behavior of the inhibitor/metal interface. X-ray diffraction scanning electron microscopy (SEM) and electron diffraction X-ray spectroscopy are mainly used for the investigation of morphological changes of the metal surface after applying a similar inhibitor. Apart from all these, recent computer calculation-based theories are applied to explain the molecular structure, electron-donating sites, and adsorption behavior of inhibitors, among which density functional theory (DFT), molecular dynamic (MD) and Monte Carlo (MC) simulation are prominent. Fig. 4–1 demonstrates the developments on synthetic environment-friendly corrosion inhibitors in the last 5 years.

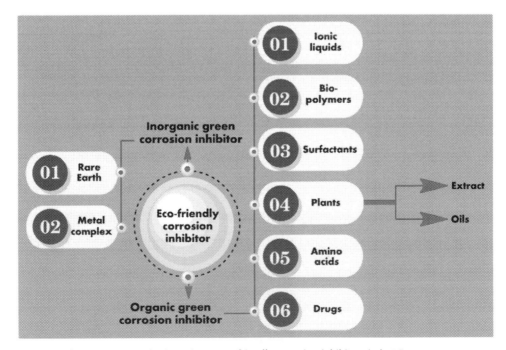

FIGURE 4–1 Developments on synthetic environment-friendly corrosion inhibitors in last 5 years.

4.2 Green corrosion inhibitors

In the recent environment, synthetic corrosion inhibitors have been constrained globally due to the increasing awareness of harsh environmental regulations and public health. For this reason, recently the attention of scientists and chemical engineers was attracted toward the synthetic development of such corrosion inhibitors which are environmentally friendly, less toxic, and inexpensive and do not involve a complicated process when making and using [22,23]. Drug molecules and plant extracts are commonly used extensively as green corrosion inhibitors because their molecules have electron-donating abilities and sites present. They are absolutely safe to use and they can be easily obtained and used. They are biodegradable and also sustainable [24–27]. Ionic liquid is in this category, which behaves like a green potential corrosion inhibitor [28,29]. Most of the synthetic organic molecules involve very complex multiple approaches from making to use and are toxic threats to the environment but there are also some synthetic organic molecules which are less toxic, water-soluble, eco-friendly, and cheap. Depositing them on metal surface protects them from further corrosion when applied in corrosive media. Heterocyclic compounds and polymers commonly come in the form of synthetic green corrosion inhibitors of which indole [30], triazole [31], benzimidazole [32], thiols [21], imidazole [33], pyrimidine [34], etc. and their derivative are of prime interest. Hence they have been used extensively for the last few years as green and sustainable corrosion inhibitors for metal and material protection. Inhibitors can be anodic, cathodic, and mixed type on the basis of their adsorption (Fig. 4–2).

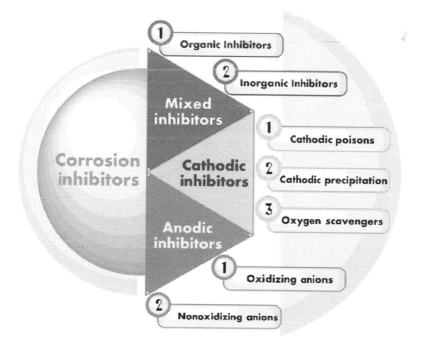

FIGURE 4–2 Types of corrosion inhibitors on the basis of their adsorption modes.

4.3 Synthesis methods of synthetic green corrosion inhibitors

Bicomponent, monocomponent, and multiple component reaction methods are commonly used to make corrosion inhibitors. Prior investigation has shown that bicomponent and multiple component reactions whose synthetic procedure involve multiple steps are not considered green routes. The main reason is the time, materials, and toxic route taken, while a one-pot multicomponent reaction is considered green chemistry. The main reason for this is that they seem to have less material, time, and energy and the yield of synthesized product is also good. Generally, MCRS are an advantage over the other synthetic routes due to its time saving, high yield, easy purification of product, and low waste production [35,36].

4.4 Experimental investigation on synthetic green corrosion inhibitors: Literature survey

The experimental techniques such as gravimetric analysis and electrochemical investigation are very important efficient tools to know about the fundamental characteristic of corrosion inhibitors like the nature of adsorption, inhibition efficiency, corrosion potential, and type of adsorption on to the metal surface by inhibitor molecule. Generally, with the help of weight loss measurement, it can be found in the lab and does not require any sophisticated and expensive instruments. An electrochemical system is usually a three electrode-based system, in which the metal for which corrosion inhibition is to be done is used as a working electrode. According to John et al. EIS reveals the higher %η of 94.24 at 200 ppm for AMTDA inhibitor molecules [37]. Also Dogdag revealed that Epoxy resin derivatives, namely ER2 and ER1 showed 95.6% and 98.1% respectively at 10^{-3} M [38]. Dogdag showed that as per PDP analysis phosphazene compound (HMC) had the %η of 99% at 10^{-3} M [39]. El Ibrahimi applied experimental approaches in which WL and EIS showed that the correct sequence of %η is given in the order: MPBD > PBD > DMBD, which may be due to presence of electron donating groups [40]. Alaoui studied triazepine carboxylate derivatives, namely Cl−Me−CO2Et, Me−CN, and Cl−Me−CN, and found that Cl−Me−CN showed the %η of 99% at 10^{-3} M [41]. According to EIS EOTPEH and EOTPE showed the %η greater than 90% at both corrosive media [42]. Rbaa studied MC simulation and showed the %η of 93.3 and 97.6 for Q-X and Q-T, respectively [43]. Mo applied a Schiff base, namely N-isonicotinamido-3-methoxy-4-hydroxybenzalaldimine (IM)-based vanillin, and isoniazid green corrosion inhibitors showed excellent inhibitor efficiency as per EIS [44]. Abdallah used phenyl sulfonylethanone derivatives Compound-1 and Compound-2 that showed the %η of 85.5 and 83.5 at 5×10^{-4} M, respectively [45]. Abd El-Raouf studied the inhibition efficiency of four bis(coumarins) compounds; inhibitor-IV showed the highest %η of 93.1 at 150 ppm [46]. Chugh et al.'s EIS study revealed the inhibition efficiency for thiadiazole compounds was in the order PMTTA > PATT > PMTA > PTA, in which PMTTA showed %η of 90.4 at 125 ppm [47]. Douche studied 8-hydroxyquinline derivatives, namely HM1, HM2, and HM3, and found the %η to be 90.00, 87.00, and 82.00, respectively, as per WL measurement [48]. According to Lgaz et al. EIS shows that HDZ1 showed the highest %η among all studied inhibitors [49]. As per Yadav et al. EIS analysis reported the %η of 95.7 for BIHT and 91.9 for MIHT. Their higher value was attributed to the presence of heteroatoms (N, O, S) and aromatic rings from which inhibiters strongly interact with Fe(110) surface [50]. According to Verma et al. EIS results showed APQD-4 inhibitor to have %η of 98.30% at 20 mg L^{-1}, the highest among all studied inhibitors [51]. Bokati et al. revealed the single and combined effect of 14-benzotriazole (BTA) toward copper and mild steel corrosion by applying SEM, EIS, and MD simulation. They found a good synergistic effect toward corrosion mitigation [52]. Farahati showed that 4-(pyridin-3-yl) thiazol-2-amine exhibited %η of 94.00% at 1 mM concentration according to EIS analysis [53]. Tan et al. found that the %η values of FFM and DFD were 97.5 and 98.9, respectively, at 5 mM concentration as per electrochemical analysis [54]. Table 4−1 summarizes the molecular structure of inhibitors, metals, and electrolyte, applied methods, and outcomes of experimental investigations.

4.5 Computational simulations on synthetic green corrosion inhibitors: Literature survey

In the corrosive aqueous electrolytic medium, the study of the adsorption behavior of the inhibitor between the inhibitor/metal interfaces is a very important phenomenon. The use of recent theoretically based calculations has proved to be a very important tool in the field of corrosion science. The main reason is the high accuracy and the time saved. These most important theoretically based calculations do not require any chemical nor are they repeated like a lab practical, that is, it is a completely green chemistry-based calculation [55,56]. Under theoretical calculation, DFT-based quantum chemical calculation and molecular simulation under MD simulation and MC simulation are prominent.

Usually, DFT-based calculations explain the reacting and interacting site of the inhibitor molecules. Different parameters, such as E_{HOMO} (highest occupied molecular orbital), E_{LOMO} (lowest unoccupied molecular orbital), dipole moment (μ), and global hardness and softness can be calculated from DFT. Based on this, an electron-donating and -accepting site of the organic inhibitor molecule is detected, which explains the interaction and adsorption between the metal inhibitors. In general, the higher the E_{HOMO} value and the lower the E_{LUMO} value, the more prominently will the molecule adsorb above the metal surface [57]. Ma et al. applied azole derivatives, namely BTA, TAZ, and TTLYK, and they proved to be efficient corrosion inhibitors on copper wiring [58]. Jmiai et al. showed that in alginate biopolymers E_{HOMO} value increased as its unit increased [59]. N-thiazolyl-2-cyanoacetamide derivatives showed the increasing E_{HOMO} values to be Inh1 > Inh2 > Inh3 [60]. According to Chaouiki et al. DFT calculation reveals −8.4983 E_{HOMO} and 1.1842 E_{LUMO} for the most effective inhibitor, which is AA-1 [61]. Belghitia et al. proved that according to DFT calculations among four hydrazine derivatives (HZ1−HZ4) HZ2 showed the highest E_{HOMO} value [62]. Fig. 4−3 demonstrates the optimized HOMO and LUMO structures of all studied inhibitors, revealing the high electron density on electron-rich sites (HOMO).

Table 4–1 Molecular structure of inhibitors, metals, and media, methods applied, and outcomes of experimental investigations.

S. no.	Molecular structure of organic molecules	Metal and electrolyte	Applied methods	Outcomes	References
1		MS, 1 N HCl	EIS, DFT, MC, and SEM	John et al. EIS reveals the higher %η of 94.24 at 200 ppm for AMTDA inhibitor molecules among all	[39]
2		MS, 1 M HCl	EIS, SEM, DFT, and MC	Dogdag epoxy resin derivatives namely ER2 and ER1 showing 95.6% and 98.1% respectively at 10^{-3} M	[40]
3		CS, 3 N NaCl	WL, EIS, DFT, and MC	Dogdag as per PDP analysis Phosphazene compound (HMC) showing the %η of 99% at 10^{-3} M	[41]
4		Iron, 1 M HCl	DFT and MC-SAA	Brahim WL and EIS shows that the correct sequence of %η is given as the order: MPBD > PBD > DMBD which may be due to presence of electron donating groups	[42]

(*Continued*)

Table 4–1 (Continued)

S. no.	Molecular structure of organic molecules	Metal and electrolyte	Applied methods	Outcomes	References
5		MS, 1 M HCl	EIS, AFM, DFT, and MC	Alaoui Triazepine carboxylate derivatives namely Cl–Me–CO2Et, Me–CN and Cl–Me–CN studied extensively in which Cl–Me–CN showing the %η of 99% at 10^{-3} M	[43]
6		MS, 1 M HCl and 0.5 M H SO	EIS, FTIR, EDX, FESEM, MC, and DFT	According to EIS EOTPEH and EOTPE showing the %η greater than 90% at both corrosive media	[44]
7		MS, 1 M HCl	EIS, PDP, SEM, JV–vs, MC, and DFT	Rbaa MC simulation showing the %η of 93.3 and 97.6 for Q-X and Q-T, respectively	[45]
8		Cu, NaC	EIS, XRD, EDS, DFT, and MC	Mp A Schiff base namely N-isonicotinamido-3-methoxy-4-hydroxybenzalaldimine (IM) based Vanillin and isoniazid green corrosion inhibitors showing excellent inhibitor efficiency as per EIS	[46]

(Continued)

Table 4–1 (Continued)

S. no.	Molecular structure of organic molecules	Metal and electrolyte	Applied methods	Outcomes	References
9		Al, 0.5 M H_2SO_4	EIS, EFM, SEM, DFT, and MC	Abdallah Phenyl sulfonylethanone derivatives (PSED) Compound-1 and Compound-2 showing the %η of 85.5 and 83.5 at 5×10^{-4} M, respectively	[47]
10		CS, 0.5 M H_2SO_4	PDP, EIS, DFT, and MD	El-Raouf Among four bis (coumarins) compounds, inhibitor-IV showing the highest %η of 93.1 at 150 ppm	[48]
11		MS, 1 M HCl	EIS, PDP, AFM, SEM, XRD, FTIR, DFT, and MD	Chugh EIS study reveals the inhibition efficiency for thiadiazole compounds was found to the order PMTTA > PATT > PMTA > PTA, in which PMTTA showing %η of 90.4 at 125Chughppm	[49]

(Continued)

Table 4–1 (Continued)

S. no.	Molecular structure of organic molecules	Metal and electrolyte	Applied methods	Outcomes	References
12		MS, 1 M HCl	EIS, WL, MD, and DFT	Douche 8-hydroxyquinline derivatives namely HM1, HM2 AND HM3 showing the %η of 90.00, 87.00, and 82.00, respectively as per WL measurement	[50]
13		MS, 1 M HCl	DM, DFT, WL, EIS, PDP, and MD	Lgaz EIS shows that HDZ1 showing highest %η among all studied inhibitors	[51]
14		N80 steel, 15% HCl	EIS, SEM, EDX, AFM, XPS, UV–vis, MD, and DFT	Yadav et al. EIS analysis reported the %η of 95.7 for BIHT and 91.9 for MIHT. Their higher value attributed to presence of heteroatoms (N, O, S) and aromatic rings from that inhibiters strongly interact with Fe(110) surface	[52]

(Continued)

Table 4–1 (Continued)

S. no.	Molecular structure of organic molecules	Metal and electrolyte	Applied methods	Outcomes	References
15		MS, 1 M HCl	EIS, WL, SEM, DFT, and MD	Verma et al. EIS result of APQD-4 inhibitor showing the %η of 98.30% at 20 mg L^{-1} highest among all studied inhibitors	[53]
16		MS, Cu, 1 M HCl	PDP, EIS, SEM, DFT, and MD	Bokati et al. studied single and combined effect of 14-benzotriazole (BTA) toward copper and mild steel corrosion by applying SEM, EIS, and MD simulation. They found the good synergistic effect toward corrosion mitigation	[54]
17		Cu, 1 M HCl	EIS, SEM, AFM DFT, and MD	Farahati 4-(pyridin-3-yl) thiazol-2-amine (PyTA) exhibited %η of 94.00% at 1 mM concentration according to EIS analysis	[55]
18		Cu, 0.5 M H$_2$SO$_4$	EIS, XPS, AFM, SEM, DFT, and MD	Tan The %η values of FFM and DFD are 97.5 and 98.9 respectively at 5 mM Concentration as per electrochemical analysis	[56]

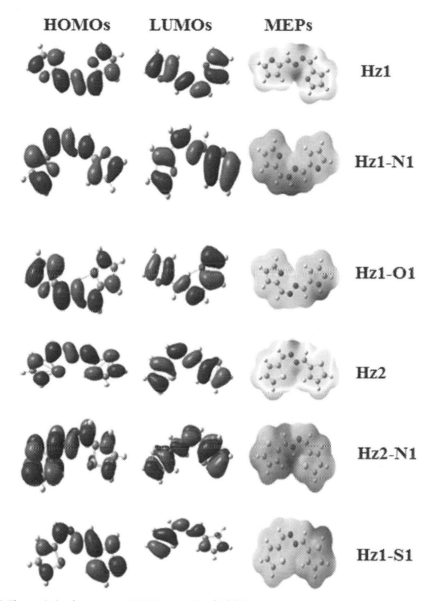

FIGURE 4–3 The optimized structures, HOMOs, LUMOs of inhibitors HZ1-HZ4. *HOMO*, Highest occupied molecular orbital; *LUMO*, lowest occupied molecular orbital. Source: *M.E. Belghitia, S. Echihic, A. Dafalib, Y. Karzazih, M. Bakassed, H. Elalaoui-Elabdallaouie, et al., Computational simulation and statistical analysis on the relationship between corrosion inhibition efficiency and molecular structure of some hydrazine derivatives in phosphoric acid on mild steel surface, Appl. Surf. Sci. 491 (2019) 707–722. Copyright 2019@Elsevier.*

Hsissou et al. showed that according to DFT calculations DGDCBA epoxy polymer showed excellent inhibition properties [63]. Fig. 4–4 shows the optimized HOMO and LUMO structures of DGDCBA molecules in which the highest electron density is located on the electron-rich sites like benzene ring and heteroatoms.

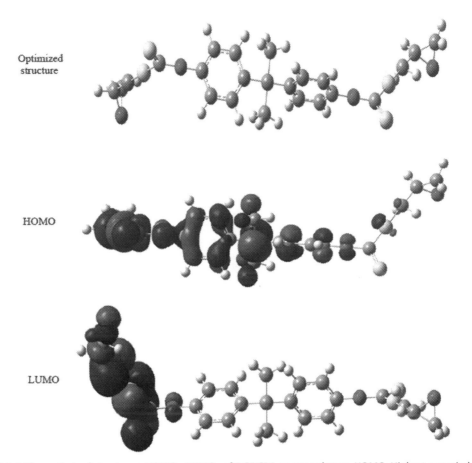

Optimized
structure

HOMO

LUMO

FIGURE 4–4 The optimized structures, HOMOs, LUMOs of DGDCBA epoxy polymer. *HOMO*, Highest occupied molecular orbital; *LUMO*, lowest occupied molecular orbital. Source: *R. Hsissou, S. Abbout, A. Berisha, M. Berradi, M. Assouag, N. Hajjaji, et al., Experimental, DFT and molecular dynamics simulation on the inhibition performance of the DGDCBA epoxy polymer against the corrosion of the E24 carbon steel in 1.0 M HCl solution, J. Mol. Struct. 1182 (2018) 340–351. Copyright 2018@Elsevier.*

Madkour et al. revealed that DFT calculations computed the high E_{HOMO} values for all studied inhibitors (BAD1–BAD-4), in which BAD1 showed the highest E_{HOMO} [64]. Saha and Banerjee used quantum chemical calculations to calculate E_{HOMO} of -6.1977, -6.1133, and -5.9400 eV for inhibitors L1, L2, and L3, respectively [65].

Based on a similar MD/MC simulation, the most important parameter energy of adsorption (E_{ads}) is known. Usually the higher the value of E_{ads}, the higher the adsorption property of that inhibitor. With the help of a simulation technique, one also learns about the stable orientation of the inhibitor molecule above the metal surface. If the orientation of the inhibitor molecule is flat, it will block the more active site above the metal surface and behave like an effective inhibitor. Generally, molecular energy reveals total energy (E_{total}), rigid energy (E_{rigid}), deformation energy (E_{def}), interaction energy ($E_{intraction}$), and binding energy ($E_{binding}$) [66−69]. MC simulation revealed the highest value of adsorption energy (E_{ads}) for Tryptophan (Try), which may be due to the presence of the indole ring [70]. Four quinazolinone derivatives QZ-CH$_3$, QZ-OH, QZ-NO$_2$, and QZ-H exhibited higher E_{ads} in which QZ-CH3 showed the highest E_{ads} of −4066.00 kJ mol^{-1} [71]. Dipyrimidin inhibitors CTDP and OPDB revealed their excellent inhibitor properties as per EIS and theoretical calculations [72]. As per MC simulation phosphorus polymer (PGEPBAP) showed the adsorption energy of −1.76 kJ mol^{-1} [73]. 8-hydroxyquinoline derivatives, namely Q-CH$_3$, Q-Br, and Q-H, were applied as potential corrosion moieties and the Q-CH3/Fe (110) system showed the highest E_{ads} value (−4257.18 kJ mol^{-1}) [20]. Arylazoazo-1,2,4-triazole derivatives (AATR 1−3) showed E_{ads} of −63.489 kJ mol^{-1} for the Cu(111)/AATR-3 system [74]. 5-Substituted-8-hydroxyquinline derivatives, namely EHQP and BHQC, exhibited the E_{ads} of −121.602 and −133.600 kJ mol^{-1}, respectively [75]. Fig. 4−5 demonstrated the side and top view of the orientation of EHQP and BHQC where a long chain is parallel to the metal surface and the heterocyclic ring is oriented perpendicular to the metal surface.

MD simulation reveals the highest binding energy (570.95 kJ mol^{-1}) and interaction energy (−570.95 kJ mol^{-1}) for quercetin [76]. An MD simulation-based calculation revealed a calculated adsorption energy (E_{ads}) of −438 and −332.0 kcal mol^{-1} for DSM and AND, respectively [77]. MD simulation revealed adsorption energies of −197.9, 202, and

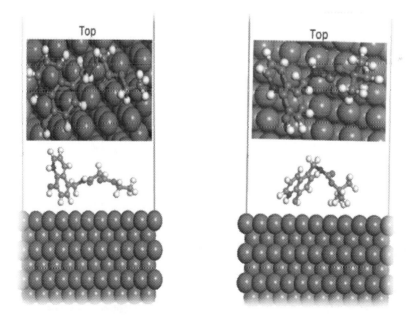

FIGURE 4–5 MC simulation of top and side views of the most stable orientation for the adsorption of EHQP (left) and BHQC (right) inhibitors on Fe (110) interface. *MC,* Monte Carlo. Source: *M. El Faydy, B. Lakhrissi, A. Guenbour, S. Kaya, F. Bentiss, I. Waradf, et al., In situ synthesis, electrochemical, surface morphological, UV−visible, DFT and Monte Carlo simulations of novel 5-substituted-8-hydroxyquinoline for corrosion protection of carbon steel in a hydrochloric acid solution, J. Mol. Liq. 280 (2019) 341−359. Copyright 2019@Elsevier.*

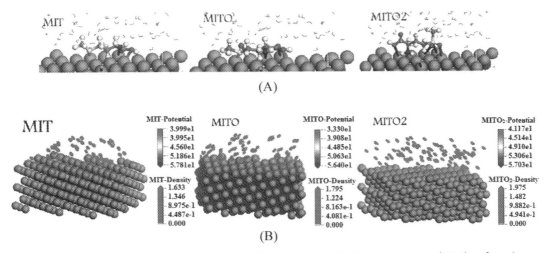

FIGURE 4–6 Adsorption mode (A) and density (B) of the L-methionine derivatives on copper(1 1 1) surface. *Source: K.F. Khaled, Corrosion control of copper in nitric acid solutions using some amino acids—a combined experimental and theoretical study, Corros. Sci. 52 (2010) 3225–3234. Copyright 2010@Elsevier.*

-235.9 kJ mol^{-1} for MITO2, MITO, and MIT, respectively [78]. Fig. 4–6 represents the adsorption mode and adsorption density of L-methionine derivatives on Cu(111) surface in which all are flat oriented on to the metal surface. Among the three benzo derivatives, BTH showed the lowest adsorption energy (-305.31 kJ mol^{-1}) and BTA showing highest adsorption energy (-356.74 kJ mol^{-1}) [79]. Table 4–2 summarizes the molecular structure of inhibitors, metals, and electrolytes, the applied methods, and outcomes of computational simulations.

4.6 Adsorption mechanism

There are usually two types of adsorption between inhibitor molecules and metal surface: (1) physisorption and (2) chemisorption [80]. Physisorption is where a corrosive anion, such as Cl$^-$, SO4^{2-}, PO4^{3-}, or NO^{3-}, forms a layer above the metal surface in the presence of high acidic medium and a protonated lone pair containing heteroatom present in the inhibitor molecule. Here there is the interaction between the protonated inhibitor molecules and deposit of the anion above the metal surface. Previous studies suggested that if the value of Gibbs adsorption energy (ΔG_{ads}) is more than -20 kJ mol^{-1}, it is usually kept under physisorption [81,82]. Whereas chemisorption is a phenomenon in which there is no direct interaction between inhibitor molecules and the metal surface. In this, the unshared and unsaturated pi-electrons present in the inhibitor molecules form a strong base by interacting with the vacant "d" orbital of the metal surface and are deposited above the metal surface as a protective thin layer. Previous studies suggested that if the value of Gibbs adsorption energy (ΔG_{ads}) is more than -40 kJ mol^{-1}, it is usually kept under chemisorption [83]. If seen, corrosion inhibition is a kind of acid–base based mechanism, in which the inhibitor(s) behaves like a Lewis base, that is, it can donate its electron, whereas a metal Lewis behaves like acid, that is, it accepts an electron.

Table 4-2 Molecular structure of inhibitors, metals, and media, simulation software applied, methods used, and outcomes of computational simulations.

S. no.	Molecular structure organic molecules	Software applied for MD/MC simulation	Metal(s)	Applied methods	Outcomes MC/MD simulations	References
1		Gaussian 9.0 Program	Cu, KOH, and HNO_3	EIS, SEM, DFT, and MC	Ma et al. Azole derivatives namely BTA, TAZ and TTLYK proved as efficient corrosion inhibitor on copper wiring as per studied techniques	[60]
2		Gaussian 9.0 Program	Cu, 1 M HCl	EIS, XRD, AFM, DFT, and MC	Jmiai et al. Alginate biopolymers E_{HOMO} value increasing as its unit increases	[61]
3		Gaussian 9.0 software	Al, NaCl	MC, MD, and DFT	N-thiazolyl-2-cyanoacetamide derivatives showing the increasing E_{HOMO} values as Inh1 > Inh2 > Inh3	[62]
4		Gaussian 9.0 Program	N'S	EIS, DFT, WL, MD, and SEM	Chaouiki et al. DFT calculation reveals the −8.4983 HOMO and 1.1842 LUMO for most effective inhibitor that is AA-1	[63]
5		Gaussian (09W) software	E2< CS, 1 M HCl	EIS, WL, DFT, and MD	Hsssou et al. According to DFT calculation DGDCBA epoxy polymer showing excellent inhibition property	[65]

(Continued)

Table 4–2 (Continued)

S. no.	Molecular structure organic molecules	Software applied for MD/MC simulation	Metal(s)	Applied methods	Outcomes MC/MD simulations	References
6		Gaussian 9.0 Program	Iron, HNO_3 and NaOH	WL, EIS, DFT, and MD	Madkour et al. DFT calculation computed the high HOMO values for all studied inhibitors (BAD1–BAD-4) in which BAD1 showing the highest E_{HOMO} as compared to others	[66]
7		ORCA program package, 2.7.0 version	MS, 1 M HCl	WL, EIS, FESEM, AFM, MD, and DFT	Saha et al. Quantum chemical calculations calculated the −6.1977, −6.1133, and −5.9400 eV HOMO for inhibitors L1, L2, and L3 respectively	[67]
8		Material Studio 6.0 software (Accelrys)	Fe (110), 1 M HCl	EIS, SEM, UV–vis, MC, and DFT	Four quinazolinone derivatives QZ-CH_3 QZ-OH QZ-NO_2 and QZ-H exhibited higher E_{ads} in which QZ-CH_3 showed the highest E_{ads} −4066.00 kJ mol^{-1}	[73]
9		Material Studio	Fe (110), 15% HCl	EIS, FESEM, AFM, DFT, and MC	Dipyrimidin inhibitors CTDP and OPDB reveals the excellent inhibitor properties as per EIS and theoretical calculations	[74]

(Continued)

Table 4–2 (Continued)

S. no.	Molecular structure organic molecules	Software applied for MD/MC simulation	Metal(s)	Applied methods	Outcomes MC/MD simulations	References
10		Material Studio 2017 software	Fe (110), 1 M HCl	EIS, WL, ESP SEM, RDF, MD, MC, and DFT	As per MC simulation phosphorus polymer (PGEPBAP) showing the adsorption energy of -1.76 kJ mol^{-1}	[75]
11		Material Studio 7.0 software, adsorption locator module (Accelrys, San Diego, CA, USA).	Fe (110), 1 M HCl	EIS, SEM, UV–vis. MC, and DFT	5-Substituted-8-hydroxyquinline derivatives namely EHQP and BHQC exhibited the E_{ads} of -121.602 and -133.600 kJ mol^{-1}, respectively	[78]
12		Material Studio software	Fe (110), HCl	DFT and MD	MD simulation reveals the highest binding energy (570.95 kJ mol^{-1}) and interaction energy (-570.95 kJ mol^{-1}) for quercetin	[79]

(Continued)

Table 4–2 (Continued)

S. no.	Molecular structure organic molecules	Software applied for MD/MC simulation	Metal(s)	Applied methods	Outcomes MC/MD simulations	References
13		Forcite module from Accelnys Inc.	Cu (111), NaCl	SEM, AFM, EIS, DFT, and MD	MD simulation-based calculation showing the calculated adsorption energy (E_{ads}) of −438 and −332.0 kcal mol^{-1} for DSM and AND, respectively	[80]
14		Materials Studio	Cu (111), 1 M HNO$_3$	MD and DFT	MD simulation reveals the adsorption energies of −197.9, 202 and −235.9 kJ mol^{-1} for MITO2, MITO, and MIT, respectively	[81]
15		Material Studio, COMPASS	Cu(110), 3.5 wt.% NaCl	EIS, SEM, XPS, DFT and MD	Among the three benzo derivatives, BTH showed lowest adsorption energy (−305.31 kJ mol^{-1}) and BTA showed highest adsorption energy (−356.74 kJ mol^{-1})	[82]

4.7 Conclusion

The above discussion shows that green synthetic corrosion inhibitors are predominantly applied toward corrosion mitigation of metals and alloys in various acidic and basic corrosive solutions. They are used predominantly due to their potential behavior, solubility in water, and stability in corrosive media at different temperature. Except for some synthetic organic molecules, they follow the green route from formation of the organic molecule to application. The discussion suggests that inhibitors used as corrosion mitigation are mainly heterocyclic compounds in whose ring pi-electrons and lone-pair electrons are present. Experimental and theoretical calculations have been used mainly to explore the potential properties of these inhibitors. Both experimental and theoretical investigations show excellent agreement toward corrosion prevention by inhibitors. Finally, the ongoing discussion concludes that a pot synthesis multicomponent approach reaction associated with ultrasound facilitated the synthesis of green sustainable corrosion inhibitors.

Acknowledgment

Dr. Dakeshwar Kumar Verma greatly acknowledges the Principal Govt. Digvijay College Rajnandgaon, Chhattisgarh for providing lab and basic instrumental facilities.

Conflict of interest

The authors declared no conflict of interest

Useful links

1. https://www.journals.elsevier.com/corrosion-science#: ~ :text = Corrosion%20Science%20prov ides%20a%20medium,metallic%20and%20non%2Dmetallic%20corrosion.&text = It%20forms% 20an%20important%20link,of%20corrosion%20and%20degradation%20phenomena.
2. https://www.corrosionjournal.org/?utm_term = corrosion%20journals%20list&utm_campaign = Corrosion + Journal + Subscriptions&utm_source = adwords&utm_medium = > ppc&hsa_ acc = 7392379026&hsa_cam = 6447992041&hsa_grp = 81133065987&hsa_ad = 409613849897 &hsa_src = g&hsa_tgt = kwd-296475749429&hsa_kw = corrosion%20journals%20list&hsa_mt = b&hsa_net = adwords&hsa_ver = 3&gclid = CjwKCAjw2dD7BRASEiwAWCtCbzz4Ap0isqM 0xS8-HGw3fOpSGaKJ-GGRpmg0w3_QP_-yAibHgKObZxoCZZkQAvD_BwE.
3. https://pubs.rsc.org/en/journals/journalissues/ra#!issueid ─ ra010059&type = current&issnon line = 2046-2069.

References

[1] G. Koch, 1 - Cost of corrosion, in: A.M. El-Sherik (Ed.), Trends in Oil and Gas Corrosion Research and Technologies, Woodhead Publishing, Boston, 2017, pp. 3–30. Available from: https://doi.org/10.1016/ B978-0-08-101105-8.00001-2.

[2] L.O. Olasunkanmi, E.E. Ebenso, Experimental and computational studies onpropanone derivatives of quinoxalin-6-yl-4,5-dihydropyrazole as inhibitors of mild steel corrosion in hydrochloric acid, J. Colloid Interface Sci. 561 (2020) 104−116. Available from: https://doi.org/10.1016/j.jcis.2019.11.097.

[3] B. Tan, S. Zhang, Y. Qiang, L. Guo, L. Feng, C. Liao, et al., A combined experimental and theoretical study of the inhibition effect of three disulfide-based flavouring agents for copper corrosion in 0.5 M sulfuric acid, J. Colloid Interface Sci. 526 (2018) 268−280. Available from: https://doi.org/10.1016/j.jcis.2018.04.092.

[4] K. Vinothkumar, M.G. Sethuraman, Corrosion inhibition ability of electropolymerised composite film of 2-amino-5-mercapto-1,3,4-thiadiazole/TiO$_2$ deposited over the copper electrode in neutral medium, Mater. Today Commun. 14 (2018) 27−39. Available from: https://doi.org/10.1016/j.mtcomm.2017.12.007.

[5] D. Verma, F. Khan, Corrosion inhibition of high carbon steel in phosphoric acid solution by extract of black tea, Adv. Res. 5 (4) (2015) 1−9. Available from: https://doi.org/10.9734/air/2015/18723.

[6] D.K. Verma, Density functional theory (DFT) as a powerful tool for designing corrosion inhibitors in aqueous phase, Advanced Engineering Testing, InTech, 2018. Available from: https://doi.org/10.5772/intechopen.78333.

[7] M. Yildiz, H. Gerengi, M.M. Solomon, E. Kaya, S.A. Umoren, Influence of 1-butyl-1-methylpiperidinium tetrafluoroborate on St37 steel dissolution behavior in HCl environment, Chem. Eng. Commun. 205 (4) (2018) 538−548.

[8] B.E.A. Rani, B.B.J. Basu, Green inhibitors for corrosion protection of metals and alloys: an overview, Int. J. Corros. 2012 (2012) 380217. Available from: https://doi.org/10.1155/2012/380217.

[9] C. Lai, X. Guo, J. Wei, B. Xie, L. Zou, X. Li, et al., Investigation on two compounds of O,O′-dithiophosphate derivatives as corrosion inhibitors for Q235 steel in hydrochloric acid solution, Open Chem. 15 (1) (2017) 263−271.

[10] J. Wang, Y. Lin, A. Singh, W. Liu, Investigation of some porphyrin derivatives as inhibitors for corrosion of N80 steel at high temperature and high pressure in 3.5% NaCl solution containing carbon dioxide, Int. J. Electrochem. Sci. 13 (12) (2018) 11961−11973.

[11] P.P. Kumari, P. Shetty, S.A. Rao, D. Sunil, T. Vishwanath, Synthesis, characterization and anticorrosion behaviour of a novel hydrazide derivative on mild steel in hydrochloric acid medium, Bull. Mater. Sci. 43 (1) (2020) 46. Available from: https://doi.org/10.1007/s12034-019-1995-x.

[12] D.K. Verma, A. Al Fantazi, C. Verma, F. Khan, A. Asatkar, C.M. Hussain, et al., Experimental and computational studies on hydroxamic acids as environmental friendly chelating corrosion inhibitors for mild steel in aqueous acidic medium, J. Mol. Liq. 314 (2020) 113651. Available from: https://doi.org/10.1016/j.molliq.2020.113651.

[13] D. Kumar Verma, E.E. Ebenso, M.A. Quraishi, C. Verma, Gravimetric, electrochemical surface and density functional theory study of acetohydroxamic and benzohydroxamic acids as corrosion inhibitors for copper in 1 M HCl, Results Phys. 13 (2019) 102194. Available from: https://doi.org/10.1016/j.rinp.2019.102194.

[14] Z. Salarvand, M. Amirnasr, M. Talebian, K. Raeissi, S. Meghdadi, Enhanced corrosion resistance of mild steel in 1 M HCl solution by trace amount of 2-phenyl-benzothiazole derivatives: experimental, quantum chemical calculations and molecular dynamics (MD) simulation studies, Corros. Sci. 114 (2017) 133−145. Available from: https://doi.org/10.1016/j.corsci.2016.11.002.

[15] A. Dutta, S.K. Saha, U. Adhikari, P. Banerjee, D. Sukul, Effect of substitution on corrosion inhibition properties of 2-(substituted phenyl) benzimidazole derivatives on mild steel in 1 M HCl solution: a combined experimental and theoretical approach, Corros. Sci. 123 (2017) 256−266.

[16] Y. El Aoufir, H. Lgaz, H. Bourazmi, Y. Kerroum, A. Ramli, A. Guenbour, et al., Quinoxaline derivatives as corrosion inhibitors of carbon steel in hydrochloridric acid media: electrochemical, DFT and montecarlo simulations studies, J. Mater. Environ. Sci. 7 (12) (2016) 4330−4347.

[17] A. Singh, K.R. Ansari, A. Kumar, W. Liu, C. Songsong, Y. Lin, Electrochemical, surface and quantum chemical studies of novel imidazole derivatives as corrosion inhibitors for J55 steel in sweet corrosive environment, J. Alloys Compd. 712 (2017) 121−133.

[18] M. Ellouz, N.K. Sebbar, H. Elmsellem, H. Steli, I. Fichtali, M.M. Mohamed Abdelahi, et al., Inhibitive properties and quantum chemical studies of 1,4-benzothiazine derivatives on mild steel corrosion in acidic medium, J. Mater. Environ. Sci. 7 (8) (2016) 2806−2819.

[19] J. Haque, V. Srivastava, D.S. Chauhan, H. Lgaz, M.A. Quraishi, Microwave-induced synthesis of chitosan Schiff bases and their application as novel and green corrosion inhibitors: experimental and theoretical approach, ACS Omega 3 (2018) 5654−5668. Available from: https://doi.org/10.1021/acsomega.8b00455.

[20] M. Rbaa, M. Galai, A.S. Abousalem, B. Lakhrissi, M.E. Touhami, I. Warad, et al., Synthetic, spectroscopic characterization, empirical and theoretical investigations on the corrosion inhibition characteristics of mild steel in molar hydrochloric acid by three novel 8-hydroxyquinoline derivatives, Ionics 26 (2019) 503. Available from: https://doi.org/10.1007/s11581-019-03160-9.

[21] J. Liu, Y. Zhou, C. Zhou, H. Lu, 1-Phenyl-1H-tetrazole-5-thiol as corrosion inhibitor for Q235 steel in 1 M HCl medium: combined experimental and theoretical researches, Int. J. Electrochem. Sci. 15 (3) (2020) 2499−2510. Available from: https://doi.org/10.20964/2020.03.76.

[22] R.V. Percival, C.H. Schroeder, A.S. Miller, J.P. Leape, Environmental Regulation: Law, Science, and Policy, Wolters Kluwer Law & Business, 2015.

[23] R. Sanghi, V. Singh, Green Chemistry for Environmental Remediation, John Wiley & Sons, 2012.

[24] D.K. Verma, F. Khan, Green approach to corrosion inhibition of mild steel in hydrochloric acid medium using extract of spirogyra algae, Green. Chem. Lett. Rev. 9 (1) (2016) 52−60.

[25] D.K. Verma, F. Khan, I. Bahadur, M. Salman, M.A. Quraishi, E.E. Ebenso, et al., Inhibition performance of Glycine max, *Cuscuta reflexa* and Spirogyra extracts for mild steel dissolution in acidic medium: density functional theory and experimental studies, Results Phys. 10 (2018) 665−674.

[26] D.K. Verma, F. Khan, Corrosion inhibition of mild steel in hydrochloric acid using extract of glycine max leaves, Res. Chem. Intermed. 42 (2016) 3489−3506.

[27] D. Verma, F. Khan, Corrosion inhibition of mild steel by using sulpha drugs in phosphoric acid medium: a combined experimental and theoretical approach, Am. Chem. Sci. J. 14 (3) (2016) 1−8. Available from: https://doi.org/10.9734/acsj/2016/26282.

[28] M.E. Mashuga, L.O. Olasunkanmi, A.S. Adekunle, S. Yesudass, M.M. Kabanda, E.E. Ebenso, Adsorption, thermodynamic and quantum chemical studies of 1-hexyl 3 methylimidazolium based ionic liquids as corrosion inhibitors for mild steel in HCl, Materials 8 (2015) 3607−3632.

[29] O. Olivares-Xometl, C. López-Aguilar, P. Herrastí-González, N.V. Likhanova, I. Lijanova, R. Martínez-Palou, et al., Adsorption and corrosion inhibition performance by three new ionic liquids on API 5L X52 steel surface in acid media, Ind. Eng. Chem. Res. 53 (2014) 9534−9543.

[30] C. Verma, M.A. Quraishi, E.E. Ebenso, I.B. Obot, A. El Assyry, 3-Amino alkylated indoles as corrosion inhibitors for mild steel in 1M HCl: experimental and theoretical studies, J. Mol. Liq. 219 (2016) 647−660.

[31] S. Weiss, J. Jakobs, T. Reemtsma, Discharge of three benzotriazole corrosion inhibitors with municipal wastewater and improvements by membrane bioreactor treatment and ozonation, Environ. Sci. Technol. 40 (2006) 7193−7199.

[32] Obot, N. Obi-Egbedi, Theoretical study of benzimidazole and its derivatives and their potential activity as corrosion inhibitors, Corros. Sci. 52 (2010) 657−660.

[33] D. Zhang, Y. Tang, S. Qi, D. Dong, H. Cang, G. Lu, The inhibition performance of long-chain alkyl-substituted benzimidazole derivatives for corrosion of mild steel in HCl, Corros. Sci. 102 (2016) 517−522.

[34] J. Haque, K. Ansari, V. Srivastava, M. Quraishi, I. Obot, Pyrimidine derivatives as novel acidizing corrosion inhibitors for N80 steel useful for petroleum industry: a combined experimental and theoretical approach, J. Ind. Eng. Chem. 49 (2017) 176−188.

[35] J.R. Donald, R.R. Wood, S.F. Martin, Application of a sequential multicomponent assembly process/ Huisgen cycloaddition strategy to the preparation of libraries of 1,2,3-triazole-fused 1, 4-benzodiazepines, ACS Comb. Sci. 14 (2012) 135–143.

[36] L.K. Ransborg, M. Overgaard, J. Hejmanowska, S. Barfüsser, K.A. Jørgensen, è. Albrecht, Asymmetric formation of bridged benzoxazocines through an organocatalytic multicomponent dienamine-mediated one-pot cascade, Org. Lett. 16 (2014) 4182–4185.

[37] S. John, A. Joseph, T. Sajini, A. James Jose, Corrosion inhibition properties of 1,2,4-hetrocyclic systems: electrochemical, theoretical and Monte Carlo simulation studies, Egypt J. Pet. 26 (2017) 721–732.

[38] O. Dagdag, Z. Safi, H. Erramli, N. Wazzan, I.B. Obot, E.D. Akpan, et al., Anticorrosive property of heterocyclic based epoxy resins on carbon steel corrosion in acidic medium: electrochemical, surface morphology, DFT and Monte Carlo simulation studies, J. Mol. Liq. 287 (2019) 110977.

[39] O. Dagdag, A. El Harfi, M. El Gouri, Z. Safi, R.T.T. Jalgham, N. Wazzan, et al., Anticorrosive properties of Hexa (3-methoxy propan-1,2-diol) cyclotri-phosphazene compound for carbon steel in 3% NaCl medium: gravimetric, electrochemical, DFT and Monte Carlo simulation studies, Heliyon 5 (2019) e01340. Available from: https://doi.org/10.1016/j.heliyon.2019.e01340.

[40] B. El Ibrahimi, Atomic-scale investigation onto the inhibition process of three 1,5-benzodiazepin-2-one derivatives against iron corrosion in acidic environment, Colloid Interface Sci. Commun. 37 (2020) 100279.

[41] K. Alaoui, M. Ouakki, A.S. Abousalem, H. Serrar, M. Galai, S. Derbali, et al., Molecular dynamics, Monte-Carlo simulations and atomic force microscopy to study the interfacial adsorption behaviour of some triazepine carboxylate compounds as corrosion inhibitors in acid medium, J. Bio-Tribo-Corrosion 5 (1) (2019). Available from: https://doi.org/10.1007/s40735-018-0196-2.

[42] H.M. Abd El-Lateef, Z.A. Abdallah, M.S. Mohamed Ahmed, Solvent-free synthesis and corrosion inhibition performance of ethyl 2-(1,2,3,6-tetrahydro-6-oxo-2-thioxopyrimidin-4-yl)ethanoate on carbon steel in pickling acids: experimental, quantum chemical and Monte Carlo simulation studies, J. Mol. Liq. 296 (2019) 111800.

[43] M. Rbaa, A.S. Abousalem, Z. Rouifi, R. Benkaddour, P. Dohare, M. Lakhrissi, et al., Synthesis, antibacterial study and corrosion inhibition potential of newly synthesis oxathiolan and triazole derivatives of 8-hydroxyquinoline: experimental and theoretical approach, Surf. Interfaces 2020 (2020) 100468. Available from: https://doi.org/10.1016/j.surfin.2020.100468.

[44] S. Mo, L.J. Li, H.Q. Luo, N.B. Li, An example of green copper corrosion inhibitors derived from flavor and medicine: vanillin and isoniazid, J. Mol. Liq. 242 (2017). Available from: https://doi.org/10.1016/j.molliq.2017.07.081.

[45] Y.M. Abdallah, Electrochemical studies of phenyl sulphonyl ethanone derivatives compounds on corrosion of aluminum in 0.5 M H$_2$SO$_4$ solutions, J. Mol. Liq. 219 (2016) 709–719.

[46] M. Abd El-Raouf, E.A. Khamis, M.T.H. Abou Kana, N.A. Negm, Electrochemical and quantum chemical evaluation of new bis(coumarins) derivatives as corrosion inhibitors for carbon steel corrosion in 0.5M H$_2$SO$_4$, J. Mol. Liq. 255 (2017) 341–353. Available from: https://doi.org/10.1016/j.molliq.2018.01.148.

[47] B. Chugh, A.K. Singh, S. Thakur, B. Pani, H. Lgaz, I.-M. Chung, et al., Comparative investigation of corrosion-mitigating behavior of thiadiazole-derived bis-schiff bases for mild steel in acid medium: experimental, theoretical, and surface study, ACS Omega 5 (23) (2020) 13503–13520. Available from: https://doi.org/10.1021/acsomega.9b04274.

[48] D. Douche, H. Elmsellem, E. Anouar, L. Guo, B. Hafez, B. Tüzün, et al., Anti-corrosion performance of 8-hydroxyquinoline derivatives for mild steel in acidic medium: gravimetric, electrochemical, DFT andmolecular dynamics simulation investigations, J. Mol. Liq. 308 (2020) 113042.

[49] H. Lgaz, R. Salghi, S. Masroor, et al., Assessing corrosion inhibition characteristics of hydrazone derivatives on mild steel in HCl: insights from electronic-scale DFT and atomic-scale molecular dynamics, J. Mol. Liq. 308 (2020) 112998. Available from: https://doi.org/10.1016/j.molliq.2020.112998.

[50] M. Yadav, T.K. Sarkar, I.B. Obot, Carbohydrate compounds as green corrosion inhibitor: electrochemical, XPS, DFT and molecular dynamics simulation studies, RSC Adv. (2016). Available from: https://doi.org/10.1039/C6RA24026G.

[51] C. Verma, L.O. Olasunkanmi, I.B. Obot, E.E. Ebenso, M.A. Quraishi, 5-Arylpyrimido-[4,5-b]quinoline-diones as new and sustainable corrosion inhibitors for mild steel in 1 M HCl: a combined experimental and theoretical approach, RSC Adv. 6 (2016) 15639.

[52] K. Sabet Bokati, C. Dehghanian, S. Yari, Corrosion inhibition of copper, mild steel and galvanically coupled copper-mild steel in artificial sea water in presence of 1H-benzotriazole, sodium molybdate and sodium phosphate, Corros. Sci. 126 (2017) 272–285. Available from: http://doi.org/10.1016/j.corsci.2017.07.009.

[53] R. Farahati, H. Behzadi, S.M. Mousavi-Khoshdel, A. Ghaffarinejad, Evaluation of corrosion inhibition of 4-(pyridin-3-yl) thiazol-2-amine for copper in HCl by experimental and theoretical studies, J. Mol. Struct. 1205 (2020) 127658. Available from: https://doi.org/10.1016/j.molstruc.2019.127658.

[54] B. Tan, S. Zhang, Y. Qiang, W. Li, H. Liu, C. Xu, et al., Insight into the corrosion inhibition of copper in sulfuric acid via two environmentally friendly food spices: combining experimental and theoretical methods, J. Mol. Liq. 286 (2019) 110891.

[55] Obot, D. Macdonald, Z. Gasem, Density functional theory (DFT) as a powerful tool for designing new organic corrosion inhibitors. Part 1: an overview, Corros. Sci. 99 (2015) 1–30.

[56] C. Verma, H. Lgaz, D.K. Verma, E.E. Ebenso, I. Bahadur, M.A. Quraishi, Molecular dynamics and Monte Carlo simulations as powerful tools for study of interfacial adsorption behavior of corrosion inhibitors in aqueous phase: a review, J. Mol. Liq. 260 (2018) 99–120.

[57] J. Vosta, J. Eliasek, Study on corrosion inhibition from aspect of quantum chemistry, Corros. Sci. 11 (1971) 223–229.

[58] T. Maa, B. Tana, Y. Xua, D. Yina, G. Liua, N. Zenga, et al., Corrosion control of copper wiring by barrier CMP slurry containing azole inhibitor: combination of simulation and experiment, Colloids Surf. A 599 (2020) 124872.

[59] Jmiai, B. El Ibrahimi, A. Tara, S. El Issami, O. Jbara, L. Bazzi, Alginate biopolymer as green corrosion inhibitor for copper in 1 M hydrochloric acid: experimental and theoretical approaches, J. Mol. Struct. 1157 (2017) 408–417. Available from: https://doi.org/10.1016/j.molstruc.2017.12.060.

[60] X.Y. Zhang, Q.X. Kang, Y. Wang, Theoretical study of N-thiazolyl-2-cyanoacetamide derivatives as corrosion inhibitor for aluminum in alkaline environments, Comput. Theor. Chem. 1131 (2018) 25–32. Available from: https://doi.org/10.1016/j.comptc.2018.03.026.

[61] A. Chaouiki, H. Lgaz, R. Salghi, M. Chafiq, H. Oudda, Shubhalaxmi, et al., Assessing the impact of electron-donating-substituted chalcones on inhibition of mild steel corrosion in HCl solution: experimental results and molecular-level insights, Colloids Surf. A: Physicochem. Eng. Asp. 588 (2019) 124366. Available from: https://doi.org/10.1016/j.colsurfa.2019.124366.

[62] M.E. Belghitia, S. Echihic, A. Dafalib, Y. Karzazib, M. Bakassed, H. Elalaoui-Elabdallaouie, et al., Computational simulation and statistical analysis on the relationship between corrosion inhibition efficiency and molecular structure of some hydrazine derivatives in phosphoric acid on mild steel surface, Appl. Surf. Sci. 491 (2019) 707–722.

[63] R. Hsissou, S. Abbout, A. Berisha, M. Berradi, M. Assouag, N. Hajjaji, et al., Experimental, DFT and molecular dynamics simulation on the inhibition performance of the DGDCBA epoxy polymer against the corrosion of the E24 carbon steel in 1.0 M HCl solution, J. Mol. Struct. 1182 (2018) 340–351. Available from: https://doi.org/10.1016/j.molstruc.2018.12.030.

[64] L.H. Madkour, S. Kaya, L. Guo, C. Kaya, Quantum chemical calculations, molecular dynamic (MD) simulations and experimental studies of using some azo dyes as corrosion inhibitors for iron. Part 2: Bis–azo dye derivatives, J. Mol. Struct. 1163 (2018) 397–417. Available from: https://doi.org/10.1016/j.molstruc.2018.03.013.

[65] S.K.K. Saha, P. Banerjee, Newly synthesized Schiff base molecules as efficient corrosion inhibitors for mild steel in 1 M HCl medium: an experimental, density functional theory and molecular dynamics simulation study, Mater. Chem. Front. (2018). Available from: https://doi.org/10.1039/C8QM00162F.

[66] L. Guo, S. Kaya, I.B. Obot, X. Zheng, Y. Qiang, Toward understanding the anticorrosive mechanism of some thiourea derivatives for carbon steel corrosion: a combined DFT and molecular dynamics investigation, J. Colloid Interface Sci. 506 (2017) 478−485.

[67] L. Guo, S. Zhang, T. Lv, W. Feng, Comparative theoretical study on the corrosion inhibition properties of benzoxazole and benzothiazole, Res. Chem. Intermed. 41 (2015) 3729−3742.

[68] M. Sikine, H. Elmsellem, Y.K. Rodi, Y. Kadmi, M. Belghiti, H. Steli, et al., Experimental, Monte Carlo simulation and quantum chemical analysis of 1, 5-di(prop-2-ynyl)-benzodiazepine-2,4-dione as new corrosion inhibitor for mild steel in 1 M hydrochloric acid solution, J. Mater. Environ. Sci. 8 (2017) 116−133.

[69] B. El Ibrahimi, L. Bazzi, Theoretical evaluation of some A-amino acids for corrosion inhibition of copper in acidic medium: DFT calculations, Monte Carlo simulations and QSPR studies, J. King Saud. Univ. 32 (2018) 163−171.

[70] B. El Ibrahimi, A. Jmiai, K. El Mouaden, R. Oukhrib, A. Soumoue, S. El Issami, et al., Theoretical evaluation of some a-amino acids for corrosion inhibition of copper in acidic medium: DFT calculations, Monte Carlo simulations and QSPR studies, J. King Saud. Univ. Sci. 32 (2020) 163−171.

[71] N. Errahmany, M. Rbaa, A.S. Abousalem, et al., Experimental, DFT calculations and MC simulations concept of novel quinazolinone derivatives as corrosion inhibitor for mild steel in 1.0M HCl medium, J. Mol. Liq. 312 (2020) 113413. Available from: https://doi.org/10.1016/j.molliq.2020.113413.

[72] V. Saraswat, M. Yadav, I.B. Obot, Investigations on eco-friendly corrosion inhibitors for mild steel in acid environment: electrochemical, DFT and Monte Carlo simulation approach, Colloids Surf. A: Physicochem. Eng. Asp. 599 (2020) 124881. Available from: https://doi.org/10.1016/j.colsurfa.2020.124881.

[73] R. Hsissoua, S. Abbout, R. Seghiri, M. Rehiouic, A. Berishad, H. Erramlic, et al., Evaluation of corrosion inhibition performance of phosphorus polymer for carbon steel in [1 M] HCl: computational studies (DFT, MC and MD simulations), J. Mater. Res. Technol. 9 (3) (2020) 2691−2703.

[74] L.H. Madkour, S. Kaya, I.B. Obot, Computational, Monte Carlo simulation and experimental studies of some arylazotriazoles (AATR) and their copper complexes in corrosion inhibition process, J. Mol. Liq. 260 (2017) 351−374. Available from: https://doi.org/10.1016/j.molliq.2018.01.055.

[75] M. El Faydy, B. Lakhrissi, A. Guenbour, S. Kaya, F. Bentiss, I. Waradf, et al., In situ synthesis, electro-chemical, surface morphological, UV−visible, DFT and Monte Carlo simulations of novel 5-substituted-8-hydroxyquinoline for corrosion protection of carbon steel in a hydrochloric acid solution, J. Mol. Liq. 280 (2019) 341−359.

[76] K.O. Sulaiman, A.T. Onawole, O. Faye, D.T. Shuaib, Understanding the corrosion inhibition of mild steel by selected green compounds using chemical quantum based assessments and molecular dynamics simulations, J. Mol. Liq. 279 (2019) 342−350.

[77] Y. Zhou, S. Xu, L. Guo, S. Zhang, H. Lu, Y. Gong, et al., Evaluating two new Schiff bases synthesized on the inhibition of corrosion of copper in NaCl solutions, RSC Adv. 5 (2015) 14804.

[78] K.F. Khaled, Corrosion control of copper in nitric acid solutions using some amino acids—a combined experimental and theoretical study, Corros. Sci. 52 (2010) 3225−3234.

[79] H. Huang, X. Guo, The relationship between the inhibition performances of three benzo derivatives and their structures on the corrosion of copper in 3.5 wt.% NaCl solution, Colloids Surf. A 598 (2020) 124809.

[80] X. Li, S. Deng, T. Lin, X. Xie, G. Du, 2-Mercaptopyrimidine as an effective inhibitor for the corrosion of cold rolled steel in HNO_3 solution, Corros. Sci. 118 (2017) 202−216.

[81] D.M. Gurudatt, K.N. Mohana, Synthesis of new pyridine based 1, 3, 4-oxadiazole derivatives and their corrosion inhibition performance on mild steel in 0.5 M hydrochloric acid, Ind. Eng. Chem. Res. 53 (2014) 2092−2105.

[82] M. Mahdavian, S. Ashhari, Corrosion inhibition performance of 2-mercaptobenzimidazole and 2-mercaptobenzoxazole compounds for protection of mild steel in hydrochloric acid solution, Electrochim. Acta 55 (2010) 1720−1724.

[83] K.R. Ansari, M.A. Quraishi, A. Singh, Isatin derivatives as a non-toxic corrosion inhibitor for mild steel in 20% H_2SO_4, Corros. Sci. 95 (2015) 62−70.

5

Experimental methods of inhibitors assessment

K.R. Ansari[1], Ambrish Singh[2], M.A. Quraishi[1], Tawfik A. Saleh[3]

[1]CENTER OF RESEARCH EXCELLENCE IN CORROSION, RESEARCH INSTITUTE, KING FAHD UNIVERSITY OF PETROLEUM AND MINERALS, DHAHRAN, SAUDI ARABIA [2]SCHOOL OF NEW ENERGY AND MATERIALS, SOUTHWEST PETROLEUM UNIVERSITY, CHENGDU, CHINA [3]DEPARTMENT OF CHEMISTRY, KING FAHD UNIVERSITY OF PETROLEUM AND MINERALS, DHAHRAN, SAUDI ARABIA

Chapter outline

Environmentally Sustainable Corrosion Inhibitors. DOI: https://doi.org/10.1016/B978-0-323-85405-4.00009-4

5.1 Introduction

A corrosion inhibitor introduced into the aggressive solution undergoes adsorption onto the selected surface of metal. For the complete understanding of the corrosion inhibitor adsorption and corresponding inhibition efficiency of inhibitor film, it is important to go through specific experimental methods. The methods consist of an evaluation of inhibition efficiency of the corrosion inhibitor, elucidating adsorption mechanism methods, different electrochemical methods, characterization of surface methods, etc. The corrosion evaluation methods may be categorized into many types depending upon the principle of operation, methods, and data obtained. In general, the primary reference for these tests is based on ASTM standards [1].

The first method category is immersion (weight loss) where information about the inhibitor concentration was gathered, in which the inhibitor provides the maximum inhibition efficiency [2−4]. At the present time, other different methods have been advanced but the weight loss method is the simplest and most trusted method for evaluation of inhibitor performance. The second category consists of electrochemical methods such as electrochemical impedance spectroscopy, electrochemical frequency modulation (EFM), electrochemical frequency modulation trend (EFMT), and polarization resistance. These methods give information on real-time electrochemical behavior of corrosion and are nondestructive [5,6]. The other electrochemical method is potentiodynamic polarization, which is a destructive method and after performing this experiment the metal surface is no longer of use. The electrochemical methods give measurable information about the adsorption of corrosion inhibitor and additionally, the electrochemistry occurring over the corroding metal surface. The third method category evaluates the surfaces of the corroded and protected metal samples. This method includes scanning electron microscopy (SEM), atomic force microscopy (AFM), scanning electrochemical microscopy (SECM), X-ray photoelectron spectroscopy (XPS), contact angle measurements, etc. These methods are very useful for providing direct information on the inhibitor structure as well as the adsorbed inhibitor composition elemental status over the surface of the metal [7−11]. Additionally, these experimental techniques assist in explaining inhibitor mechanism adsorption and inhibitor−metal interaction.

In the present chapter, we try to describe the commonly used experimental methods for the examination of corrosion inhibitor performance. The abovementioned methods are explained using typical examples.

5.2 Method of weight loss

Weight loss is the simplest, most affordable, and most commonly used method for the evaluation of the corrosion inhibition efficiency of a corrosion inhibitor. Initially, the metal samples are cut into desired coupon sizes and are polished using different grit abrasive papers. The rinsed, cleaned coupons are kept in a desiccator. Then the coupon samples are immersed into the aggressive medium and after a known time (6/12/24 hours) samples are removed and washed as per the ASTM standard [12−15]. The weight loss difference before and after the immersion gives the average weight loss [16].

The speed of the metal destruction in the given specific corrosive environment is known as the corrosion rate (C_R). The corrosion rate (mm year^{-1}) is estimated as per Eq. (5.1):

$$C_R = \frac{8.76 \times 10^4 \times \Delta m}{s \times t \times \rho}$$

(5.1)

where C_R is corrosion rate (mm year^{-1}), Δm is weight loss (g), ρ is steel density (g cm^{-3}), A is steel exposed area (cm^2), and t is testing duration (hours).

The application of corrosion rate is used for the estimation of surface coverage (θ) and the inhibition efficiency ($\eta\%$):

$$\theta = \frac{C_{R(a)} - C_{R(p)}}{C_{R(a)}}$$

(5.2)

$$\eta\% = \frac{C_{R(a)} - C_{R(p)}}{C_{R(a)}} \times 100$$

(5.3)

where $C_{R(a)}$ and $C_{R(p)}$ are corrosion rates without and with corrosion inhibitor, respectively.

Alternatively, the corrosion rate (C_R) in mpy (Eq. 5.4) and mg cm^{-2} h^{-1} (Eq. 5.5) can also be calculated as follows:

$$C_R = \frac{3.45 \times 10^3 \times \Delta m}{\rho At}$$

(5.4)

$$C_R = \frac{\Delta m}{At}$$

(5.5)

5.2.1 Concentration of inhibitor

The method of weight loss is used to analyze the inhibition efficiency ($\eta\%$) and corrosion rate (C_R). Fig. 5−1 represents the $\eta\%$ and C_R variation with an increasing concentration of inhibitor [17].

The inhibitor addition leads to an increase in $\eta\%$ and a decrease in C_R values, respectively. After reaching the maximum values the increase in the concentration does not produce any significant change in the values. Thus the value is supposed to be the optimum inhibitor concentration. The increase in $\eta\%$ and decrease in C_R with increasing inhibitor concentration suggests the formation of an inhibitor protective layer that covers more surface area at an optimum concentration and minimizes the contact of a corrosive medium with metal active sites. This supports the adsorption of inhibitor molecules.

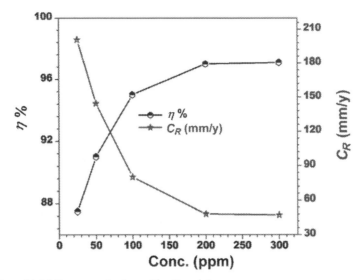

FIGURE 5–1 Variation of inhibitor concentration with η% and C_R.

5.2.2 Temperature effect and parameters of activation

Temperature plays an important role in terms of the kinetics of corrosion reaction behavior of corroding metal by changing the adsorption strength of inhibiting compounds. It is a well-known fact that an increase in temperature can break weak electrostatic adsorption bonds between the metal and the corrosion inhibitor [18]. This process mainly occurs by an increase in the kinetic energy of the corrosion inhibitor by an increase in temperature that makes the molecules desorb from the metal surface, which in turn lowers the performance of the corrosion inhibitor [19]. However, whenever the increase in temperate causes no change/increase in corrosion inhibitor performance then the adsorption is said to be chemical adsorption. In this, chemical/coordinate bonds have formed between the metal and corrosion inhibitor.

The dependency of corrosion inhibitor performance with temperature can be justified using the Arrhenius and the transition state equations [20]:

$$C_R = A \exp\left(\frac{-E_a}{RT}\right) \tag{5.6}$$

$$C_R = \frac{RT}{Nh} \exp\left(\frac{\Delta S*}{R}\right) \exp\left(-\frac{\Delta H*}{RT}\right) \tag{5.7}$$

where E_a, T, A, and R are energy of activation, temperature, preexponential factor, and gas constant, respectively; N, h, ΔH^*, and ΔS^* are Avogadro number, Planck's constant, enthalpy, and entropy of activation, respectively.

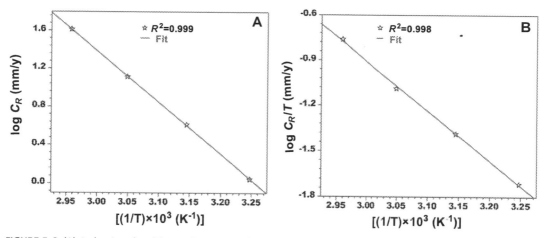

FIGURE 5–2 (A) Arrhenius plot. (B) Transition-state plot.

Fig. 5–2A shows an Arrhenius plot. The values of E_a are calculated from the slope of the plot, that is, $E_a/2.303R$. A plot of log C_R/T versus $1/T$ (Fig. 5–2B) provides a straight line with a slope of $\Delta H^*/2.303R$ and intercept of log $(R/Nh) + \Delta S^*/2.303R$, known as the transition-state plot, which allows the calculation of ΔH^* and ΔS^*.

The higher value of E_a in the presence of an inhibitor as compared to the absence of an inhibitor reveals an increment in the thickness of double layer thickness, which ultimately increases the barrier of activation energy of the corrosion reaction, and thus reduces the corrosion process [21]. In general, the increment in the activation energy with the addition of inhibitor represents the physical nature of corrosion inhibition adsorption and this is the initial stage of the corrosion inhibitor adsorption [22]. However, chemical adsorption of inhibitor is represented by the lower E_a. But, the adsorption type cannot be judged only on the basis of E_a values because the adsorption process is competitive in nature and it is taking place by the replacement of the preadsorbed water molecules which also require some activation energy [23].

The positive value of ΔH^* reveals the slower dissolution of metal [24]. The increasing value of ΔH^* in the presence of an inhibitor suggests that the energy barrier that governed the corrosion reaction has increased. In the same way, an increase in the values of ΔS^* with the addition of corrosion inhibitor reveals the adsorption of the inhibitor over the metal surface [25]. The inhibitor adsorption is replacing preadsorbed water molecules. Thus an increase in entropy of solvent reflects the increase in the value of ΔS^* [25].

5.3 Parameters of adsorption

5.3.1 Isotherms

The interactive nature of the corrosion inhibitor and metal surface were analyzed using various kinds of adsorption isotherms models [26]. The selection isotherm plays a key role in the understanding of the adsorptive nature of corrosion inhibitors. The parameters obtained from the fitting of the isotherm are used for the calculation of standard free energy of adsorption, which helps to determine whether the corrosion reaction is spontaneous or not and also whether inhibitor adsorption is physical or chemical [27]. Some adsorption isotherm models are explained briefly in this section:

1. Langmuir isotherm model: the isotherm assumes that a single inhibitor layer has form over the metal surface with the number of localized equilibrium sites fixed. It also assumes that there is no lateral interaction between the adsorbed and free inhibitor molecules. The below equation describes the Langmuir isotherm model [28]:

$$\frac{C_{inh}}{\theta} = \frac{1}{K_{ads}} + C_{inh} \tag{5.8}$$

where C_{inh} and K_{ads} are concentration and equilibrium constant. The typical Langmuir isotherm graph is given in Fig. 5−3

2. Temkin isotherm model: the equation used for the calculation of this model is as follows:

$$\exp(-2a\theta) = K_{ads}C \tag{5.9}$$

where C = inhibitor concentration, θ = fraction of site occupied by inhibitor molecules, and a = interaction parameter between adsorbing species and the metal surface [29]. The Temkin isotherm is plotted using the values of surface coverage (θ) versus log C (Fig. 5−4).

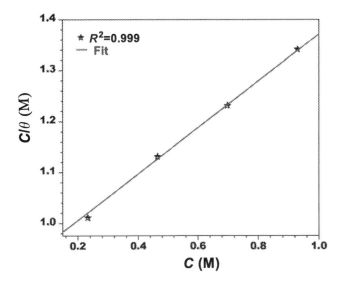

FIGURE 5–3 Langmuir isotherm for the adsorption.

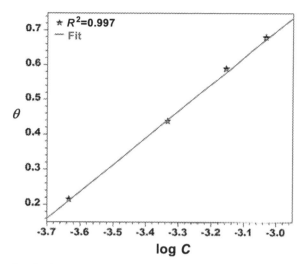

FIGURE 5–4 Temkin adsorption isotherm.

3. Frumkin isotherm model: the below equation represents the Frumkin adsorption isotherm model:

$$KC = \frac{\theta}{1-\theta}e^{-f\theta} \tag{5.10}$$

where f is the lateral interaction parameter among adsorbed inhibitor molecules and the positive or negative values of f represent the attraction or repulsion, respectively. An example of the Frumkin isotherm model is depicted in Fig. 5–5.

4. Flory–Huggins isotherm: the below equation represents this model (Fig. 5–6) [30]:

$$\frac{\theta}{x(1-\theta)^x} = K_{ads}C \tag{5.11}$$

5. El-Awady isotherm model: the El-Awady isotherm model is given by Eq. (5.12) (Fig. 5–7) [31]:

$$\log\left(\frac{\theta}{1-\theta}\right) = \log K + y\log C_{inh} \tag{5.12}$$

where $K_{ads} = K^{1/y}$ and y represents the occupied number of molecules of inhibitors at reactive centers. Generally, a value of $1/y$ less or greater than one corresponds to multilayer adsorption and more than one site occupation, respectively.

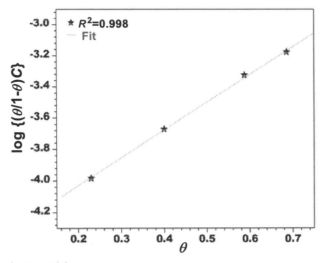

FIGURE 5–5 Frumkin isotherm model.

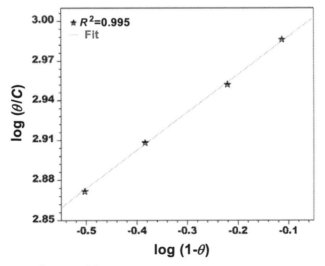

FIGURE 5–6 Flory–Huggins isotherm model.

5.3.2 Energy of adsorption

The isotherm models help in the calculation of standard free energy of adsorption (ΔG^0_{ads}) and adsorption equilibrium constant (K_{ads}). The K_{ads} and ΔG^0_{ads} are related mathematically as per the below equations:

[32]:

$$\Delta G_{ads} = -2.303RT\log(55.55 \times K_{ads}) \tag{5.13}$$

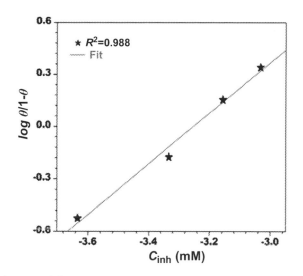

FIGURE 5–7 El-Awady isotherm model.

$$\Delta G_{ads} = -2.303RT\log(10^6 \times K_{ads})$$ (5.14)

where ΔG_{ads}, T, R, and $55.55/10^6$ are standard adsorption free energy, temperature, gas constant, concentration of water molecules, respectively.

In general, the negative values of ΔG^0_{ads} represent the spontaneity of the adsorption process. If the value of ΔG^0_{ads} is -20 kJ mol L or lower then the adsorption is physical in nature. However, if the value is -40 kJ mol L or higher then the adsorption is chemical in nature [33]. But, if the value is in-between -20 and -40 kJ mol L, then combined adsorption, that is, physical and chemical, is observed. Additionally, we can also calculate parameters like adsorption enthalpy, and entropy [34].

The heat of adsorption (Q_{ads}) was calculated using the below-mentioned equation [35]:

$$Q_{ads} = 2.303\left[\log\left(\frac{\theta_2}{1-\theta_2}\right) - \log\left(\frac{\theta_1}{1-\theta_1}\right)\right] \times \left(\frac{T_1 T_2}{T_2 - T_1}\right)$$ (5.15)

where θ_1 and θ_2 are the surface coverage at temperatures T_1 and T_2 (65°C), respectively. The positive values of Q_{ads} represent the chemical adsorption mechanism [35].

5.4 Electrochemical techniques

5.4.1 Open circuit potential curves

Fig. 5–8 represents open circuit potential curves in the absence and presence of different concentrations of inhibitor. It could be observed that in 30 min the curves achieved steady

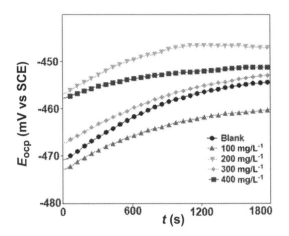

FIGURE 5–8 Open circuit potential without and with different concentration of inhibitor.

state. Adsorption of inhibitor molecules onto metal surface with inhibitor addition was represented by curves shifting.

5.4.2 Electrochemical impedance spectroscopy

An impedance technique was used for the measurement of capacitance of a double-layer (C_{dl}) and the resistance of charge transfer (R_{ct}). In this technique an alternate current (AC) is applied. The application of a complex number helps to resolve the impedance into two parts, that is,

$$\text{Real part} = |Z'| = |Z\cos\theta| \tag{5.16}$$

$$\text{Imaginary part} = |Z''| = |Z\sin\theta| \tag{5.17}$$

The total resistance for the AC flow is given by Z,

$$Z = Z' - Z'' \tag{5.18}$$

where,

$$Z' = R_s + \frac{R_{ct}}{1 + \omega^2 C_{dl} R_{ct}^2} \tag{5.19}$$

$$Z'' = \frac{\omega C_{dl} R_{ct}^2}{1 + \omega C_{dl}^2 R_{ct}^2} \tag{5.20}$$

From the above equations, a plot was generated between Z' and Z'' that looks like a semi-circle and that intersect the real axis at higher and lower frequencies. The intersection at higher and lower frequencies represents R_s, and $R_s + R_{ct}$, that is, R_p. The double-layer capacitance is calculated as per Eq. 5.21:

$$\omega(Z''_{max}) = \frac{1}{C_{dl} \times R_{ct}} \qquad (5.21)$$

Inhibition efficiency by using impedance measurements was carried out by using the formula:

$$\eta\% = \frac{R_{ct(inh)} - R_{ct(blank)}}{R_{ct(inh)}} \times 100 \qquad (5.22)$$

where $R_{ct(inh)}$ and $R_{ct\ (blank)}$ are inhibitor-containing and inhibitor-free resistance of charge. A general example of fitted Nyquist plots without and with different concentrations of inhibitor in an acidic medium at 308K temperature is shown in Fig. 5−9A. This figure represents a depressed semicircle which is due to the surface imperfection and inhomogeneity. The semicircle's appearance also represents the capacitive behavior both in the absence and in presence of an inhibitor. As can be seen from the figure, as the inhibitor concentration increases, the semicircle size increases, which is due to the increase in charge transfer resistance (R_{ct}) of the metal. The equivalent circuit is represented in Fig. 5−9B. The equivalent circuit was modeled using elements of constant phase (*CPE*), resistance in charge transfer (R_{ct}), and resistance in solution (R_s). The double-layer capacitance (C_{dl}) and thickness of the adsorbed layer were calculated using the below equations [36]:

$$C_{dl} = \left(Y_o R_{ct}^{1-n}\right)^{1/n} \qquad (5.23)$$

The plots of phase angle are represented in Fig. 5−9C. The phase angle plots suggest the increasing values of phase angles with increasing inhibitor concentrations at an intermediate frequency, which indicate that inhibitor molecules are adsorbed over the metal surface and inhibit the dissolution of metal.

5.4.3 Potentiodynamic polarization

The technique is for the measurement of anodic and cathodic polarization. Fig. 5−10 represents the anodic and cathodic reactions as metal dissolution and hydrogen evolution, respectively. It is notable that at high applied current the hydrogen evolution process becomes identical. The Tafel regions were extrapolated for the determination of polarization parameters (Fig. 5−10). The hydrogen evolution and metal dissolution rates are equal at corrosion potential, providing the value of corrosion current density. The Tafel constant can also be calculated from the anodic and cathodic Tafel curves.

This technique governs the kinetics of the corrosion process and could be explained on the basis of the below points.

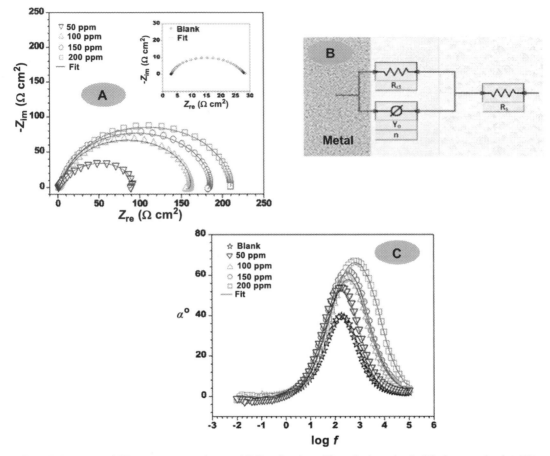

FIGURE 5–9 A general EIS measurements images (A) Nyquist plots, (B) equivalent circuit, (C) phase angle plot. *EIS*, Electrochemical impedance spectroscopy.

5.4.3.1 Kinetics of corrosion

Consider an electrochemical reaction, where the reaction takes place at the metal–solution interface and the rate of reaction is proportional to current–potential dependence. Overvoltage (polarization) η is the potential change, $E-E_r$, from the equilibrium half-cell electrode potential E_r, caused by a net surface reaction rate for the half-cell reaction and was introduced by Nernst and Caspari. The dependence of η on the current-density for a hydrogen evolution reaction has been shown to be $\eta = a + b \log i$ by Tafel. Butler gave the kinetic treatment of a reversible electrode in which the concepts of the partial anodic and cathodic currents are related to η through an exponential equation.

5.4.3.1.1 Activation controlled corrosion

Activation polarization is generally caused by a slow electrode reaction. The reaction at the electrode requires activation energy in order to proceed. The most important example of activation controlled corrosion reaction is that of hydrogen ion reduction at cathode.

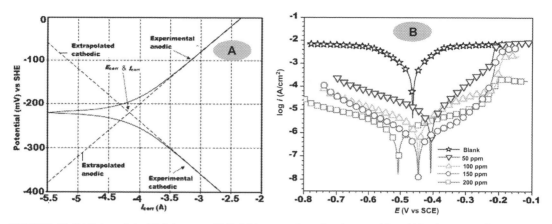

FIGURE 5–10 (A) Extrapolated Tafel curves. (B) Tafel for corrosion of carbon steel in acid in absence and presence of inhibitor.

The relationship between current and potential for a corroding system in which anodic reaction is metal dissolution and cathodic reaction is hydrogen evolution can be derived by the application of electrochemical kinetic theory.

For the metal dissolution reaction,

$$M \rightarrow M^{+2} + 2e^- \tag{5.24}$$

$$i_c = i_c^o \left\{ \exp\left(\frac{\alpha_c F}{RT}(E - E_c^{\,r})\right) - \exp\left(\frac{-\beta_c F}{RT}(E - E_c^{\,r})\right) \right\} \tag{5.25}$$

where $E_c^{\,r}$ is the reversible potential of cathodic dissolution reaction, i_c^o is the exchange current density for cathodic reaction, and α_c, β_c are transfer coefficients of reduction reaction.

Normally the corrosion potential (E_{corr}) will be far away from the equilibrium potential of the reversible reaction. Hence the contribution from the deposition reaction of metal dissolution and the reduction reaction is negligible. Therefore the net current of the mixed electrode system is given as follows:

$$i = i_a - i_c \tag{5.26}$$

$$= i_a^o \left\{ \exp\left(\frac{\alpha_a F}{RT}(E - E_a^{\,r})\right) - \exp\left(\frac{-\beta_a F}{RT}(E - E_a^{\,r})\right) \right\} \tag{5.27}$$

At corrosion potential $E = E_{corr}, i = 0$

i.e., $\quad i_{corr} = i_a^o \left(\exp\frac{\alpha_a F}{RT}(E_{corr} - E_a^{\,r}) \right)$ (for anodic reaction) $\tag{5.28}$

i.e., $\quad i_{corr} = i_c^o \left(\exp\frac{\alpha_c F}{RT}(E_{corr} - E_c^{\,r}) \right)$ (for cathodic reaction) $\tag{5.29}$

Substituting the terms E_a^r and E_c^r in terms of E_{corr}

$$i = i_{corr} \left\{ \exp\left(\frac{\alpha_a F}{RT}(E - E_{corr}) \right) - \exp\left(\frac{-\beta_c F}{RT}(E - E_{corr}) \right) \right\} \qquad (5.30)$$

The above equation can be rewritten in terms of Tafel slopes β_a and β_c as

$$i = i_{corr} \left\{ \exp\left(\frac{2.3(E - E_{corr})}{\beta_a} \right) - \exp\left(\frac{-2.3(E - E_{corr})}{\beta_c} \right) \right\} \qquad (5.31)$$

Since $E - E_{corr} = \eta$

$$i = i_{corr} \left\{ \exp\left(\frac{2.3\eta}{\beta_a} \right) - \exp\left(\frac{-2.3\eta}{\beta_c} \right) \right\} \qquad (5.32)$$

The above expression forms the basis of measuring corrosion rate by electrochemical method.

5.4.3.1.2 Diffusion controlled reaction

Concentration polarization or diffusion overpotential is the potential difference of a cathode in the absence and presence of an external current. The corrosion process in neutral media consists of metal dissolution reaction as anodic and oxygen reduction as a cathodic reaction. In such cases,

$$i_a = i_{corr} \left(\exp \frac{\alpha_a F}{RT}(E - E_{corr}) \right) \qquad (5.33)$$

$$i_c = i_d = \frac{\eta F D C_b}{\delta} \qquad (5.34)$$

where D is diffusion coefficient, δ is diffusion layer thickness, and C_b is the concentration of reduced species.

$$i = i_a - i_c \qquad (5.35)$$

$$i_a = i_{corr} \left(\exp \frac{\alpha_a F}{RT}(E - E_{corr}) \right) - i_d \qquad (5.36)$$

i_d = limiting diffusion current density.

At corrosion potential $E = E_{corr}$, $i = 0$, therefore $i_{corr} = i_d$

It follows that the i_d is the most significant parameter in the corrosion reaction in which the cathodic reaction is diffusion controlled and any factor that increases i_d will increase the corrosion rate.

The extrapolation of the Tafel curves helps in the determination of kinetic parameters like i_{corr}, E_{corr}, β_a, β_c, and $\eta_{PDP\%}$ [37]:

$$\eta_{PDP\%} = \frac{i_{corr} - i_{corr(inh)}}{i_{corr}} \times 100 \qquad (5.37)$$

where i_{corr} = corrosion current density without inhibitor and $i_{corr(inh)}$ = corrosion current density with inhibitor. In general, if the shift in E_{corr} is less than 85 mV, that is, in a positive direction as compared to without inhibitor, then the inhibitor is anodic-type. However, if the shift is greater than 85 mV, that is, in a negative direction, then the inhibitor is cathodic-type. Nevertheless, if the shift in both is more/less than 85 mV, then the inhibitor is said to be mixed type. But there are conflicts among the scientists above this type of shifting. According to this community, a prominent shift in either anodic or cathodic direction helps to decide the action of the inhibitor is anodic or cathodic.

5.4.4 Electrochemical frequency modulation trend/electrochemical frequency modulation trend

The EFMT and EFM both are nondestructive and very sensitive electrochemical technique that helps in the calculation of various parameters. The below equations were used for kinetic parameters calculation [5]:

$$i_{corr} = \frac{i_\omega^2}{\sqrt{48(2i_\omega i_{3\omega} - i_{2\omega})}}$$ (5.38)

$$\text{Causality factor (2)} = \frac{i_{\omega_2 \pm \omega_1}}{i_{2\omega_1}} = 2.0$$ (5.39)

$$\text{Causality factor (2)} = \frac{i_{\omega_2 \pm \omega_1}}{i_{3\omega_1}} = 3.0$$ (5.40)

It helps in the simultaneous measurement of corrosion rate and corrosion current. Fig. 5−11 represents the EFMT in the absence and presence of inhibitor to the corrosive medium. It is observed from Fig. 5−11 that the values of corrosion rate and corrosion current are very high. However, the addition of inhibitor causes a significant reduction in corrosion rate and corrosion current value [38]. This observation supports the protective nature of inhibitor molecules. Similarly, in EFM the value of corrosion current decreases with the addition of corrosion inhibitor as compared to a blank (Fig. 5−12).

The causality factors provide the validation of the EFMT measurements. The theoretical values of causality factors, that is, CF-2 and CF-3 are 2.0 and 3.0, respectively. If the experimentally obtained values are close to the theoretical value then the obtained results are in good agreement.

5.4.5 Linear polarization resistance

It is also called as LPR in short. The polarization resistance of a material is defined as the $\Delta E/\Delta i$ slope of a potential−current density curve at the free corrosion potential (Fig. 5−13), yielding the polarization resistance R_p that can be itself related to the corrosion current (i_{corr}) with the help of the Stern−Geary approximation in Eq. (1.45) [39].

$$R_p = \frac{B}{i_{corr}} = \frac{(\Delta E)}{(\Delta i)_{\Delta E \to 0}}$$ (5.41)

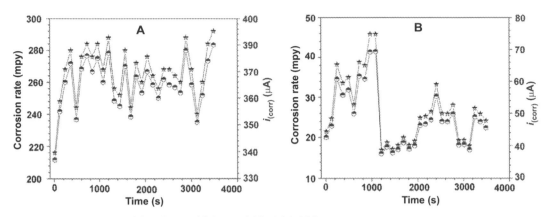

FIGURE 5–11 EFMT curves (A) without inhibitor and (B) with inhibitor.

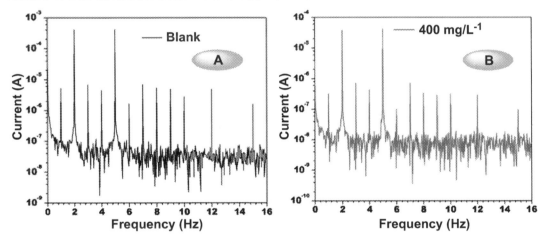

FIGURE 5–12 EFM curves (A) without inhibitor and (B) with inhibitor.

where R_p is the polarization resistance, i_{corr} is the corrosion current, B is an empirical polarization resistance constant that can be related to the anodic (b_a) and cathodic (b_c) Tafel slopes with Eq. (1.46).

$$B = \frac{b_a \times b_c}{2.3(b_a + b_c)} \tag{5.42}$$

The corrosion inhibition efficiency can be calculated as given below:

$$\eta_{LPR\%} = \frac{R_p^{inh} - R_p}{R_p^{inh}} \times 100 \tag{5.43}$$

where the terms R_p and R_p^{inh} represent the polarization resistance without and with the inhibitor, respectively.

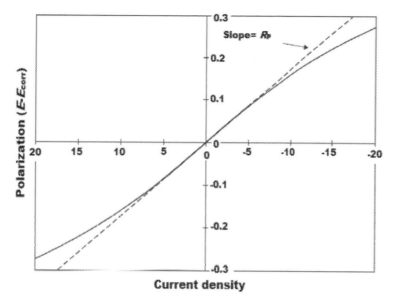

FIGURE 5–13 Hypothetical linear polarization plot.

5.5 Surface analytical techniques

Surface characterization techniques are commonly used for understanding the corrosion inhibition mechanism and interaction among the surface of metals and corrosion inhibitor molecules [40]. The most commonly used techniques are discussed in brief:

5.5.1 Scanning electron microscopy

SEM is the most commonly used technique for understanding the surface morphology of metal. It can provide the images of the surface whose magnification could range from $10 \times$ to $10^6 \times$. In SEM, a fine probe is used for focusing the electrons in a rectangular pattern of parallel scanning lines for scanning the metal sample's surface [41]. In the SEM, the metal samples like carbon steel, Zn, Cu, etc. are dipped in the corrosive solution without and with inhibitor for an appropriate period of time (3–24 hours). After that, the samples are taken out, washed with distilled water, and analyzed. It was observed that the metal sample without inhibitor was severely damaged due to the direct contact of metal and corrosive solution. However, the metal sample with inhibitor was smooth with very little damage due to the formation of a protective film of inhibitor molecules over the metal surface that could act as a barrier between the metal surface and the corrosive solution (Fig. 5–14). Although no quantitative information about the surface is provided, smoothness could be obtained using the SEM technique.

5.5.2 Energy-dispersive X-ray spectroscopy

The energy-dispersive X-ray spectroscopy (EDX) provides the chemical composition of the adsorbed inhibitor film and thus provides some quantitative data. The EDX data are

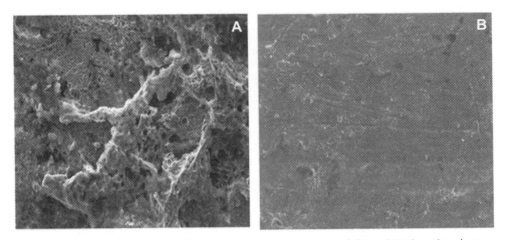

FIGURE 5–14 SEM images of carbon steel (A) without inhibitor and (B) with inhibitor. *SEM*, Scanning electron microscopy.

presented as the energy vs. intensity (counts) plots and the X-rays characterized the respective elements from where they come, such as from Fe, Cu, Al, etc. If the inhibitor layer is formed over the metal surface then the respective peaks of elements like C, O, N, S, P, etc., could also appear that supports corrosion inhibitor adsorption onto the metal surface (Fig. 5–15) [41].

The second way of EDX representation is in the form of a table that consists of atomic weight % of the different elements. The limitation of EDX analysis is that it cannot provide any information about the metal and corrosion inhibitor interaction.

5.5.3 Water contact angle

The adsorption of organic inhibitor molecules over the metal surface causes changes to the wetting properties, and the degree of wettability could be determined by measuring the water contact angle (WCA) of the surface [42]. Thus if the corrosion inhibitor film is hydrophilic or hydrophobic can be differentiated using WCA measurement. The measurement of WCA is done using the tensiometer that is popularly known as a contact angle meter. This instrument has various modes for the estimation of WCA but studying the wettability of inhibitor films over the metal surface sessile drop method is most commonly used. In the sessile method, a water drop (5 mL) is applied over the metal sample surface and WCA is measured immediately after locating. The drops' images are analyzed and a contact angle with a precision of ± 0.1 degree is calculated depending upon the shapes of the drops. Fig. 5–16 represents the WCA without and with the addition of inhibitor molecules. Here, it is noticeable that without additional inhibitor the WCA is much less. However, with the inhibitor addition, the value of WCA is increased, which corresponds to the decrease in the metal surface wettability due to the presence of a hydrophobic film of organic corrosion inhibitor molecules [43].

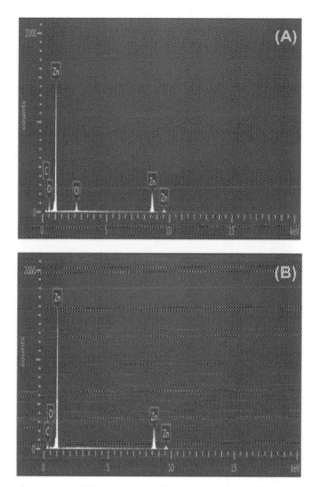

FIGURE 5–15 EDX images of carbon steel (A) without inhibitor and (B) with inhibitor.

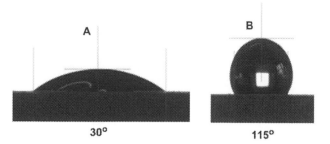

FIGURE 5–16 WCA images of carbon steel (A) without inhibitor and (B) with inhibitor. *WCA*, Water contact angle.

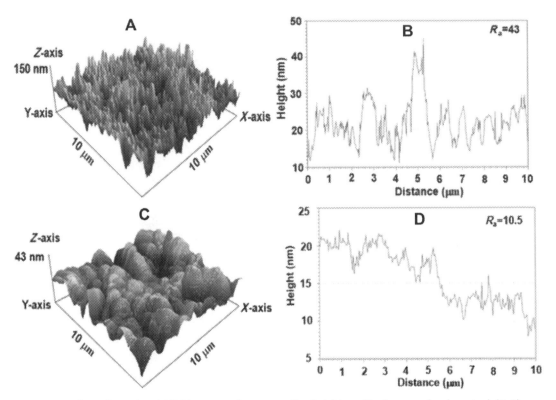

FIGURE 5–17 Three-dimensional AFM images and corresponding height profile diagram of carbon steel: (A, B) without inhibitor and (C, D) with inhibitor.

5.5.4 Atomic force microscopy

The AFM technique is similar to the SEM where in both we get the surface roughness. However, in AFM we can get some quantitative data. As in SEM, the metal samples are dipped into the corrosive solution in the presence and absence of corrosion inhibitor for the appropriate time. The measurement could be done in contact, semicontact (tapping), or noncontact modes, respectively. However, due to the rough and damaged metal surface, the last modes are more preferred. The advantage of AFM is that it provides the values of surface roughness (i.e., root-mean-square roughness, average roughness, etc.) along with the surface profile images (Fig. 5−17). Thus with the observation of roughness values it is easy to differentiate the performance of the corrosion inhibitor with respect to the blank. It also helps to compare the inhibition performance among the series of corrosion inhibitors. However, the AFM technique gives no information about the chemical composition of the adsorbed inhibitor film.

5.5.5 X-ray photoelectron spectroscopy

XPS is the most commonly used surface analysis technique for the estimation of materials composition [44]. The basic principle of XPS is a photoelectric effect phenomenon. The energy of X-rays is strong enough to kick out the electrons from the core orbitals. Depending upon the binding energy of the electrons, a different energy is used for the removal and thus an energy spectrum is formed. The X-rays can easily penetrate deep into the materials and can remove the electrons up to the first 4 nm thickness. Thus the commonly formed layer of inhibitor films is of a few nm and thus can be easily probed. The mathematical relation between the binding energy (E_B) of an electron and the kinetic energy (E_k) of the ejected photoelectron are as follows:

$$E_B = h\upsilon - (E_k + \phi) \tag{5.44}$$

where $h\upsilon$ and ϕ are X-ray energy and spectrometer work function.

The XPS spectra of adsorbed inhibitor over carbon steel are represented in Fig. 5−18. The Fe_{2p} spectrum consists of two peaks at the binding energy of 711.25 eV ($Fe_{2p3/2}$) and 724.61 eV ($Fe_{2p1/2}$). These peaks correspond to iron oxide, that is, Fe_2O_3, FeOOH, and FeO (OH) [45]. The $Fe_{2p3/2}$ peak has two satellite peaks at 711.13 and 713.54 eV [46]. The $Fe_{2p1/2}$ also consists of one satellite peak at 725.37 eV [46]. Also, one satellite peak at 718.01 eV was observed [46]. The spectrum of N1s (Fig. 5−6B) consists of three peaks. The peaks at 398.46 and 399.16 eV correspond to C-N and C-N-Fe coordination, respectively [9]. The peak at 400.11 eV is protonated nitrogen ($=N^+$-) [9]. The C 1s spectra show five peaks. The peak from 284.38 to 285.1 cV represents C-C and C-H bonds [47]. The peak at 285.9 and 288.1 eV are attributed to C-O-C, C = N, and O-C = O, respectively [48]. The O1s deconvoluted spectra have four peaks. The peak at 529.8−530.8 eV represents O^{2-} and it represents Fe^{3+} in the form of Fe_2O_3 and Fe_3O_4 [10]. The peak at 531.8 eV corresponds to OH^- and it represents the presence of iron oxides, that is, FeOOH [10]. The peak at 532.9 eV corresponds to oxygen molecules adsorbed in the form of water [49]. XPS results confirm the inhibitor adsorption onto the carbon steel surface.

5.5.6 X-ray diffraction

X-Ray diffraction (XRD) falls into the category of nondestructive techniques for the investigation of a varying range of materials [50]. In this method, monochromatic X-rays with constructive interference and a crystalline sample are used. The XRD instrument produces X-rays via the cathode ray tube, which generates the monochromatic radiation. These are passed toward the testing sample. The interaction between the testing sample and the incident X-ray beam generates constructive interference that satisfies the Bragg's law condition [51]:

$$n\lambda = 2d \sin\theta \tag{5.45}$$

where n, λ, d, and θ are integers, X-ray wavelength, interplanar spacing, and angle of diffraction. Since every sample has a unique "d" spacing that helps to identify the characteristic of that sample.

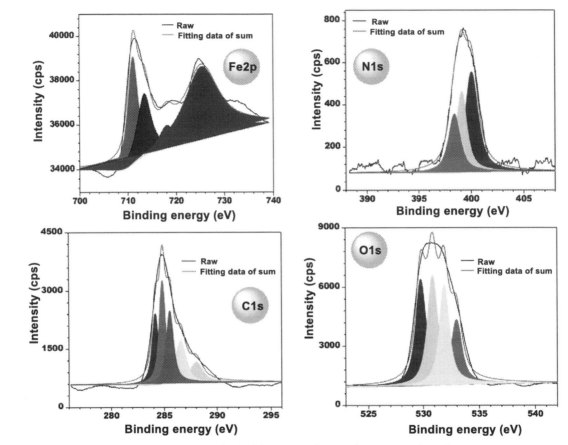

FIGURE 5–18 XPS spectra images for adsorbed inhibitor on carbon steel. *XPS*, X-ray photoelectron spectroscopy.

Fig. 5−19 represents the XRD images of a corrosion product that is present over the surface of carbon steel samples. In the absence of a corrosion inhibitor different peaks at $2\theta = 33°$, $40°$, $44°$, $48°$, $51°$, $52°$, and $66°$ are observed that correspond to iron oxide (Fig. 5−19A). However, only iron peaks are observed in the XRD patterns in the presence of an inhibitor (Fig. 5−19B). The absence of iron oxide peaks with the addition of an inhibitor suggests the formation of an inhibitor protective film over the metal surface.

5.5.7 Scanning electrochemical microscopy

SECM is a versatile and powerful technique for the investigation of the local electrochemical activity that is ongoing over the corroding metal surface using an ultramicroelectrode tip. Here, the tip of the ultramicroelectrode is moved over the surface of the studied metal sample using the precise positioning systems. The pictorial presentation of SECM in feedback mode for the corrosion of carbon steel in 3.5% NaCl solution in the presence and absence of

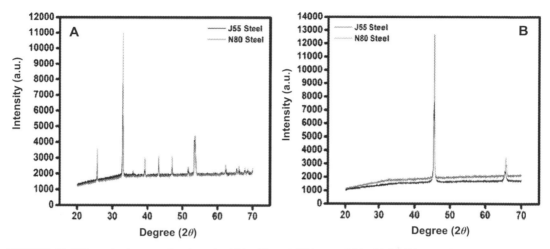

FIGURE 5–19 XPS spectra images of carbon steel (a) without inhibitor and (b) with inhibitor. *XPS*, X-ray photoelectron spectroscopy.

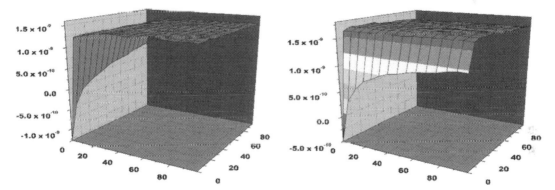

FIGURE 5–20 SECM images (A) without inhibitor and (B) with inhibitor. *SECM*, Scanning electrochemical microscopy.

corrosion inhibitor is shown in Fig. 5–20A and B. According to Fig. 5–20A, in the absence of a corrosion inhibitor, the value of the corrosion current increases as the microtip probe comes close to the carbon steel surface, which suggests that the metal surface is conductive in nature [52]. However, in the presence of an inhibitor, the value of the corrosion current decreases as the probe comes close to a metal surface (Fig. 5–20B), which suggests that the surface metal becomes insulating due to the adsorption and formation of the inhibitor film [53].

5.6 Conclusion

In the present chapter different techniques have been applied for the successful study of metallic corrosion inhibition. In practical cases, satisfactory results were obtained from weight loss and electrochemical methods for the determination of the protection ability of metal. However, in order to understand the proper mechanism of corrosion inhibitor performance surface analytical techniques, including AFM, SEM, and XPS, should be used. The application of surface techniques provides the exact nature of the adsorbed inhibitor film composition, thickness, and types of bonds between the inhibitor molecules and metal surface.

Acknowledgment

The authors acknowledge the financial support provided by the King Fahd University of Petroleum and Minerals (KFUPM), Kingdom of Saudi Arabia, under the Deanship of Scientific Research (DSR) project number DF191051.

Useful links

https://onlinelibrary.wiley.com/doi/abs/10.1002/9781118015438.ch4
https://link.springer.com/chapter/10.1007/978-1-4757-4825-3_6
https://www.sciencedirect.com/topics/engineering/surface-corrosion

References

[1] R. Baboian, Corrosion Tests and Standards: Application and Interpretation, ASTM Manual Series 20, ASTM International, 2005.

[2] D.S. Chauhan, K. Ansari, A. Sorour, M. Quraishi, H. Lgaz, R. Salghi, Thiosemicarbazide and thiocarbohydrazide functionalized chitosan as ecofriendly corrosion inhibitors for carbon steel in hydrochloric acid solution, Int. J. Biol. Macromol. 107 (2018) 1747−1757.

[3] K.R. Ansari, D.S. Chauhan, M.A. Quraishi, T.A. Saleh, Bis(2-aminoethyl)amine-modified graphene oxide nanoemulsion for carbon steel protection in 15% HCl: effect of temperature and synergism with iodide ions, J. Colloid Interface Sci. 564 (2020) 124−133.

[4] A. Singh, K.R. Ansari, M.A. Quraishi, S. Kaya, Theoretically and experimentally exploring the corrosion inhibition of N80 steel by pyrazol derivatives in simulated acidizing environment, J. Mol. Struct. 1206 (2020) 127685.

[5] B. Onyeachu, D.S. Chauhan, K.R. Ansari, I. Obot, M.A. Quraishi, A.H. Alamri, Hexamethylene-1,6-bis (NeD-glucopyranosylamine) as a novel corrosion inhibitor for oil and gas industry: electrochemical and computational analysis, N. J. Chem. 43 (2019) 7282−7293.

[6] A. Singh, K.R. Ansari, M.A. Quraishi, H. Lgaz, Y. Lin, Synthesis and investigation of pyran derivatives as acidizing corrosion inhibitors for N80 steel in hydrochloric acid: theoretical and experimental approaches, J. Alloys Compd. 762 (2018) 347−362.

[7] A. Singh, K.R. Ansari, M.A. Quraishi, S. Kaya, L. Guo, Aminoantipyrine derivatives as a novel eco-friendly corrosion inhibitors for P110 steel in simulating acidizing environment: experimental and computational studies, J. Nat. Gas Sci. Eng. 83 (2020) 103547.

[8] E. Kang, K. Neoh, K. Tan, The intrinsic redox states in polypyrrole and polyaniline: a comparative study by XPS, Surf. Interface Anal. 19 (1992) 33−37.

[9] M.A. Mazumder, H.A. Al-Muallem, M. Faiz, S.A. Ali, Design and synthesis of a novel class of inhibitors for mild steel corrosion in acidic and carbon dioxide-saturated saline media, Corros. Sci. 87 (2014) 187−198.

[10] W. Temesghen, P. Sherwood, Analytical utility of valence band X-ray photoelectron spectroscopy of iron and its oxides, with spectral interpretation by cluster and band structure calculations, Anal. Bioanal. Chem. 373 (2002) 601−608.

[11] S.S. Jamalia, S.E. Moulton, D.E. Tallman, M. Forsyth, J. Weber, G.G. Wallace, Applications of scanning electrochemical microscopy (SECM) for local characterization of AZ31 surface during corrosion in a buffered media, Corros. Sci. 86 (2014) 93−100.

[12] S. Wade, Y. Lizama, Clarke's solution cleaning used for corrosion product removal: efects on carbon steel substrate, in: Corrosion and Prevention 2015, Conference Paper e 050, Australasian CorrosionnAssociation (2015).

[13] ASTM, G1e03: Standard Practice for Preparing, Cleaning, and Evaluating Corrosion Test Specimens, ASTM International (2004).

[14] A. Kina, J. Ponciano, Inhibition of carbon steel CO_2 corrosion in high salinity solutions, Int. J. Electrochem. Sci. 8 (2013) 12600 12612.

[15] M. Finsgar, J. Jackson, Application of corrosion inhibitors for steels in acidic media for the oil and gas industry: a review, Corros. Sci. 86 (2014) 17−41.

[16] ASTM, G4-01, Standard Guide for Conducting Corrosion Tests in Field Applications, ASTM International (2001).

[17] K.R. Ansari, M.A. Quraishi, A. Singh, Chromenopyridin derivatives as environmentally benign corrosion inhibitors for N80 steel in 15% HCl, J. Assoc. Arab. Univ. Basic Appl. Sci. 22 (2017) 45−54.

[18] V. Rajeswari, D. Kesavan, M. Gopiraman, P. Viswanathamurthi, Physicochemical studies of glucose, gellan gum, and hydroxypropyl cellulosedinhibition of cast iron corrosion, Carbohydr. Polym. 95 (2013) 288−294.

[19] X. Luo, C. Ci, J. Li, K. Lin, S. Du, H. Zhang, et al., 4-aminoazobenzene modified natural glucomannan as a green eco-friendly inhibitor for the mild steel in 0.5 M HCl solution, Corros. Sci. 151 (2019) 132−142.

[20] S.B. Aoun, On the corrosion inhibition of carbon steel in 1 M HCl with a pyridinium-ionic liquid: chemical, thermodynamic, kinetic and electrochemical studies, RSC Adv. 7 (2017) 36688−36696.

[21] K. Ansari, M. Quraishi, A. Singh, S. Ramkumar, I.B. Obote, Corrosion inhibition of N80 steel in 15% HCl by pyrazolone derivatives: electrochemical, surface and quantum chemical studies, RSC Adv. 6 (2016) 24130−24141.

[22] R. Solmaz, Investigation of adsorption and corrosion inhibition of mild steel in hydrochloric acid solution by 5-(4-Dimethylaminobenzylidene) rhodanine, Corros. Sci. 79 (2014) 169−176.

[23] L.M. Vracar, D. Drazic, Adsorption and corrosion inhibitive properties of some organic molecules on iron electrode in sulfuric acid, Corros. Sci. 44 (2002) 1669−1680.

[24] K. Ansari, M. Quraishi, Bis-Schiff bases of isatin as new and environmentally benign corrosion inhibitor for mild steel, J. Ind. Eng. Chem. 20 (2014) 2819−2829.

[25] B. Ateya, B. El-Anadouli, F. El-Nizamy, The adsorption of thiourea on mild steel, Corros. Sci. 24 (1984) 509−515.

[26] A. Fateh, M. Aliofkhazraei, A. Rezvanian, Review of corrosive environments for copper and its corrosion inhibitors, Arab. J. Chem. 13 (2017) 481−544. Available from: https://doi.org/10.1016/j.arabjc.2017.05.021.

[27] F. El-Hajjaji, M. Messali, A. Aljuhani, M. Aouad, B. Hammouti, M. Belghiti, et al., Pyridazinium-based ionic liquids as novel and green corrosion inhibitors of carbon steel in acid medium: electrochemical and molecular dynamics simulation studies, J. Mol. Liq. 249 (2018) 997−1008.

[28] P. Singh, A. Singh, M. Quraishi, Thiopyrimidine derivatives as new and effective corrosion inhibitors for mild steel in hydrochloric acid: electrochemical and quantum chemical studies, J. Taiwan Inst. Chem. Eng. 60 (2016) 588−601.

[29] E. Oguzie, Y. Li, F. Wang, Corrosion inhibition and adsorption behavior of methionine on mild steel in sulfuric acid and synergistic effect of iodide ion, J. Colloid Interface Sci. 310 (2007) 90−98.

[30] P.J. Flory, Thermodynamics of high polymer solutions, J. Chem. Phys. 10 (1942) 51−61.

[31] R. Karthikaiselvi, S. Subhashini, J. Assoc. Arab. Univ. Basic Appl. Sci. 16 (2014) 74−82.

[32] A. Singh, K.R. Ansari, D.S. Chauhan, M.A. Quraishi, S. Kaya, Sustain. Chem. Pharm. 16 (2020) 100257.

[33] D.K. Yadav, D. Chauhan, I. Ahamad, M. Quraishi, Electrochemical behavior of steel/acid interface: adsorption and inhibition effect of oligomeric aniline, RSC Adv. 3 (2013) 632−646.

[34] P. Mourya, S. Banerjee, R.B. Rastogi, M.M. Singh, Inhibition of mild steel corrosion in hydrochloric and sulfuric acid media using a thiosemicarbazone derivative, Ind. Eng. Chem. Res. 52 (2013) 12733−12747.

[35] E.E. Oguzie, Corrosion inhibition of aluminium in acidic and alkaline media by Sansevieria trifasciata extract, Corros. Sci. 49 (2007) 1527−1539.

[36] L. Fragoza-Mar, O. Olivares-Xometl, M.A. Domínguez-Aguilar, E.A. Flores, P. Arellanes-Lozada, F. Jiménez-Cruz, Corrosion inhibitor activity of 1,3-diketone malonates for mild steel in aqueous hydrochloric acid solution, Corros. Sci. 61 (2012) 171−184.

[37] M. Hosseini, S.F. Mertens, M. Ghorbani, M.R. Arshadi, Asymmetrical Schiff bases as inhibitors of mild steel corrosion in sulphuric acid media, Mater. Chem. Phys. 78 (2003) 800−808.

[38] A. Singh, X. Dayu, E. Ituen, K.R. Ansari, M.A. Quraishi, S. Kaya, et al., J. Mater. Res. Technol. 9 (2020) 5161−5173. Available from: https://doi.org/10.1016/j.jmrt.2020.03.033.

[39] R. Grauer, P.J. Moreland, G. Pini, A Literature Review of Polarisation Resistance Constant (B) Values for the Measurement of Corrosion Rate, NACE International, Houston, Texas, 1982.

[40] A. Grosvenor, B. Kobe, M. Biesinger, N. McIntyre, Investigation of multiplet splitting of Fe 2p XPS spectra and bonding in iron compounds, Surf. Interface Anal. 36 (2004) 1564−1574.

[41] J.I. Goldstein, D.E. Newbury, J.R. Michael, N.W. Ritchie, J.H.J. Scott, D.C. Joy, Scanning electron microscopy and X-ray microanalysis, Springer, 2017.

[42] A. Singh, Y. Lin, K. Ansari, M. Quraishi, E.E. Ebenso, S. Chen, et al., Electrochemical and surface studies of some Porphines as corrosion inhibitor for J55 steel in sweet corrosion environment, Appl. Surf. Sci. 359 (2015) 331−339.

[43] M. Murmu, S.K. Saha, N.C. Murmu, P. Banerjee, Effect of stereochemical conformation into the corrosion inhibitive behaviour of double azomethine based Schiff bases on mild steel surface in 1 mol/L HCl medium: an experimental, density functional theory and molecular dynamics simulation study, Corros. Sci. 146 (2019) 134−151.

[44] D. Briggs, Handbook of X-Ray Photoelectron Spectroscopy, in: C.D. Wanger, W.M. Riggs, L.E. Davis, J.F. Moulder, G.E. Muilenberg (Eds.), Perkin-Elmer Corp., Physical Electronics Division, Eden Prairie, Minnesota, USA, 1979.

[45] S.J. Roosendaal, B. van Asselen, J.W. Elsenaar, A.M. Vredenberg, F.H.P.M. Habraken, The oxidation state of Fe(100) after initial oxidation in O_2, Surf. Sci. 442 (1999) 329−337.

[46] L. Yuan, Y. Wang, R. Cai, Q. Jiang, J. Wang, B. Boquan Li, et al., The origin of hematite nanowire growth during the thermal oxidation of iron, Mater. Sci. Eng. B 177 (2012) 327−336.

[47] F. Bentiss, C. Jama, B. Mernari, H. El Attari, L. El Kadi, M. Lebrini, et al., Corrosion control of mild steel using 3,5-bis(4-methoxyphenyl)-4-amino-1,2,4-triazole in normal hydrochloric acid medium, Corros. Sci. 51 (2009) 1628−1635.

[48] I. Arukalam, I. Madufor, E. Oguzie, Inhibition of mild steel corrosion in sulfuric acid medium by hydroxyethyl cellulose, Chem. Eng. Commun. 202 (1) (2014) 112–122.

[49] K. Babić-Samardžija, C. Lupu, N. Hackerman, A.R. Barron, A. Luttge, Inhibitive properties and surface morphology of a group of heterocyclic diazoles as inhibitors for acidic iron corrosion, Langmuir 21 (2005) 12187–12196.

[50] H. Stanjek, W. Häusler, Basics of X-ray diffraction, Hyperfine Interact. 154 (2004) 107–119.

[51] S.A. Speakman, Basics of X-Ray Diffraction, Massachusetts Institute of Technology, 2016. Available from: http://prism.mit.edu/xray/.

[52] J. Izquierdo, L. Nagy, J.J. Santana, G. Nagy, R.M. Souto, A novel microelectrochemical strategy for the study of corrosion inhibitors employing the scanning vibrating electrode technique and dual potentiometric/amperometric operation in scanning electrochemical microscopy: application to the study of the cathodic inhibition by benzotriazole of the galvanic corrosion of copper coupled to iron, Electrochim. Acta 58 (2011) 707–716.

[53] N. Aouinaa, et al., Initiation and growth of a single pit on 316L stainless steel: Influence of $SO_4\,2-$ and ClO_4- anions, Electrochim. Acta 104 (2013) 274–281.

Sustainable corrosion inhibitors: candidates and characterizations

6

Corrosion inhibitors for basic environments

Mine Kurtay Yildiz, Mesut Yildiz

CORROSION RESEARCH LABORATORY, DEPARTMENT OF MECHANICAL ENGINEERING, FACULTY OF ENGINEERING, DUZCE UNIVERSITY, DUZCE, TURKEY

Chapter outline

6.1 Introduction

Corrosion is the main reason for the degradation of metals used in many parts of our lives [1]. Moreover, corrosion is also common in other material grades such as ceramic, plastic, and rubber. Corrosion which is the main cause of material failures, as all environments are to a certain extent corrosive, is a major economic problem for society [2].

A basic environment is encountered in most of the industrially significant applications (alkaline etching, batteries, boiler feedwater systems, fuel cells, etc.). Basic environments are less corrosive compared to acidic environments as oxide/hydroxide layers form on the metal surface in these mediums. Pit corrosion is observed as a result of the dissolution of these layers [3].

Inhibitors, which effectively reduce the corrosion rate of metals without significantly reacting with the components of the environment when added to a corrosive medium at low concentrations, are a substance or combination of substances [4]. Corrosion inhibitors contribute to extend the corrosion initiation time or decrease the rate in the propagation phase [5]. Further, they are gaining in popularity as they increase the use of materials by forming a protective layer on the metal surface [6]. It is predicted that the global corrosion inhibitors market, which was US$ 5.99 billion in 2015, will show significant growth due to the rising

Environmentally Sustainable Corrosion Inhibitors. DOI: https://doi.org/10.1016/B978-0-323-85405-4.00008-2

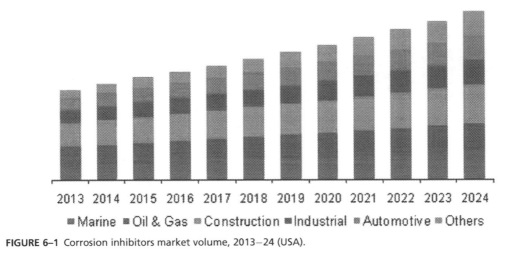

FIGURE 6–1 Corrosion inhibitors market volume, 2013–24 (USA).

construction sector, metal processing, and power generation (Fig. 6–1). Particularly, its widespread use in construction, mining, power generation and water treatment sectors will increase the demand for products in the coming years [7].

Conventional corrosion inhibitors are divided into three categories: organic, inorganic, and hybrid [8]. Studies show that many inorganic and organic compounds can successfully reduce the corrosion rate when incorporated into the corrosive environment even at low concentrations [9–13]. Although inorganic inhibitors have long been applied to protect against corrosion [14–18], they have been forbidden in many countries because of their toxicity [19–21]. Therefore, it has been observed that the use of nontoxic, environment-friendly, biodegradable, cheap, and effective molecules, such as organic inhibitors, amines, aldehydes, and plant extracts, has increased [22–27]. Due to the oxide/hydroxide layers formed on the metal surface, the inhibitors used in alkaline environments are difficult to adhere to the surface, so fewer inhibitors are used in these environments.

6.2 Corrosion inhibitors for various metals in a basic environment

Recent research has focused on using environment-friendly and nontoxic inhibitors instead of toxic inorganic and organic inhibitors. These nontoxic, odorless, colorless, harmless to humans and the environment, cheap, easily soluble in water and the environment in which they are used, corrosion inhibitors are generally easily obtained from plants, natural polymers, and sources such as amino acids [28–30]. Inhibitors are used in basic environments to stop the corrosion of metals such as steel, aluminum, copper, and zinc, which are known to form amphoteric oxides.

6.2.1 Corrosion inhibitors for steel in a basic environment

Steel, which is one of the most used metals in the industry, has gained great importance in recent years due to its width of usage area and durability. Steel, which has many uses such as the production of equipment that is of great importance in industrial processes, for example, heating boilers, machine parts, oil transportation pipes, and the construction sector, has become one of the most important metals today [31,32].

One of the most important factors determining the corrosion behavior is the pH value. pH also affects the resistance of the material to stress corrosion cracking and pitting corrosion. A potential–pH diagram showing the effect of pH on the corrosion of iron in aqueous environment was created by taking into account the reactions of iron with aqueous environments between ions and oxides (Fig. 6–2). Although steel passivates in an alkaline environment, when exposed to aggressive ions (e.g., chloride ions), a decrease in pH (<10) causes the layer to deteriorate [33].

In Fig. 6–2 differences are observed in the corrosion behavior of steel depending on the potential and pH. The regions where the metal is expected to undergo acid or alkali corrosion are defined as the "corrosion area," while regions with hydroxide or oxide formation are called the "passive area" in the diagram.

The reactions in which steels exposed to alkali and oxygen-rich electrolytes become rust are shown below [34]:

$$Fe^{2+} + 2OH^- \rightarrow Fe(OH)_2 \tag{6.1}$$

$$4Fe(OH)_2 + O_2 + 2H_2O \rightarrow 4Fe(OH)_3 \tag{6.2}$$

$$2Fe(OH)_3 \rightarrow Fe_2O_3 \cdot H_2O + 2H_2O \tag{6.3}$$

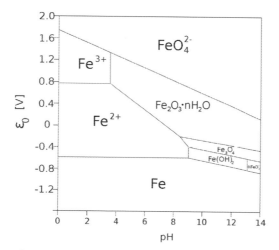

FIGURE 6–2 Pourbaix diagrams for Fe-H$_2$O.

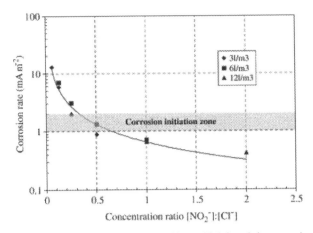

FIGURE 6–3 Relationship between the concentration ratio of $[NO_2^-]:[Cl^-]$ and the corrosion rate.

In a study, the property of calcium nitrite as an inhibitor was investigated by using a polarization method and its impact on compressive strength of concrete was also evaluated. It has been observed that the calcium nitrite-based inhibitor significantly reduces the corrosion rate of the steel in the chloride-contaminated mortar (Fig. 6–3). It has been observed that calcium nitrite increases the compressive strength at 28 days but decreases it in the long term (900 days) [35].

In the study investigating the effect of calcium nitrite, sodium nitrite, and bitter leaf (*Vernonia amygdalina*) extract on the corrosion behavior of reinforced steel in concrete exposed to tap water and 3.5% NaCl, it has been determined that calcium nitrite and bitter leaf extract decrease the corrosion rate of steel compared to sodium nitrite, and bitter leaf extract can be used as a more effective inhibitor against reinforcement corrosion [36]. In another study, the effect of nitrite ions was investigated on the corrosion behavior of steel embedded into concrete. The results obtained showed that the nitrite ions have an important effect on the polarization ratios of the anode and cathode [37]. Table 6–1 gives corrosion inhibitors for steel in basic media.

6.2.2 Corrosion inhibitors for aluminum in a basic environment

Aluminum and its alloys are widely used in packaging, aerospace engineering, automotive, and other machinery and engineering industries because of their light weight and high strength compared to their low density (Fig. 6–4) [48,49]. In addition, Al alloys are known for their superior resistance to oxidation thanks to their passivation property in the presence of oxygen [50].

In basic media, the dissolution of aluminum proceeds as

$$Al + 4OH^- \rightarrow Al(OH)_4^- + 3e^- \tag{6.4}$$

Table 6-1 Corrosion inhibitors for steel.

Name of inhibitor	Metal	Environment	Techniques	IE%, inhibitor efficiency	Reference
Moringa leaves (ML) extract	Carbon steel	3 M KOH	FTIR, WL, PDP, EIS, SEM	86.9%, 85.1%, 85.9%	[38]
Dipotassium hydrogen phosphate	Carbon steel A106	Mildly and highly alkaline solutions (pH 8 and 12)	EIS, PDP, WL, SEM, Raman spectroscopy	97%	[39]
MgAl-NO₂ layered double hydroxides (LDHs)	Reinforcing steel	Simulated carbonated concrete pore solutions	EIS, SEM, XRD, FTIR	Cl-free: 92 % Cl-contaminated: 75%	[40]
Brahmi (Bacopa monnieri) and Henna	Low-carbon steel	0.5 M NaOH	WL, PDP	80%	[41]
Deoxyribonucleic acid (DNA)	Reinforced steel	Simulated concrete pore solutions (SCPS)	LPR, EIS, XPS,	94,41%	[42]
Maize gluten meal extract	Mild steel	SCPS	EIS, PDP, SEM-EDS, ATR-FTIR	88.10%	[43]
Ginger extract	Carbon steel	SCPS	PDP, LP, EIS, ATR-FTIR, XPS,	–	[44]
Na₂HPO₄ (DSP) and benzotriazole (BTA)	Rebar	Carbonated concrete pore solutions	EIS, LPR, PDP, XPS, SEM	89%	[45]
Triethanolammonium dodecylbenzene sulfonate (TDS)	Q235 carbon steel	SCPS	PDP, EIS, Mott–Schottky plots, OM	90.91%	[46]
Polycarboxylate derivatives	ST12 carbon steel	SCPS	EIS, PDP, FTIR, SEM, AFM	98.55%	[47]

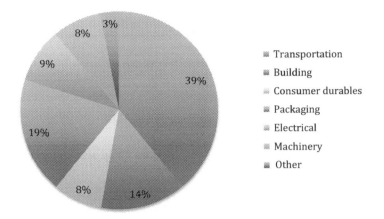

FIGURE 6–4 Aluminum consumption in the United States.

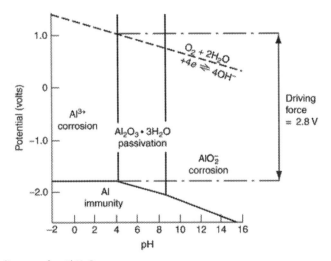

FIGURE 6–5 Pourbaix diagrams for Al-H$_2$O.

and the oxide layer dissolves as

$$Al(OH)_3 + OH^- \rightarrow Al(OH)_4^- \tag{6.5}$$

In Fig. 6–5, Al is stable at negative potentials and corrosion is not expected. In these conditions the metal is immune to corrosion and this region is called the "immunity area." The zone where the metal has a corrosion tendency is called the "corrosion area." In addition,

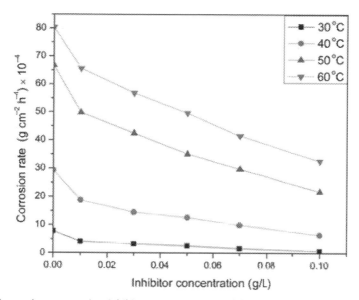

FIGURE 6–6 Plot of corrosion rate against inhibitor concentration at different temperatures.

the area where a corrosion-resistant passive layer is formed on the surface is defined as the "passive area."

Various applications such as alkaline etching and aluminum air batteries, are exposed to alkaline solutions [51–53]. For instance, aluminum and its alloys tend to corrode when exposed to basic solution, particularly potassium and sodium hydroxides [54]. It was shown that three polyacrylic acids with different molecular weights inhibit alkali corrosion of aluminum in weak basic media. It is stated that these polymers have an inhibitory effect by being adsorbed on the metal surface. It appears that the inhibition effects of all polymers used for aluminum corrosion are excellent at pH 8 [55].

In a study examining humic acids, the corrosion prevention potential of humic acid extracted from cow dung was investigated for aluminum alloy in 0.1 M NaOH solution. The results showed that humic acid was a great corrosion inhibitor for this metal in basic solution. It has been found that the corrosion rate increases with the increase in humic acid concentration, but reduces with the increase in temperature (Fig. 6–6) [56]. Table 6–2 gives corrosion inhibitors for aluminum in basic mediums.

6.2.3 Corrosion inhibitors for zinc in a basic environment

Zinc, ranking fourth among metals in production and consumption worldwide, is widely used in many industrial areas (Fig. 6–7) [67]. Zinc is used in the production of many alloys

Table 6–2   Corrosion inhibitors for aluminum.

Name of inhibitor	Metal	Environment	Techniques	IE%, inhibitor efficiency	Reference
Maleic acid, DL-malic acid, succinic acid, DL-tartaric acid, citric acid, tricarballylic acids, and DL-serine	AA5754	Bicarbonate buffer	g-DEIS, EIS, Cyclic polarization, SEM, EDS, XPS	>99%	[57]
Cetyl trimethyl ammonium bromide (CTAB)	Aluminum	2 M NaOH	PDP	99.68%	[58]
Citric acid	AA7020	5 M NaOH	g-DEIS, XPS, SEM	99.4%	[59]
Kalmegh leaf extract	Aluminum alloy	1 M NaOH	EIS, PDP, SEM, AFM	82.45%	[60]
Aromatic carboxylic acids	Al-1050	1 M NaOH	EIS, PDP, SEM	79, 71 and 68%	[61]
Urea and thiourea	Aluminum alloy	5 M KOH	Hydrogen evolution, EIS, PDP, FT-IR, AFM	51.5% and 57.0%	[62]
2-Aminobenzene-1, 3-dicarbonitriles	Al-1060	0.5 M NaOH	WL, OCP, PDP, SEM, EDS	94.50%	[63]
Mangifera indica (MI), *Moringa oleifera* (MO), and *Terminalia arjuna* (TA)	Aluminum alloy	1 M NaOH	EIS, PDP, SEM, AFM	MO: 85.3%	[64]
Rauwolfia serpentina (RS), *Cannabis sativa* (CS), *Annona squamosa* (AS), *Adhatoda vasica* (AV), and *Cymbopogon citratus* (CC)	Aluminum alloy	1 M NaOH	EIS, PDP, LPR, SEM, AFM	RS: 97%	[65]
Piper longum seed extracts	Al-1060	1 M NaOH	EIS, WL, PDP	94%	[66]

and compounds in the industry because it is chemically active and can easily alloy with other metals. It is especially used in galvanizing to protect steel against corrosion [68].

A potential–pH diagram of zinc is given in Fig. 6–8. Zinc dissolving in its ions in an acidic environment has a stable structure at negative potentials. Zinc corrodes over a wide pH range, because it forms oxyanions in basic conditions and cations up to pH 8.

The dissolution of zinc is a successive process of two steps of a charge-transfer process in noncomplexing and neutral solutions.

$$Zn \rightarrow Zn^+_{ad} + e^- \tag{6.6}$$

$$Zn^+_{ad} \rightarrow Zn^{2+}_{ad} + e^- \tag{6.7}$$

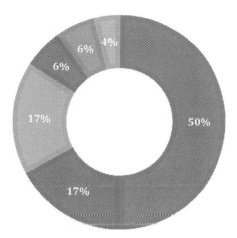

Galvanizing   Alloys   Brass/bronze   Semi-manufacturers   Chemicals   Miscellaneous

FIGURE 6–7 Zinc, global uses, 2018.

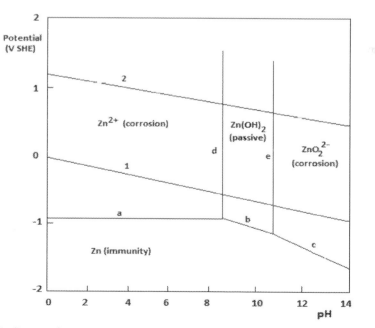

FIGURE 6–8 Pourbaix diagrams for Zn-H$_2$O.

Table 6–3 Corrosion inhibitor for zinc.

Name of inhibitor	Metal	Environment	Techniques	*IE*%, inhibitor concentration	Reference
Polyoxyethylene (40) nonylphenyl ether	Zinc	7.0 M KOH	PDP, EIS, SEM, FT-IR	98%	[69]
Sodium eperuate	Zinc	0.1 M NaCl	PDP, EIS	92%	[70]
Cetyltrimethylammonium bromide (CTMAB) and poly(ethylene glycol)-400 (PEG-400)	Zn and Zn − Ni alloys	8 mol/L KOH	PDP, EIS, SEM	96%	[71]
Polyethylene glycols	Zinc	7 M KOH	PDP, volumetric and gravimetric, SEM	−	[72]
$[Fe(CN)_6]^{3-}$	Zinc and Zn−0.5Ni alloy	7 M KOH	PDP, EIS, SEM, EDX	75%	[73]
Dodecyltrimethylammonium bromide	Zinc	7.0 M KOH	PDP, EIS, SEM	80%	[74]
Poly(ethylene glycol) 600 (PEG) and imidazole (IMZ)	Zinc	3 mol L^{-1} KOH	WL, PDP, SEM	80%	[75]
Polyethylene glycols (PEG400, PEG 600)	Zinc	8 M KOH	PDP, EIS, SEM	93.8 (PEG 600) 90.4 (PEG 400)	[76]
Polyoxyethylene alkyl phosphate ester acid form (GAFAC RA600) and polyethylene glycol (PEG)	Zinc	8.5 M KOH	PDP, LP, SEM	−	[77]
CTAB, FC-129, FC-135, FC-170C	Zinc	30% KOH	Hydrogen evolution reaction, cyclic voltammetry	−	[78]

In its simplest form, the formation and dissolution of zinc hydroxide film can be represented as:

$$Zn + 2OH^- \rightarrow Zn(OH)_2 + 2e^- \tag{6.8}$$

$$Zn(OH)_2 + xOH^- \rightarrow Zn(OH)_{(2+x)}^{(x-)} \tag{6.9}$$

Table 6−3 gives corrosion inhibitors for zinc in basic mediums.

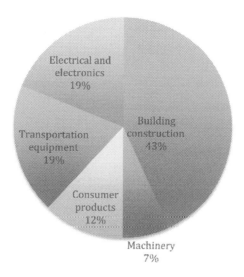

FIGURE 6–9 Uses of copper in United States, 2017.

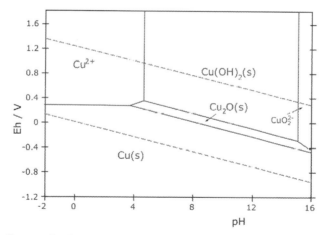

FIGURE 6–10 Pourbaix diagrams for Cu-H$_2$O.

6.2.4 Corrosion inhibitors for copper in a basic environment

Copper, which is a very important element in terms of its physical and chemical properties, is frequently used in electricity, electronics, industry, construction, transportation, industrial hardware, chemistry, jewelry, and paint industries (Fig. 6–9) [79].

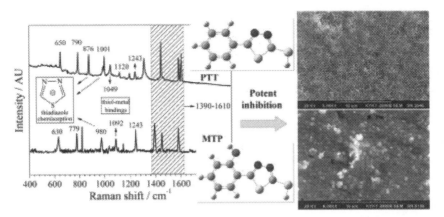

FIGURE 6–11 Raman microspectroscopy of PTT (upper line) and MTP (bottom line) adsorbed on copper surface.

Copper is resistant to the effects of the many chemicals and atmosphere, but is known to be susceptible to corrosion in aggressive environments (Fig. 6–10) [80–82].

Copper dissolves at low and high pH values. The corrosion rate is fundamentally determined by the formation of oxide layers over the metal surface. First, copper dissolves to Cu^+ ions: The corrosion reaction of copper can be formulated by the following reactions:

$$Cu \rightarrow Cu^+ + e^- \tag{6.10}$$

$$2Cu^+ + H_2O \rightarrow Cu_2O + 2H^+ \tag{6.11}$$

$$1/2Cu_2O + 1/2H_2O \rightarrow CuO + OH^- + e^- \tag{6.12}$$

$$Cu^{2+} + H_2O \rightarrow CuO + 2H^+ \tag{6.13}$$

In the study in which the inhibition of benzotriazole against copper corrosion was investigated in a static environment (100–700 rpm), it was observed that the corrosion rate increased due to the increase in speed, but the weight loss decreased four to five times with the addition of inhibitor [83]. The inhibition effect of two new nontoxic thiadiazole derivatives on copper was investigated in 3.5% NaCl solution. The presence of inhibitors and increased concentration have been shown to greatly reduce the corrosion rate (Fig. 6–11). It has also been observed that it suppresses the charge transfer process by adsorption on the copper surface [84].

Methionine has been investigated on the corrosion behavior of copper in 3.5% NaCl solution at different temperatures (298K–328K). It was found that as an organic corrosion inhibitor, methionine can prevent corrosion of copper in 3.5% NaCl solution by forming a protective film. The inhibition effect of methionine increases with decreasing concentration and has been found to act as a mixed type inhibitor [85]. Table 6–4 gives corrosion inhibitors for copper in basic mediums.

Table 6-4 Corrosion inhibitors for copper.

Name of inhibitor	Metal	Environment	Techniques	IE%, inhibitor efficiency	Reference
Tobacco extracted from discarded cigarettes (NDC)	Pure copper and pure zinc	Artificial seawater	WL, EIS, PDP, UV–vis spectroscopy, SEM, AFM, FTIR	96.8% (copper) 98.2% (zinc)	[86]
4-Amino-5-methyl-4H-1,2,4-triazole-3-thiol (AMTT)	Pure copper	3.5% NaCl	EIS, PDP, Cyclic voltammetry, AFM, FTIR	>94%	[87]
Thiadiazole (TDA), benzotriazole (BTA), mercaptobenzimidazole (MBIMD), benzimidazole (BIMD) and mercaptobenzothiazole (MBT)	Pure copper	Sodium borate solution	EIS, PDP	MBT > TDA > MBIMD ≈ BIMD > BTA	[88]
Cysteine	Copper	0.5 M Na_2SO_4	PDP	88.16%	[89]
Benzotriazole	Copper	Borax buffer solutions	OCP, PDP	—	[90]
Benzotriazole	Cu37Zn brass	Sodium tetraborate	Cyclic voltammetry, OCP	—	[91]
Benzotriazole	CuAlNiSi alloy	Sodium tetraborate (borax)	Cyclic voltammetry, OCP	—	[92]
Tetrazole, benzotriazole (BTA), mercaptobenzothiazole (MBT), imidazole (IMD), benzimidazole (BIMD), and mercaptobenzimidazole (MBIMD)	Copper	0.1 M NaOH	EIS, cyclic voltammetry	MBT > BIMD ≈ MBIMD > BTA ≈ IMD	[93]
Cysteine	Cu37Zn	0.5 mol dm^{-3} Na_2SO_4	PDP, OCP, chronoamperometric	85.1%	[94]
Benzotriazole	Cu24Zn5Al alloy	Sodium tetraborate	PDP, OCP	—	[95]

6.3 Conclusion

- Corrosion is a natural process that causes the properties of materials to deteriorate.
- As a result of corrosion, the atom loses one or more electrons, forming an oxidized metal.
- Inhibitors are a great way to prevent corrosion due to their ease of use and application in a wide variety of industries.
- For the success of inhibitor applications, principles such as the type of the metal, the composition of solution, the temperature, the pH of the environment, and the material design should be considered in the selection of the inhibitor.
- Environment-friendly green inhibitors should be preferred to toxic inhibitors due to the latter's negative effects on humans and the environment.

References

[1] B. Valdez, M. Schorr, R. Zlatev, M. Carrillo, M. Stoytcheva, L. Alvarez, et al., Chapter 2: Corrosion control in industry, environmental and industrial corrosion - practical and theoretical aspects, 2012, pp. 19−54.

[2] P.R. Roberge, Corrosion Engineering Principles and Practice, McGraw-Hill, New York, 2008.

[3] M. Kurtay, H. Gerengi, Y. Kocak, The effect of caffeine molecule on the physico-chemical properties of blended cement, Constr. Build. Mater. 255 (2020) 119394.

[4] G. Palanisamy, Corrosion Inhibitors, InTech Open, 2019, pp. 1−24. Available from: http://doi.org/10.5772/intechopen.80542.

[5] F. Bolzoni, A. Brenna, G. Fumagalli, S. Goidanich, L. Lazzari, M. Ormellese, et al., Experiences on corrosion inhibitors for reinforced concrete, Int. J. Corros. Scale Inhibition 3 (4) (2014) 254−278.

[6] M. Erbil, Korozyon İlkeler-Önlemler, 1st Ed., Korozyon Derneği Yayınevi, Ankara, Türkiye, 2012.

[7] Corrosion Inhibitors Market Size, Share & Trends Analysis Report By Product (Organic, Inorganic), By Type (Water Based, Oil Based), By End Use, By Region, And Segment Forecasts. < https://www.grandviewresearch.com/industry-analysis/corrosion-inhibitors-market >. 2020, pp. 2020−2027.

[8] S. Mohammadi, F.B. Ravari, A. Dadgarinezhad, Improvement in corrosion inhibition efficiency of molybdate-based inhibitors via addition of nitroethane and zinc in stimulated cooling water, Int. Sch. Res. Not. 2012 (2012) 1−9.

[9] M.L.S. Rivetti, J.S.A. Neto, N.S.A. Júnior, D.V. Ribeiro, Corrosion Inhibitors for Reinforced Concrete, Corrosion Inhibitors, Principles and Recent Applications, InTech Open, 2018, pp. 35−58.

[10] I.B. Obot, M.M. Solomon, S.A. Umoren, R. Suleiman, M. Elanany, N.M. Alanazi, et al., Progress in the development of sour corrosion inhibitors: past, present, and future perspectives, J. Ind. Eng. Chem. 79 (2019) 1−18.

[11] B.P. Maliekkal, J.T. Kakkassery, V.R. Palayoo, Efficacies of sodium nitrite and sodium citrate−zinc acetate mixture to inhibit steel rebar corrosion in simulated concrete interstitial solution contaminated with NaCl, Int. J. Ind. Chem. 9 (2) (2018) 105−114.

[12] P.B. Raja, S. Ghoreishiamiri, M. Ismail, Natural Corrosion Inhibitors for Steel Reinforcement in Concrete − A Review, 22, World Scientific Publishing Company, 2015, pp. 68−75.

[13] Corrosion of metals and alloys—Basic terms and definitions, ISO 8044, The International Organization for Standardization, 2015.

[14] M. Balonis, G. Sant, O.B. Isgor, Mitigating steel corrosion in reinforced concrete using functional coatings, corrosion inhibitors, and atomistic simulations, Cem. Concr. Compos. 101 (2019) 15−23.

[15] J.K. Das, B. Pradhan, Effect of cation type of chloride salts on corrosion behaviour of steel in concrete powder electrolyte solution in the presence of corrosion inhibitors, Constr. Build. Mater. 208 (2019) 175−191.

[16] A.H. Noorbakhsh Nezhad, A. Davoodi, E.M. Zahrani, R. Arefinia, The effects of an inorganic corrosion inhibitor on the electrochemical behavior of superhydrophobic micro-nano structured Ni films in 3.5% NaCl solution, Surf. Coat. Technol. 395 (2020) 125946.

[17] T. Bellezze, D. Timofeeva, G. Giuliani, G. Roventi, Effect of soluble inhibitors on the corrosion behaviour of galvanized steel in fresh concrete, Cem. Concr. Res. 107 (2018) 1−10.

[18] H.S. Lee, H.M. Yang, J.K. Singh, S.K. Prasad, B. Yoo, Corrosion mitigation of steel rebars in chloride contaminated concrete pore solution using inhibitor: an electrochemical investigation, Constr. Build. Mater. 173 (2018) 443−451.

[19] Q. Liu, Z. Song, H. Han, S. Donkor, L. Jiang, W. Wang, et al., A novel green reinforcement corrosion inhibitor extracted from waste *Platanus acerifolia* leaves, Constr. Build. Mater. 260 (2020) 119695.

[20] F. Cao, J. Wei, J. Dong, W. Ke, The corrosion inhibition effect of phytic acid on 20SiMn steel in simulated carbonated concrete pore solution, Corros. Sci. 100 (2015) 365−376.

[21] H. Lee, H. Ryu, W. Park, M. Ismail, Comparative study on corrosion protection of reinforcing steel by using amino alcohol and lithium nitrite inhibitors, Materials 8 (2015) 251−269.

[22] K. Zhang, W. Yang, X. Yin, Y. Chen, Y. Liu, J. Le, et al., Amino acids modified konjac glucomannan as green corrosion inhibitors for mild steel in HCl solution, Carbohydr. Polym. 181 (2018) 191−199.

[23] M.A. Asaad, M. Ismail, P.B. Raja, N.H.A. Khalid, Rhizophora apiculata as eco-friendly inhibitor against mild steel corrosion in 1M HCl, Surf. Rev. Lett. 24 (2017) 1850013.

[24] P. Muthukrishnan, B. Jeyaprabha, P. Prakash, Adsorption and corrosion inhibiting behavior of *Lannea coromandelica* leaf extract on mild steel corrosion, Arab. J. Chem. 10 (2017) 2343−2354.

[25] B. El Ibrahimi, A. Jmiai, L. Bazzi, S. El Issami, Amino acids and their derivatives as corrosion inhibitors for metals and alloys, Arab. J. Chem. 13 (2020) 740−771.

[26] M. Yadav, T.K. Sarkar, T. Purkait, Amino acid compounds as eco-friendly corrosion inhibitor for N80 steel in HCl solution: electrochemical and theoretical approaches, J. Mol. Liq. 212 (2015) 731−738.

[27] S. Jiang, S. Gao, L. Jiang, M.Z. Guo, Y. Jiang, C. Chen, et al., Effects of deoxyribonucleic acid on cement paste properties and chloride-induced corrosion of reinforcing steel in cement mortars, Cem. Concr. Compos. 91 (2018) 87−96.

[28] G.L.F. Mendonça, S.N. Costa, V.N. Freire, P.N.S. Casciano, A.N. Koreia, P. Lima-Neto, Understanding the corrosion inhibition of carbon steel and copper in sulphuric acid medium by amino acids using electrochemical techniques allied to molecular modelling methods, Corros. Sci. 115 (2017) 41−55.

[29] N. Etteyeb, X.R. Nóvoa, Inhibition effect of some trees cultivated in arid regions against the corrosion of steel reinforcement in alkaline chloride solution, Corros. Sci. 112 (2016) 471−482.

[30] M.M. Solomon, H. Gerengi, S.A. Umoren, N.B. Essien, U.B. Essien, E. Kaya, Gum Arabic-silver nanoparticles composite as a green anticorrosive formulation for steel corrosion in strong acid media, Carbohydr. Polym. 181 (2018) 43−55.

[31] M. Yildiz, H. Gerengi, M.M. Solomon, E. Kaya, S.A. Umoren, Influence of 1-butyl-1-methylpiperidinium tetrafluoroborate on St37 steel dissolution behavior in HCl environment, Chem. Eng. Commun. 205 (4) (2018) 538−548.

[32] H. Gerengi, M.M. Solomon, S.A. Umoren, H.I. Ugras, M. Yildiz, P. Slepski, Improved performance of 1-ethyl-3-methylimidazolium tetrafluoroborate at steel/HCl interface by iodide ions, J. Bio- Tribo-Corros. 4 (12) (2018) 1−11.

[33] H. Verbruggen, H. Terryn, I. De Graeve, Inhibitor evaluation in different simulated concrete pore solution for the protection of steel rebars, Constr. Build. Mater. 124 (2016) 887−896.

[34] F. Pruckner, Corrosion and Protection of Reinforcement in Concrete Measurements and Interpretation, University of Vienna Faculty of Natural Sciences and Mathematics, Vienna, 2001.

[35] K.Y. Ann, H.S. Jung, H.S. Kim, S.S. Kim, H.Y. Moon, Effect of calcium nitrite-based corrosion inhibitor in preventing corrosion of embedded steel in concrete, Cem. Concr. Res. 36 (2006) 530−535.

[36] D.G. Eyu, H. Esah, C. Chukwuekezie, J. Idris, I. Mohammad, Effect of green inhibitor on the corrosion behaviour of reinforced carbon steel in concrete, RPN J. Eng. Appl. Sci. 8 (5) (2013) 326−332.

[37] Z. Cao, M. Hibino, H. Goda, Effect of nitrite inhibitor on the macrocell corrosion behavior of reinforcing steel, J. Chem. 2015 (2015) 1−15.

[38] V.C. Anadebe, C.S. Okafor, O.D. Onukwuli, Electrochemical, molecular dynamics, adsorption studies and anti-corrosion activities of Moringa Leaf biomolecules on carbon steel surface in alkaline and acid environment, Chem. Data Collect. 28 (2020) 100437.

[39] A. Mohagheghi, R. Arefinia, Corrosion inhibition of carbon steel by dipotassium hydrogen phosphate in alkaline solutions with low chloride contamination, Constr. Build. Mater. 187 (2018) 760−772.

[40] X. Jinxia, W. Jiafeng, M. Guoxu, T. Qipin, Effect of MgAl-NO$_2$ LDHs inhibitor on steel corrosion in chloride-free and contaminated simulated carbonated concrete pore solutions, Corros. Sci. 176 (2020) 108940.

[41] N.H.J.A. Hasan, H.J. Alaradi, Z.A.K.A. Mansor, A.H.J.A. Shadood, The dual effect of stem extract of Brahmi (Bacopamonnieri) and Henna as a green corrosion inhibitor for low carbon steel in 0.5M NaOH solution, Case Stud. Constr. Mater. 11 (2019) e00300.

[42] C. Chen, L. Jiang, M.-Z. Guo, P. Xu, L. Chen, J. Zha, Effect of sulfate ions on corrosion of reinforced steel treated by DNA corrosion inhibitor in simulated concrete pore solution, Constr. Build. Mater. 228 (2019) 116752.

[43] Z. Zhang, H. Ba, Z. Wu, Sustainable corrosion inhibitor for steel in simulated concrete pore solution by maize gluten meal extract: electrochemical and adsorption behavior studies, Constr. Build. Mater. 227 (2019) 117080.

[44] Y. Liu, Z. Song, W. Wang, L. Jiang, Y. Zhang, M. Guo, et al., Effect of ginger extract as green inhibitor on chloride-induced corrosion of carbon steel in simulated concrete pore solutions, J. Clean Prod. 214 (2019) 298−307.

[45] D. Wang, J. Ming, J. Shi, Enhanced corrosion resistance of rebar in carbonated concrete pore solutions by Na$_2$HPO$_4$ and benzotriazole, Corros. Sci. 174 (2020) 108830.

[46] Y. Zhao, T. Pan, X. Yu, D. Chen, Corrosion inhibition efficiency of triethanolammonium dodecylbenzene sulfonate on Q235 carbon steel in simulated concrete pore solution, Corros. Sci. 158 (2019) 108097.

[47] A.S. Fazayel, M. Khorasani, A.A. Sarabi, The effect of functionalized polycarboxylate structures as corrosion inhibitors in a simulated concrete pore solution, Appl. Surf. Sci. 441 (2018) 895−913.

[48] H. Gerengi, G. Bereket, M. Kurtay, A morphological and electrochemical comparison of the corrosion process of aluminum alloys under simulated acid rain conditions, J. Taiwan Inst. Chem. Eng. 58 (2016) 509−516.

[49] C. Woodford, Aluminum. < https://www.explainthatstuff.com/aluminum.html >. (Accessed 24 September 2020).

[50] D.G. Ladha, N.K. Shah, Z. Ghelichkhah, I.B. Obot, F. Khorrami Dehkharghani, J.-Z. Yao, et al., Experimental and computational evaluation of illicium verum as a novel eco-friendly corrosion inhibitor for aluminium, Mater. Corros., 69, 2018, pp. 125−139.

[51] K. Xhanari, M. Finšgar, Organic corrosion inhibitors for aluminum and its alloys in chloride and alkaline solutions: a review, Arab. J. Chem. 12 (8) (2019) 4646−4663.

[52] D. Egan, C.P. De León, R. Wood, R. Jones, K. Stokes, F. Walsh, Developments in electrode materials and electrolytes for aluminium−air batteries, J. Power Sources 236 (2013) 293−310.

[53] Y. Ma, X. Zhou, G. Thompson, P. Skeldon, Surface texture formed on AA2099 Al−Li−Cu alloy during alkaline etching, Corros. Sci. 66 (2013) 292−299.

[54] M.A. Quraishi, D.S. Chauhan, V.S. Saji, 7-Heterocyclic corrosion inhibitors for alkaline environments, Heterocyclic Organic Corrosion Inhibitors: Principles and Applications, Elsevier, 2020, pp. 195−209.

[55] M.A. Amin, S.S. AbdEI-Rehim, E.E.F. El-Sherbini, O.A. Hazzazi, M.N. Abbas, Polyacrylic acid as a corrosion inhibitor for aluminium in weakly alkaline solutions. Part I: weight loss, polarization, impedance EFM and EDX studies, Corros. Sci. 51 (3) (2009) 658−667.

[56] S.A. Umoren, E.I. Inam, A.A. Udoidiong, I.B. Obot, U.M. Eduok, K.-W. Kim, Humic acid from livestock dung: ecofriendly corrosion inhibitor for 3SR aluminum alloy in alkaline medium, Chem. Eng. Commun. 202 (2015) 206−216.

[57] J. Wysocka, M. Cieslik, S. Krakowiak, J. Ryl, Carboxylic acids as efficient corrosion inhibitors of aluminium alloys in alkaline media, Electrochim. Acta 289 (2018) 175−192.

[58] A.M. Abdel-Gaber, E. Khamis, H. Abo-Eldahab, S. Adeel, Novel package for inhibition of aluminium corrosion in alkaline solutions, Mater. Chem. Phys. 124 (1) (2010) 773−779.

[59] J. Wysocka, S. Krakowiak, J. Ryl, Evaluation of citric acid corrosion inhibition efficiency and passivation kinetics for aluminium alloys in alkaline media by means of dynamic impedance monitoring, Electrochim. Acta 258 (2017) 1463−1475.

[60] N. Chaubey, V.K. Singh, M.A. Quraishi, Electrochemical approach of Kalmegh leaf extract on the corrosion behavior of aluminium alloy in alkaline solution, Int. J. Ind. Chem. 8 (2017) 75−82.

[61] A.R. Madram, F. Shokri, M.R. Sovizi, H. Kalhor, Aromatic carboxylic acids as corrosion inhibitors for aluminium in alkaline solution, Portugaliae Electrochim. Acta 34 (6) (2016) 395−405.

[62] Z. Moghadam, M. Shabani-Nooshabadi, M. Behpour, Electrochemical performance of aluminium alloy in strong alkaline media by urea and thiourea as inhibitor for aluminium-air batteries, J. Mol. Liq. 242 (2017) 971−978.

[63] C. Verma, P. Singh, I. Bahadur, E.E. Ebenso, M.A. Quraishi, Electrochemical, thermodynamic, surface and theoretical investigation of 2-aminobenzene-1,3-dicarbonitriles as green corrosion inhibitor for aluminum in 0.5M NaOH, J. Mol. Liq. 209 (2015) 767−778.

[64] N. Chaubey, V.K. Singh, M.A. Quraishi, Corrosion inhibition performance of different bark extracts on aluminium in alkaline solution, J. Assoc. Arab. Univ. Basic Appl. Sci. 22 (1) (2017) 38−44.

[65] N. Chaubey, D.K. Yadav, V.K. Singh, M.A. Quraishic, A comparative study of leaves extracts for corrosion inhibition effect on aluminium alloy in alkaline medium, Ain Shams Eng. J. 8 (4) (2017) 673−682.

[66] A. Singh, I. Ahamad, M.A. Quraishi, Piper longum extract as green corrosion inhibitor for aluminium in NaOH solution, Arab. J. Chem. 9 (2) (2016) S1584−S1589.

[67] Minerals and Metals Facts, Zinc Facts. < https://www.nrcan.gc.ca/our-natural-resources/minerals-mining/zinc-facts/20534 >. (Accessed 27 November 2019).

[68] M. Lebrini, F. Suedile, P. Salvina, C. Roos, A. Zarrouk, C. Jama, et al., Bagassa guianensis ethanol extract used as sustainable eco-friendly inhibitor for zinc corrosion in 3% NaCl: electrochemical and XPS studies, Surf. Interfaces 20 (2020) 100588.

[69] M.A. Deyab, Application of nonionic surfactant as a corrosion inhibitor for zinc in alkaline battery solution, J. Power Sources 292 (2015) 66−71.

[70] M.C. Li, M. Royer, D. Stien, A. Lecante, C. Roos, Inhibitive effect of sodium eperuate on zinc corrosion in alkaline solutions, Corros. Sci. 50 (7) (2008) 1975−1981.

[71] I.M. Abdel-Lateef, M. Elrouby, Synergistic inhibition effect of poly(ethylene glycol) and cetyltrimethylammonium bromide on corrosion of Zn and Zn−Ni alloys for alkaline batteries, Trans. Nonferrous Met. Soc. China 30 (1) (2020) 259−274.

[72] J. Dobryszycki, S. Biallozor, On some organic inhibitors of zinc corrosion in alkaline media, Corros. Sci. 43 (7) (2001) 1309−1319.

[73] A.-R. El-Sayed, H.S. Mohran, H.M.A. El-Lateef, Inhibitive action of ferricyanide complex anion on both corrosion and passivation of zinc and zinc−nickel alloy in the alkaline solution, J. Power Sources 196 (15) (2011) 6573−6582.

[74] K. Liu, P. He, H. Bai, J. Chen, F. Dong, S. Wang, et al., Effects of dodecyltrimethylammonium bromide surfactant on both corrosion and passivation behaviors of zinc electrodes in alkaline solution, Mater. Chem. Phys. 199 (2017) 73–78.

[75] H. Zhou, Q. Huang, M. Liang, D. Lv, M. Xu, H. Li, et al., Investigation on synergism of composite additives for zinc corrosion inhibition in alkaline solution, Mater. Chem. Phys. 128 (1–2) (2011) 214–219.

[76] J.H. Wang, Y.T. Horng, Y.C. Lin, R.S. Lin, Effects of organic corrosion inhibitors on the electrochemical characteristics of zinc in alkaline media, Corrosion 64 (1) (2008) 51–59.

[77] Y. Ein-Eli, M. Auinat, D. Starosvetsky, Electrochemical and surface studies of zinc in alkaline solutions containing organic corrosion inhibitors, J. Power Sources 114 (2) (2003) 330–337.

[78] J.L. Zhu, Y.H. Zhou, C.Q. Gao, Influence of surfactants on electrochemical behavior of zinc electrodes in alkaline solution, J. Power Sources 72 (2) (1998) 231–235.

[79] Copper Uses, Resources, Supply, Demand and Production Information. https://geology.com/usgs/uses-of-copper

[80] A. Dehghani, G. Bahlakeh, B. Ramezanzadeh, M. Ramezanzadeh, Electronic/atomic level fundamental theoretical evaluations combined with electrochemical/surface examinations of Tamarindus indiaca aqueous extract as a new green inhibitor for mild steel in acidic solution (HCl 1M), J. Taiwan Inst. Chem. Eng. 102 (2019) 349–377.

[81] W. Gong, B. Xu, X. Yin, Y. Liu, Y. Chen, W. Yang, Halogen-substituted thiazole derivatives as corrosion inhibitors for mild steel in 0.5M sulfuric acid at high temperature, J. Taiwan. Inst. Chem. Eng. 97 (2019) 466–479.

[82] T. Yan, S. Zhang, L. Feng, Y. Qiang, L. Lu, D. Fu, et al., Investigation of imidazole derivatives as corrosion inhibitors of copper in sulfuric acid: combination of experimental and theoretical researches, J. Taiwan. Inst. Chem. Eng. 106 (2020) 118–129.

[83] P.F. Khan, V. Shanthi, R.K. Babu, S. Muralidharan, R.C. Barik, Effect of benzotriazole on corrosion inhibition of copper under flow conditions, J. Environ. Chem. Eng. 3 (1) (2015) 10–19.

[84] H. Tian, W. Li, K. Cao, B. Hou, Potent inhibition of copper corrosion in neutral chloride media by novel non-toxic thiadiazole derivatives, Corros. Sci. 73 (2013) 281–291.

[85] G. Kilinççeker, H. Demir, The inhibition effects of methionine on corrosion behavior of copper in 3.5% NaCl solution at pH = 8.5, Prot. Met. Phys. Chem. Surf. 49 (2013) 788–797.

[86] A. Singh, X. Dayu, E. Ituen, K. Ansari, M.A. Quraishi, S. Kaya, et al., Tobacco extracted from the discarded cigarettes as an inhibitor of copper and zinc corrosion in an ASTM standard D1141-98(2013) artificial seawater solution, J. Mater. Res. Technol. 9 (3) (2020) 5161–5173.

[87] D.S. Chauhan, M.A. Quraishi, C. Carrière, A. Seyeux, P. Marcus, A. Singh, Electrochemical, ToF-SIMS and computational studies of 4-amino-5-methyl-4H-1,2,4-triazole-3-thiol as a novel corrosion inhibitor for copper in 3.5% NaCl, J. Mol. Liq. 289 (2019) 111113.

[88] F. Altaf, R. Qureshi, S. Ahmed, Surface protection of copper by azoles in borate buffers-voltammetric and impedance analysis, J. Electroanal. Chem. 659 (2) (2011) 134–142.

[89] M.B. Petrović, M.B. Radovanović, A.T. Simonović, S.M. Milić, M.M. Antonijević, The effect of cysteine on the behaviour of copper in neutral and alkaline sulphate solutions, Int. J. Electrochem. Sci. 7 (10) (2012) 9043–9057.

[90] M.M. Antonijevic, S.M. Milic, M.D. Dimitrijevic, M.B. Petrovic, M.B. Radovanovic, A.T. Stamenkovic, The influence of pH and chlorides on electrochemical behavior of copper in the presence of benzotriazole, Int. J. Electrochem. Sci. 4 (7) (2009) 962–979.

[91] M.M. Antonijević, S.M. Milić, S.M. Šerbula, G.D. Bogdanović, The influence of chloride ions and benzotriazole on the corrosion behavior of Cu37Zn brass in alkaline medium, Electrochim. Acta 50 (18) (2005) 3693–3701.

[92] S.M. Milić, M.M. Antonijević, S.M. Šerbula, G.D. Bogdanović, Influence of benzotriazole on corrosion behaviour of CuAlNiSi alloy in alkaline medium, Corros. Eng. Sci. Technol. 43 (1) (2008) 30–37.

[93] R. Subramanian, V. Lakshminarayanan, Effect of adsorption of some azoles on copper passivation in alkaline, Corros. Sci. 44 (3) (2002) 535–554.

[94] S.M. Abd El Haleem, E.E. Abd El Aal, Electrochemical behavior of copper in alkaline-sulfide solutions, Corrosion 62 (2) (2006) 121–128.

[95] M. Radovanovic, M.P. Mihajlovic, A.T. Simonović, Snežana M. Milić, M. Antonijevic, Cysteine as a green corrosion inhibitor for Cu37Zn brass in neutral and weakly alkaline sulphate solutions, Environ. Sci. Pollut. Res. 20 (2013) 4370–4381.

7

Corrosion inhibitors for neutral environment

Ruby Aslam[1], Mohammad Mobin[1], Afroz Aslam[2], Saman Zehra[1], Jeenat Aslam[3]

[1]CORROSION RESEARCH LABORATORY, DEPARTMENT OF APPLIED CHEMISTRY, FACULTY OF ENGINEERING AND TECHNOLOGY, ALIGARH MUSLIM UNIVERSITY, ALIGARH, INDIA [2]DEPARTMENT OF CHEMISTRY, FACULTY OF SCIENCE, ALIGARH MUSLIM UNIVERSITY, ALIGARH, INDIA [3]DEPARTMENT OF CHEMISTRY, COLLEGE OF SCIENCE, YANBU, TAIBAH UNIVERSITY, AL-MADINA, SAUDI ARABIA

Chapter outline

Abbreviations

WL   Weight loss
PDP   Potentiodynamic polarization

Environmentally Sustainable Corrosion Inhibitors. DOI: https://doi.org/10.1016/B978-0-323-85405-4.00006-9

EIS   Electrochemical impedance spectroscopy
HE   Hydrogen evolution
CP   Cathodic polarization
LSV   Linear sweep voltammetry
OCP   Open circuit potential
SEM   Scanning electron microscopy
XRD   X-ray diffraction
T   Temperature
RT   Room temperature
IE   Inhibition efficiency

7.1 Introduction

Metals have a variety of favorable properties including electrical and heat conductivity, higher melting and boiling points, higher tensile strength, high mass-to-volume ratio, and ductility, and are commonly used as raw materials in industries and a variety of other areas. Chemical or electrochemical reactions occur between metals and their environments as they interact with the environment. Corrosion is the process of converting the refined metal into a more chemically stable form, such as oxide, hydroxide, or sulphide [1]. The structures and properties of different materials are influenced by this interaction. Corrosion is the irreversible degradation of materials, and it is a very expensive occurrence, making it the world's most serious problem. Corrosion is influenced by a variety of physical and chemical factors, including temperature, gases, salts, contaminants, stresses, and changes in pH and electrolyte species. Chloride ions, which were used to mimic seawater conditions because NaCl molecules occur in greater proportion than other salts, are among the most common aggressive corrosive media. There are various ways to prevent corrosion and the rates at which it can propagate in order to extend the life of metallic and alloy materials. Metal surface coatings, environmental enhancements, corrosion inhibitors, pH modifications, and cathodic or anodic protection are some of the strategies available. Several methods were used to combat metal corrosion, the use of inhibitors being among the most common [2,3]. A corrosion inhibitor is a substance which reduces or eliminates the action of the ambient environment to corrode the metal surface when introduced to an aggressive solution in small quantities [4]. These inhibitors function by developing a film over the metal surface, which entails the donation of their electrons to the metal. The efficacy of the protection process depends on the form, characteristics, and time of protection of the surface of the metal, as well as the following atmospheric conditions: temperature, humidity, and corrosive atmosphere characteristics [4]. The primary aim of the chapter is to cover the fundamentals related to the corrosion of various metals in neutral media and the inhibitors employed to combat the corrosion.

7.2 Corrosion in neutral media

In freshwater, seawater, and salt solutions metal corrosion takes place. Corrosion only happens significantly in nearly all of these environments when dissolved oxygen is available. Water solutions dissolve oxygen from the air easily and this is the source of the oxygen needed in the process of corrosion.

7.2.1 Iron

Iron rusting when exposed to a moist environment is the most common type of corrosion. This process consumes the electrons released during the oxidation reaction, where iron corrosion can undergo two separate transformations as follows [5]:

$$Fe \rightarrow Fe^{2+} + 2e^- \tag{7.1}$$

$$Fe^{2+} \; \rightarrow Fe^{3+} + e \tag{7.2}$$

It is well-known that the cathodic reaction of metals in aerated neutral solutions is the reduction of oxygen, in accordance with the Eq. (7.3),

$$2H_2O + O_2 + 4e^- \rightarrow 4OH^- \tag{7.3}$$

Nonetheless, the second transformation Eq. (7.2) in practice, is not going to occur. According to Darwish et al. [6], the dissolution of iron in concentrated NaCl solutions into ferrous cations can be understood,

$$Fe + H_2O \rightarrow Fe(OH)_{ads} + H^+ \tag{7.4}$$

$$Fe + Cl^- \rightarrow Fe(Cl^-)_{ads} \tag{7.5}$$

$$Fe(OH)_{ads} + Fe(Cl^-)_{ads} \rightarrow Fe + FeOH^+ + (Cl^-) + 2e^- \tag{7.6}$$

$$FeOH^+ + H^+ = Fe_{aq}^{2+} + H_2O \tag{7.7}$$

Furthermore, the resulting hydroxide ions from Eq. (7.3) will react with Fe^{2+} in order to form a deposit of $Fe(OH)_2$,

$$Fe + \tfrac{1}{2}O_2 + H_2O \rightarrow Fe(OH)_2 \tag{7.8}$$

The ferrous hydroxide formed converts into the final corrosion product called magnetite (Fe_3O_4) when there is an abundance of oxygen, according to the following reaction,

$$3Fe(OH)_2 + \tfrac{1}{2}O_2 \rightarrow Fe_3O_4 + 3H_2O \tag{7.9}$$

7.2.2 Copper

Copper is recognized as a noble metal which, due to the formation on its surface of a nonconductive layer of corrosion products [7,8], provides sufficient corrosion resistance in the atmosphere and in some chemical environments. However, pitting corrosion can occur on the surface of copper in the presence of oxygen and certain aggressive anions such as chloride and sulfate ions [9]. Copper corrosion and oxide layer accumulation on its surface have a negative effect on the productivity of a copper-built framework and can restrict its functionality [10,11]. In neutral solutions, the cathodic reaction is:

$$O_2 + 2H_2O + 4e^- \rightarrow 4OH^- \tag{7.10}$$

In the above environment, however, the anodic reactions are regulated independently of pH by chloride concentration. At chloride concentrations <1 M, the mechanism of copper dissolution can be expressed as [12]:

$$Cu + Cl^- \leftrightarrow CuCl + e^- \tag{7.11}$$

$$CuCl + Cl^- \rightarrow CuCl_2^- \tag{7.12}$$

Higher cuprous complexes such as $CuCl_3^{2-}$ and $CuCl_4^{2-}$ are formed when chloride concentrations <1 M [13].

In neutral solutions, the cathodic response is:

$$O_2 + 2H_2O + 4e^- \rightarrow 4OH^- \tag{7.13}$$

The use of corrosion inhibitors in these conditions is essential as no actual passivation layer can be predicted.

7.2.3 Aluminum

Aluminum has exemplary formability, high thermal and electrical conductivity, light weight ($2.7 \, g \, cm^{-3}$ density), and high reflectivity. Aluminum is comparatively cheap and almost double as plentiful as iron. Aluminum can be used under various conditions in various types of machinery. In addition, aluminum shows high corrosion resistance when exposed to the atmosphere and many aqueous environments because of the formation of a resistive oxide layer. In addition, colorless and nonpoisonous corrosion materials are formed [14]. However, the oxide layer developed at aluminum is dissolved in the presence of chloride-containing solutions and therefore the metal gets corroded. The most dangerous form of corrosion in aluminum is pitting corrosion. It exists on the metal surface, as holes and pits with irregular shapes. Depending on the type of environment to which aluminum and its alloys are exposed, the diameter and depth of the pits rely on the corrosive medium and the environmental conditions. Aluminum pitting corrosion occurs commonly in solutions of aerated chloride. The following are three reactions that occur on anodic and cathode sites:

At anode

$$Al \rightarrow Al^{3+} + 3e^- \tag{7.14}$$

$$Al^{3+} + 3H_2O \rightarrow Al(OH)_3 + 3H^+ \tag{7.15}$$

At cathode

$$AlCl_3 + 3H_2O \rightarrow Al(OH)_3 + 3HCl \tag{7.16}$$

$$3H^+ + 3e- \rightarrow 3/2\,H_2 \tag{7.17}$$

$$1/2\,O_2 + H_2O + 2e- \rightarrow 2OH^- \tag{7.18}$$

7.3 Corrosion inhibitors for various metals

7.3.1 Fe and Fe-based alloys

Several reports related to iron and alloys corrosion inhibition in neutral environment are available. In 3.5% NaCl, The inhibition activity of water-soluble carboxymethyl chitosan on carbon steel was evaluated by Macedo et al. [15]. At 80 mg L^{-1}, the eco-friendly inhibitor showed 85% inhibition efficiency. Mixed-type inhibition with an anodic predominance was shown by the inhibitor. Luo et al. [16] evaluated the anticorrosive characteristics of a combination of glucomannan and bisquaternary ammonium salt that showed 99.1% inhibition efficiency (IE) at 40 mg L^{-1} in simulated seawater.

Lakhrissi et al. [17] conducted the preparation of a series of bisglucobenzimidazolones by grafting the 6-deoxy-3-O-methyl-D-glucopyranos-6-yl group on the N-3 nitrogen atom of two benzimidazolone units connected by an alkyloxypropylene group and measured their inhibition activity in 200 mg L^{-1} NaCl on mild steel. The standard mixed type protection has been shown by all inhibitors, all displaying the IE higher than 90% at 10^{-5} M. Four azole-based corrosion inhibitors were tested on C15 grade mild steel by Finsgar et al. [18] in 3 wt.% NaCl at 25°C and 70°C. The findings have endorsed superior performance of benzotriazole (BTAH), 2-aminobenzimidazole, 2-methylimidazole, and tolyltriazole. Sharifi et al. [19] used chemically functionalize graphene oxide (GO) as corrosion inhibitors to obtain N-, S-, and P-decorated GO. At 500 mg L^{-1} concentration, the functionalized GO yielded very high (100%) inhibition efficiencies.

Singh et al. [20] have shown that four porphyrins acted as powerful N80 steel corrosion inhibitors employing experimental and theoretical approaches in a 3.5% CO_2-saturated NaCl solution. Both experimental and theoretical techniques showed that the IE of the four investigated porphyrin derivatives follows the order of IE: T4PP (91%) > TPP (88%) THP (86%) > HPTB (85%) at 200 ppm. The inhibition property of BTAH in 3.5% H_2S saturated NaCl solution was examined by Solehudin [21] using EIS, XRD, SEM, and EDX studies on API 5LX65 carbon steel. They recorded that BTAH showed the highest inhibitory activity of 93% at a 5 mM concentration.

Mobin and Aslam [22] examined the anticorrosive effects of two nonionic surfactants, namely, N-alkyl-N' glucosylethylenediamine, designated as Glu(10) and Glu(12). They reported that inhibition efficiencies increased with the increase in the concentration of

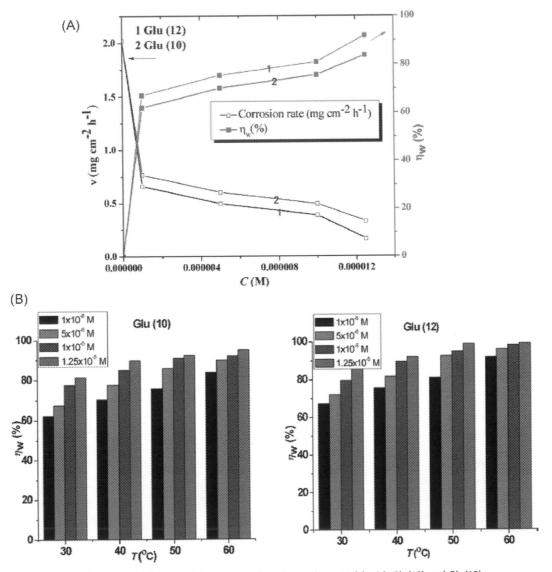

FIGURE 7–1 (A) Change of inhibition efficiency (ηw, %) and corrosion rate (v) with Glu(10) and Glu(12) concentrations at 30°C obtained by weight loss method. (B) Change of inhibition efficiencies with Glu(10) and Glu (12) concentrations at 30°C–60°C obtained by weight loss method [22].

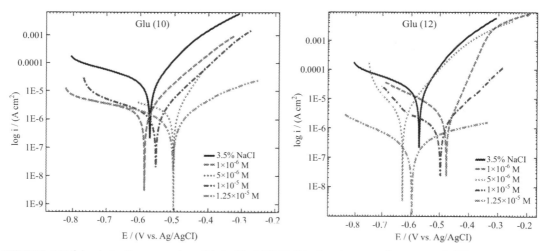

FIGURE 7–2 Tafel polarization curves for mild steel in 3.5% NaCl in the absence and presence of different concentrations of Glu(10) and Glu(12) at 30°C [22].

surfactants, hydrophobic chain length, and temperature, achieving 94.8% and 99.1% for Glu (10) and Glu(12), respectively, at 60°C (Fig. 7–1). From Fig. 7–2, it is evident that in the presence of inhibitor, the corrosion potential values have shifted to a less negative potential relative to the blank solution, that is, 3.5%NaCl.

In another report, Aslam et al. [23] examined the anticorrosive potential of nonionic sugar-based N,N'-didodecyl-N,N'-digluconamideethylenediamine gemini surfactant referred as Glu(12)-2-Glu(12) for mild steel corrosion in 3.5% NaCl at 30°C–60°C. The inhibition formulations consisting of 2.5×10^{-3} mM of Glu(12)-2-Glu(12) and 10 mM of KI (potassium iodide) demonstrated a 96.9% inhibitory efficacy at 60°C.

Fawzy et al. [24] investigated the inhibiting ability of three molecules based on surfactant synthesized from amino acids, namely sodium N-dodecyl asparagines (AS), sodium N-dodecylhistidine (HS), and sodium N-dodecyltryptophan (TS) in carbon steel dissolution in 0.5 M NaCl and 0.5 M NaOH solutions at 25°C. Inhibition efficiencies have been found to increase as the concentrations of surfactants increased, while decreasing with increasing corrosive media (NaCl and NaOH) concentration and temperature. Results obtained from the different methods showed that the efficiency of compound TS was greater than that of both AS and HS.

The pitting corrosion of 304 stainless steel was examined by Wei et al. [25] employing N-lauroylsarcosine sodium salt (NLS) as an inhibitor at neutral pH in 0.1 molar sodium chloride solutions via chemical and electrochemical methods. At the maximum concentration of 30 mM, NLS was found to improve the pitting resistance of stainless steel. Bilayer coverage is indicated by NLS adsorption on stainless steel. The inhibitor mechanism was introduced as a consequence of the inhibitory activity of the negatively charged NLS adsorption layer. The study concludes that NLS prevents stainless steel pitting corrosion in

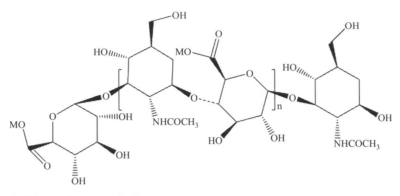

FIGURE 7–3 Eggshell chemical structure [26].

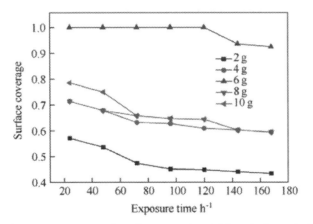

FIGURE 7–4 Degree of surface coverage versus time of exposure for stainless steel sample dipped in NaCl solution in the presence of ES [26]. *ES,* Eggshell.

NaCl solution and the performance of the inhibitor is a function of the concentration of NLS.

Sani and Papoola [26] studied eggshell (ES) powder (Fig. 7−3) as a sustainable inhibitor for N08904 austenitic stainless steel in simulated saline (3.5% NaCl) medium. The experimental data clarified the successful performance of ES with IE values of 57%−100%, at 2−10 g of tested inhibitor concentration due to stainless steel inhibition. The highest performance of ES was obtained when the concentration of the inhibitor was 6 g. The rate of corrosion decreased steadily with the presence of an inhibitor due to anion adsorption at the metal film interface.

Outcomes from Fig. 7−4, showed that 6 g of ES revealed 100% inhibition effectiveness after 24 h of exposure. At the end of the test i.e., after 120 h, 92.45% inhibitive performance in the presence of 6 g ES powder was reported.

Zargouni et al. [27] explored the impact of trisdimethylaminoselenophosphoramide (SeAP) as a corrosion inhibitor for AISI 630 stainless steel and AISI 316L in 3-wt.% NaCl using glow discharge optical emission spectroscopy (GDOES), potentiodynamic polarization (PDP), and SEM techniques. The inhibitor efficacy improves with SeAP concentration with the addition of SeAP at varying concentrations in the studied solution. It is apparent from the findings that the rise in inhibitor concentrations reduces the density of corrosion current for both modified-AISI 630 and AISI 316L. In addition, SEM and GDOES analysis suggest the presence of protective film formed by SeAP molecules on the AISI 630 and AISI 316L surface.

7.3.2 Copper

For many decades, researchers have used a wide variety of corrosion inhibitors to mitigate damage sustained in corrosive environments [28]. Currently, copper corrosion research is geared to the production of efficient and safe corrosion inhibitors to minimize chromate use.

Sherif and Park investigated the effect of 2-amino-5-ethylthio-1,3,4-thiadiazole (AETD) [29] on copper corrosion in HCl aerated solution as well as the effect of AETD [30], 2-amino-5-ethyl-1,3,4-thiadiazole (AETDA) [31], and 5-(phenyl)-4H-1,2,4-triazole 3 thiole (PTAT) [32] in NaCl solution. AETD, AETDA, and PTAT were shown to be effective mixed-type corrosion inhibitors for copper and the efficacy of inhibition improved with their concentration.

Dafali et al. [33] studied substituted uracils, namely, uracil, 5,6-dihydrouracil; 5 amino uracil; 2-thiouracil; 5-methyl-thiouracil; and dithiouracil (DTUr) as copper corrosion inhibitors in 3% NaCl. Comparison of results proved DTUr to be the best inhibitor with a maximum IE that reached 98% at 10^{-3} M. Zucchi et al. [34] studied the inhibiting action of tetrazole (T), 5-mercapto-1-methyl-tetrazole, 5-mercapto(Na salt)-1-methyl-tetrazole, 5-mercapto-1-acetic acid (Na salt)-tetrazole, 5-mercapto-1 phenyl-tetrazole, 5-phenyl-tetrazole, and 5-amino tetrazole in 0.1 M NaCl solution in the range of pH from 4 to 8 and at temperatures of 40°C and 80°C. IE of 2-ethylthio-4-(p-methoxyphenyl)-6-oxo-1,6-dihydropyrimidine-5-carbonitrile (EPD) for copper corrosion was examined by Khaled et al. [35] in a 3.5% NaCl solution. Experimental investigation revealed that 94.6% IE was shown by 10^{-3} M EPD. Two inhibitors, namely, potassium 4-(2H-benzo[d][1−3]triazole-2-yl)benzene-1,3-bis(olate) (PBTB) and potassium 2,4-di(2H-benzo[d][1−3]triazole-2-yl)benzene-1,3-bis(olate), (PDBTB), have been investigated by Huang et al. [36] to prevent copper corrosion at 298K. The order of was as follows: PDBTB (96.3%) > PBTB (84.1%). The mixed type behavior of inhibitors examined was revealed by polarization parameters. XPS spectra showed that mainly Cu(I) species were corrosion products on the copper surface. The FT-IR spectrum revealed that the C-N bond present in the studied inhibitors played a key role in chemical complexions with copper.

A series of six pyrazoles, namely, *N,N*-bis (3,5-dimethylpyrazol-1-ylmethyl) butylamine (bipy1); *N,N*-bis (3,5-dimethylpyrazol-1-ylmethyl) allylamine (bipy2); *N,N*-bis (3,5-dimethylpyrazol-1-ylmethyl) ethanolamine. (bipy3); *N,N*-bis (3,5-dimethylpyrazol-1-ylmethyl) cyclohexylamine (bipy4); *N,N*-bis (3-carbomethoxy-5-methylpyrazol-1-ylmethyl) cyclohexylamine (bipy5); and *N,N*-bis(3-carboethoxy-5-methylpyrazol-1-ylmethyl) cyclohexylamine (bipy6), were studied to inhibit Cu corrosion in 3% NaCl by Dafali et al. [37]. Among them bipy1 showed the maximum IE of around 99% at 5×10^{-4} M of inhibitor. It was observed that the inhibitors exhibited chemisorption on the Cu surface by interacting with the cyclic center of the 5 N atoms. Anticorrosive properties of 4-amino-antipyrine (AAP) was measured by Hong et al. [38] for Cu in a 3.5% NaCl solution. DFT studies have shown that the AAP molecule has a strong negative charge of N and O atoms, which promotes the adsorption of AAP on the Cu surface by donating the unshred pair of electrons of N and O atoms to the vacant d-orbitals of Cu.

The corrosion protection of copper in 3.5% NaCl solution in the presence of BTAH through a rotating cage was investigated by Khan et al. [39]. At varying speeds of $0.5-3.0 \text{ m s}^{-1}$ (100−700 rpm), the flow analysis was conducted. With regard to the increase in velocity, copper exhibited elevated corrosion activity. The corrosion attack on the copper was, however, minimized by BTAH applied to the test solution. It obviously demonstrates the adsorption of inhibitor on copper surface which resulted in reduction of the mass loss by 4−5 times. The Langmuir adsorption isotherm accompanied the adsorption of BTAH on the copper surface. Fig. 7−5 displays the photographs of the samples taken to confirm the surface condition and the existence of the corrosion substance through a Canon EOS 20D digital SLR camera. Fig. 7−5A showed clearly the corroded surface of copper solution with a brown and black film formed at 0.5 m s^{-1} tested in 3.5% NaCl.

This leads to formation of oxides or the copper chloride complex on the copper surface. However, the 3.5% NaCl solution added to the test solution by the inhibitor BTAH keeps the copper surface bright, see Fig. 7−5B, and free of visible corrosion, likely due to inhibitor adsorption. The transition of the hydrophilic surface to hydrophobic in the presence of an inhibitor without any flow effects has been further established by contact angle calculation. Otmacic and Stupnisek-Lisac [40] investigated the efficiency of seven nontoxic imidazole derivatives, namely,

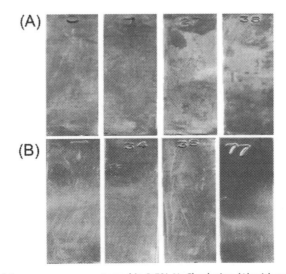

FIGURE 7–5 Photograph of the copper coupons tested in 3.5% NaCl solution (A) without and (B) with BTAH inhibitor at 0.5 m s^{-1} [39].

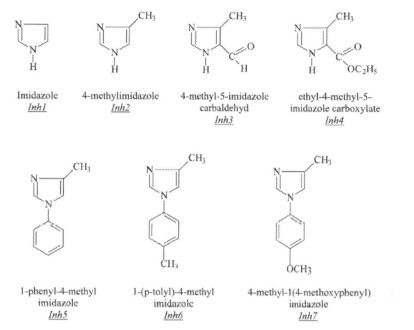

FIGURE 7 6 Organic corrosion inhibitors [40].

imidazole (Inh1), 4-methylimidazole (Inh2), 4-methyl-5-imidazolecarbaldehyde (Inh3), ethyl-4-methyl-imidazolecarboxylate (Inh4), 1-phenyl 4 methylimidazole (Inh5), 1-(p-tolyl)-4-methylimidazole (Inh6), and 4-methyl-1(4-methoxyphenyl)imidazole (Inh7) for copper in 3% NaCl solution. The structures of all seven inhibitors are given in Fig. 7–6. The results obtained revealed that the inhibitors with higher molecular weight have better inhibiting properties, especially those with a phenyl ring. And with increasing concentration and solution temperature, their IE rised, following the order: inh 5 > inh 6 > inh 4 > inh 7 > inh 3 > inh 2 > inh 1. They declared them to be mixed-type inhibitors and 4-methyl-1-phenyl imidazole was the best copper corrosion inhibitor, as confirmed by electrochemical and weight-loss methods.

Rajkumar and Sethuraman [41] reported that at different immersion times, that is, 1, 2, 6, and 12 hours, 3.5% NaCl solution containing 3-mercapto-1H-1,2,4-triazole formed a self-assembled monolayer (SAM) on the copper surface. The increase in immersion time increases the adsorption period on the copper surface of SAM monolayers, resulting in the formation of a more homogeneous and compact protective layer. Table 7–1 shows the structure of certain common organic corrosion inhibitors, test condition, test methods, and inhibition performances that have been established in chloride containing corrosive media as an effective inhibitor for copper corrosion [42–48].

7.3.3 Aluminum

The use of corrosion inhibitors for aluminum and its alloys in chloride-containing solutions published previously are discussed below.

Table 7-1 The inhibition effectiveness of different compounds as corrosion inhibitors for copper in chloride solution.

S. no.	Inhibitor's name	Inhibitor concentration	Test condition	IE (%)	Test methods	Reference
1.	(5-Methyl-[1,3,4]thiadiazol-2-ylsulfanyl)-acetic acid (4-hydroxy-3-methoxybenzylidene)-hydrazide	100 (mg L^{-1})	3.5 wt.% NaCl pH 5.5–9.5	96.7	EIS, PDP	[42]
2.	Hexa propylene glycol cyclotriphosphazene	1 mM	3.5 wt.% NaCl	96.0	EIS, PDP	[43]
3.	5-Phenyl-1,3,4-thiadiazole-2-thiol	100 mg L^{-1}	3.5 wt.% NaCl	98.2	PDP, EIS, WL	[44]
4.	2-(5-Mercapto-1,3,4-thiadiazole-2-yl)-phenol	100 mg L^{-1}	3.5 wt.% NaCl	97.7	PDP, EIS, WL	[45]
5.	Sulfathiazole (ST)	80 ppm	0.1 M NaCl, T 40°C	91.1	PDP	[45]
6.	5-Nitroindazole	0.4 mM	3 wt.% NaCl	98.9	EIS, PDP, WL	[46]
7.	1-Butyl-3-methylimidazolium bromide	0.15 mM	3 wt.% NaCl	83.1	PDP, EIS	[47]
8.	4-(4-Aminostyryl)-N,N-dimethylaniline	0.15 mM	3 wt.% NaCl	84.6	PDP, EIS	[47]
9.	2-((4-(Dimethylamino)styryl)phenylimino)methyl)	0.15 mM	3 wt.% NaCl	94.5	PDP, EIS	[47]
10.	4,6-Diamino-2-mercaptopyrimidine	2 mM	3.5 wt.% NaCl	91.2	EIS, PDP, WL	[48]

Table 7-2 The inhibition effectiveness of different compounds as corrosion inhibitors for aluminum and its alloys in chloride solution.

S. no.	Material	Inhibitor's name	Inhibitor concentration	Test condition	IE (%)	Test methods	Reference
1.	Al	Sulfathiazole	80 ppm	0.1 M NaCl, T 60°C in aerated stirred solutions	66.5	PDP, EIS	[48]
2.	Al	3-Amino-1,2,4-triazole-5-thiol	1 mM	3.5 wt.% NaCl, T RT in aerated stagnant solutions	71.7	CP, EIS	[49]
3.	Al	3-Amino-1,2,4-triazole-5-thiol	5 mM	3.5 wt.% NaCl, T RT in aerated stagnant solutions	79.8	CP, EIS	[49]
4.	Al	Monoethanolamine	8 mM	3 wt.% NaCl, CO_2, T 20°C	73.1	EIS, LSV	[50]
5.	Al	Phthalic acid	4×10^{-3} M	0.6 M NaCl, T 30°C	57.6	PDP	[51]
6.	Al	o-Phenylenediamine	4×10^{-3} M	0.6 M NaCl, T 30°C	92.3	PDP	[51]
7.	Al	Anthranilic acid	4×10^{-3} M	0.6 M NaCl, T 30°C	94.1	PDP	[51]
8.	Al	Ferrous gluconate	2.0 g mL^{-1}	0.05 M NaCl, T 28°C	99.9	PDP, EIS	[52]
9.	Al	Diisopropyl thiourea	120 ppm	3.5 wt.% NaCl	91.0	WL, HE, PDP	[53]
10.	Al	Calcium gluconate	250 ppm	60 ppm Cl$^-$, pH 11 after 1 day immersion	51.0	WL	[54]
11.	Al	Calcium gluconate + 50 ppm Zn^{2+}	250 ppm	60 ppm Cl$^-$, pH 11 after 1 day immersion	78.0	WL	[54]
12.	Al	Calcium gluconate + 50 ppm Zn^{2+}	250 ppm	60 ppm Cl$^-$, pH 11	55.1	PDP	[54]
13.	AA2024	Sodium dodecylbenzensulfonate	0.42 ppm	0.58 g L^{-1} NaCl, pH 10 after 24 h immersion	93.1	PDP, EIS	[55]
14.	AA2024	Sodium dodecylbenzensulfonate + 0.1 g L^{-1} LaCl$_3$	0.42 ppm	0.58 g L^{-1} NaCl, pH 10 after 24 h immersion	95.5	PDP, EIS	[55]
15.	AA2024	Sodium dodecylbenzensulfonate + 0.1 g L^{-1} LaCl$_3$	0.5 ppm	0.58 g L^{-1} NaCl, pH 10 after 24 h immersion	95.4	PDP	[55]
16.	AA2024	Sodium dodecylbenzensulfonate + 0.05−0.50 g L^{-1} LaCl$_3$	0.42 ppm	0.58 g L^{-1} NaCl, pH 10 after 24 h immersion	98.1	PDP	[55]
17.	AA2024–T3	Benzotriazole	5 mM	3.5 wt.% NaCl	55.5	EIS, PDP	[56]
18.	AA2024	Tryptophan	8.0 mM	3.5 wt.% NaCl after 120 h immersion	80.5	WL	[57]
19.	AA2198	Sodium salts of N-lauroylsarcosine	5 mM	0.01 M NaCl, T 25°C after 1−168 h immersion	99.9	PDP, EIS	[58]
20.	AA2198	Dodecylbenzensulfonic acid	5 mM	0.01 M NaCl, T 25°C after 1−168 h immersion	99.8	PDP, EIS	[58]
21.	AA2198	Sodium lauryl sulfate	5 mM	0.01 M NaCl, T 25°C after 1−168 h immersion	99.9	PDP, EIS	[58]
22.	AA5052	8-Aminoquinoline	0.02 mM	3 wt.% NaCl, T 25°C	91.8	WL, PDP, EIS	[59]
23.	AA5052	8-Nitroquinoline	0.02 mM	3 wt.% NaCl, T 25°C	94.7	WL, PDP, EIS	[59]
24.	AA7022	Cerium diphenyl phosphate	150 ppm	0.1 M NaCl, pH 5.5−6 after 6 days immersion	96.0	EIS	[60]
25.	AA7075	8-Hydroxyquinoline	5.52 mM	3.5 wt.% NaCl	96.0	EIS, PDP	[61]
26.	Al	3-Amino-1,2,4-triazole-5-thiol	5 mM	Naturally aerated stagnant Arabian Gulf seawater	86.1	EIS, CP	[62]
27.	ADC12	Triisopropanolamine	2.0 mM	3 wt.% NaCl, T 25°C after 24 h immersion	92.3	PDP, EIS, WL	[62]
28.	Al	Hexamine	1.2 g L^{-1}	3.65% NaCl, T RT, pH 7	47.1	PDP, OCP, WL, SEM	[64]

Xhanari and Finsgar [49] studied 1,2,4-triazole (TA), 3-amino-1,2,4-triazole (ATA), BTA, and 2-mercaptobenzothiazole (MBT) as corrosion inhibitors for 2024Al alloy in neutral chloride solution. Both anodic and cathodic corrosion reactions in the presence of inhibitors have been retarded. Compared to TA and ATA, BTA and MBT provided better performance. Balaskas et al. [50] studied a series of inhibitors including MBT, 5,7-dibromo-8-hydroxyquinoline (8-HQ), and quinaldic acid (QA) for AA2024 alloy in 0.05 M NaCl. The inhibition performance of 8-HQ and QA was better than MBT. 4-Amino-N-(1,3-thiazol-2-yl)benzenesulfonamide was tested by Zor and Sağdinc [51] as a 0.1 M NaCl inhibitor for Al which shows 81.22% effectiveness at 80 mg L^{-1}. The inhibitor displayed a mixed-type activity and interacted following Langmuir adsorption isotherm.

The results are reported in Table 7−2 with information on the inhibition efficacy of the compounds examined in chloride-containing solution for aluminum and its alloys [51−67].

7.3.4 Others metals

El-Etre et al. [68] tested the aqueous extract of henna leaves (lawsonia) as a corrosion inhibitor of carbon steel, nickel, and zinc in acid, acidic, and alkaline solutions. The extract has been found to serve as a strong corrosion inhibitor in all the tested media for the selected electrodes. The inhibition performance improved as the added concentration of extract was raised. For nickel, steel, and zinc, the observed inhibition performance was 82.88%, 91.01%, and 93.44% at 800 ppm. The inhibitor acted as a mixed type inhibitor.

Nada et al. tested [69] the binary framework of pectin−ascorbic acid as a tin corrosion inhibitor in neutral solution using WL and PDP techniques. The inhibitory effect of this inhibitor mechanism was due to the development of the tin/solution interface protective complex. Dang et al. [70] examined eco-friendly Na alginate as a corrosion inhibitor for magnesium alloy (AZ31 alloy grade) in 3.5 wt.% NaCl. The authors found that, with Na alginate concentration, corrosion inhibition increased, although a reduction in magnesium safety was also observed after prolonged immersion in the inhibitor solution. For 500 ppm Na alginate, a corrosion IE of 90% was obtained and this was due to the molecular adsorption and subsequent formation on the metal surface of a compact film (freshly produced magnesium hydroxide).

Wang et al. [71] investigated the anticorrosion properties of Ca−alginate gel filled with imidazoline quaternary ammonium salts; this composite was prepared by a piercing-solidifying method set to efficiently release this imidazoline corrosion inhibitor at the metal surface (P110 steel) soaked in CO_2-saturated 3.5 wt.% NaCl medium. The experiments showed that during their sinking phase, the synthesized capsules gradually released inhibitors and actively prevented corrosion in the oil well tube.

7.4 Conclusion

The chapter focuses on the causes and mechanisms of corrosion and its inhibition for various metals in neutral media. The current discussion showed that several kinds of corrosion inhibitors such as plant extracts, organic compounds, and their mixtures were employed to appreciably inhibit metallic corrosion in neutral solutions employing various techniques. The literature survey revealed that most of the inhibitors acted as mixed type inhibitors.

Acknowledgments

Ruby Aslam acknowledges Council of Scientific & Industrial Research, New Delhi, India for providing financial aid under Research Associate fellowship (file number-09/112(0616)2K19 EMR-I).

Conflict of interest

Authors declare that there is no conflict of interest in any way.

Author's contributions

All authors collectedly contributed in designing and write-up of the book chapter.

References

[1] M. Ladan, W.J. Basirun, S.N. Kazi, F.A. Rahman, Corrosion protection of AISI 1018 steel using Co-doped TiO_2/polypyrrole nanocomposites in 3.5% NaCl solution, Mater. Chem. Phys. 192 (2017) 361−373.

[2] A. Fattah-alhosseini, M. Noori, Corrosion inhibition of SAE 1018 carbon steel in H_2S and HCl solutions by lemon verbena leaves extract, Measurement 94 (2016) 787−793.

[3] M. Abdallah, H. Al-Tass, B.A. AL Jahdaly, A.S. Fouda, Inhibition properties and adsorption behavior of 5-arylazothiazole derivatives on 1018 carbon steel in 0.5M H_2SO_4 solution, J. Mol. Liq. 216 (2016) 590−597.

[4] E. Samiento-Bustosa, J.G. González, R.J. Uruchurtua, G. Dominguez-Patiñoac, V.M. Salinas-Bravo, Effect of inorganic inhibitors on the corrosion behavior of 1018 carbon steel in the LiBr + ethylene glycol + H_2O mixture, Corros. Sci. 50 (2008) 2296−2303.

[5] M.S. El-Sayed, R.S. Erasmus, J.D. Comins, In situ Raman spectroscopy and electrochemical techniques for studying corrosion and corrosion inhibition of iron in sodium chloride solutions, Electrochim. Acta 55 (2010) 3657−3663.

[6] N.A. Darwish, F. Hilbert, W.J. Lorenz, H. Rosswag, The influence of chloride ions on the kinetics of iron dissolution, Electrochim. Acta 18 (1973) 421.

[7] B. Duran, G. Bereket, M. Duran, Electrochemical synthesis and characterization of poly(m-phenylenediamine) films on copper for corrosion protection, Prog. Org. Coat. 73 (2012) 162−168.

[8] D. Gopi, E.S.M. Sherif, M. Surendiran, D.M.A. Sakila, L. Kavitha, Corrosion inhibition by benzotriazole derivatives and sodium dodecyl sulfate as corrosion inhibitors for copper in ground water at different temperatures, Surf. Interface Anal. 47 (2015) 618−625.

[9] K. Habib, In-situ monitoring of pitting corrosion of copper alloys by holographic interferometry, Corros. Sci. 40 (1998) 1435−1440.

[10] A.A. Attia, E.M. Elmelegy, M. El-Batouti, A.-M.M. Ahmed, Anodic corrosion inhibition in presence of protic solvents, Asian J. Chem. 28 (2016) 267.

[11] N.A. Al-Mobarak, K.F. Khaled, M.N.H. Hamed, K.M. Abdel-Azim, N.S. Abdelshafi, Corrosion inhibition of copper in chloride media by 2-mercapto-4-(p-methoxyphenyl)-6-oxo-1,6-dihydropyrimidine-5-carbonitrile: Electrochemical and theoretical study, Adv. J. Chem. 3 (2010) 233−242.

[12] H.O. Curkovic, E. Stupnisek-Lisac, H. Takenouti, The influence of pH value on the efficiency of imidazole based corrosion inhibitors of copper, Corros. Sci. 52 (2010) 398−405.

[13] H.P. Lee, K. Nobe, Kinetics and mechanisms of Cu electrodissolution in chloride media, J. Electrochem. Soc. 133 (1986) 2035.

[14] R.W. Revie, Uhlig's Corrosion Handbook, third ed., Wiley, 2011.

[15] R.G.M. de Araujo Macedo, N. do Nascimento Marques, J. Tonholo, R. de Carvalho Balaban, Water soluble carboxymethylchitosan used as corrosion inhibitor for carbon steel in saline medium, Carbohydr. Polym. 205 (2019) 371−376.

[16] X. Luo, X. Pan, S. Yuan, S. Du, C. Zhang, Y. Liu, Corrosion inhibition of mild steel in simulated seawater solution by a green eco-friendly mixture of glucomannan (GL) and bisquaternary ammonium salt (BQAS), Corros. Sci. 125 (2017) 139−151.

[17] L. Lakhrissi, B. Lakhrissi, R. Touir, M.E. Touhami, M. Massoui, E.M. Essassi, Mild steel corrosion inhibition in 200 ppm NaCl by new surfactant derivatives of bis-glucobenzimidazolones, Arab. J. Chem. 10 (2017) S3142−S3149.

[18] M. Finsgar, B. Petovar, K. Xhanari, U. Maver, The corrosion inhibition of certain azoles on steel in chloride media: electrochemistry and surface analysis, Corros. Sci. 111 (2016) 370−381.

[19] Z. Sharifi, M. Pakshir, A. Amini, R. Rafiei, Hybrid graphene oxide decoration and water-based polymers for mild steel surface protection in saline environment, J. Ind. Eng. Chem. 74 (2019) 41−54.

[20] A. Singh, Y. Lin, M.A. Quraishi, L.O. Olasunkanmi, O.E. Fayemi, Y. Sasikumar, et al., Porphyrins as corrosion inhibitors for N80 steel in 3.5% NaCl solution: electrochemical, quantum chemical, QSAR and Monte Carlo simulations studies, Molecules 20 (2015) 15122−15146.

[21] A. Solehudin, Performance of benzotriazole as corrosion inhibitors of carbon steel in chloride solution containing hydrogen sulfide, Inter. Refer. J. Eng. Sci. 1 (2012) 21−26.

[22] M. Mobin, R. Aslam, Experimental and theoretical study on corrosion inhibition performance of environmentally benign non-ionic surfactants for mild steel in 3.5% NaCl solution corrosion, Process. Saf. Environ. Prot. 114 (2018) 279−295.

[23] R. Aslam, M. Mobin, J. Aslam, H. Lgaz, Sugar based N,N′-didodecyl-N,N′ digluconamideethylenediamine gemini surfactant as corrosion inhibitor for mild steel in 3.5% NaCl solution-effect of synergistic KI additive, Sci. Rep. 8 (2018) 1−20.

[24] A. Fawzy, M. Abdallah, I.A. Zaafarany, S.A. Ahmed, I.I. Althagafi, Thermodynamic, kinetic and mechanistic approach to the corrosion inhibition of carbon steel by new synthesized amino acids-based surfactants as green inhibitors in neutral and alkaline aqueous media, J. Mol. Liq. 265 (2018) 276−291.

[25] Z. Wei, P. Duby, P. Somasundaran, Pitting inhibition of stainless steel by surfactants: an electrochemical and surface chemical approach, J. Colloid Interface Sci. 259 (2003) 97−102.

[26] O. Sanni, A.P.I. Popoola, O.S.I. Fayomi, The inhibitive study of egg shell powder on UNS N08904 austenitic stainless steel corrosion in chloride solution, Def. Technol. 14 (2018) 463−468.

[27] Y. Zargouni, W. Sassi, K. Alouani, Corrosion inhibition of aisi 316l and modified-aisi 630 stainless steel by the new organic inhibitor [(CH3)2N]3PSe in chloride media: electrochemical and physical study, Mediterr. J. Chem. 4 (2015) 105−110.

[28] A. Fateh, M. Aliofkhazraei, A.R. Rezvanian, Review of corrosive environments for copper and its corrosion inhibitors, Arab. J. Chem. 13 (2020) 481−544.

[29] E.M. Sherif, S.M. Park, Inhibition of copper corrosion in acidic pickling solutions by N-phenyl-1, 4-phenylenediamine, Electrochim. Acta 51 (2006) 4665−4673.

[30] E.M. Sherif, S.M. Park, Effects of 2-amino-5-ethylthio-1,3,4-thiadiazole on copper corrosion as a corrosion inhibitor in aerated acidic pickling solutions, Electrochim. Acta 51 (2006) 6556−6562.

[31] E.M. Sherif, S.M. Park, 2-Amino-5-ethyl-1,3,4-thiadiazole as a corrosion inhibitor for copper in 3.0% NaCl solutions, Corros. Sci. 48 (2006) 4065–4079.

[32] E.M. Sherif, A.M. Shamy, M.M. Ramla, A.O.H. Nazhawy, 5-(Phenyl)-4H-1,2,4-triazole-3-thiol as a corrosion inhibitor for copper in 3.5% NaCl solutions, Mater. Chem. Phys. 102 (2007) 231.

[33] A. Dafali, B. Hammouti, R. Mokhlisse, S. Kertit, Substituted uracils as corrosion inhibitors for copper in 3% NaCl solution, Corros. Sci. 45 (2003) 1619–1630.

[34] F. Zucchi, G. Trabanelli, M. Fonsati, Tetrazole derivatives as corrosion inhibitors for copper in chloride solutions, Corros. Sci. 38 (1996) 2019–2029.

[35] K.F. Khaled, M.N.H. Hamed, K.M. Abdel-Azim, N.S. Abdelshafi, Inhibition of copper corrosion in 3.5% NaCl solutions by a new pyrimidine derivative: electrochemical and computer simulation techniques, J. Solid State Electrochem. 15 (2011) 663–673.

[36] H. Huang, Z. Wang, Y. Gong, F. Gao, Z. Luo, S. Zhang, et al., Water soluble corrosion inhibitors for copper in 3.5wt% sodium chloride solution, Corros. Sci. 123 (2017) 339–350.

[37] A. Dafali, B. Hammouti, R. Touzani, S. Kertit, A. Ramdani, K. El Kacemi, Corrosion inhibition of copper in 3 per cent NaCl solution by new bipyrazolic derivatives, Anti-Corros. Methods Mater. 49 (2002) 96–104.

[38] S. Hong, W. Chen, H.Q. Luo, N.B. Li, Inhibition effect of 4-amino-antipyrine on the corrosion of copper in 3 wt.% NaCl solution, Corros. Sci. 57 (2012) 270–278.

[39] P.F. Khan, V. Shanthi, R.K. Babu, S. Muralidharan, R.C. Barik, Effect of benzotriazole on corrosion inhibition of copper under flow conditions, J. Environ. Chem. Eng. 3 (2015) 10–19.

[40] H. Otmacic, E. Stupnisek-Lisac, Copper corrosion inhibitors in near neutral media, Electrochim. Acta 48 (2003) 985–991.

[41] G. Rajkumar, M.G. Sethuraman, A study of copper corrosion inhibition by selfassembled films of 3-mercapto-1H-1,2,4-triazole, Res. Chem. Intermed. 42 (2016) 1809–1821.

[42] F. Ma, W. Li, H. Tian, B. Hou, The use of a new thiadiazole derivative as a highly efficient and durable copper inhibitor in 3.5% NaCl solution, Int. J. Electrochem. Sci. 10 (2015) 5862–5879.

[43] O. Dagdag, M. El Gouri, M. Galai, M. Ebn Touhami, A. Essamri, A. Elharfi, Application of hexa propylene glycolcyclotriphosphazene as corrosion inhibitor for copper in 3% NaCl solution, Der Pharm. Chem. 7 (2015) 114–122.

[44] H. Tian, W. Li, K. Cao, B. Hou, Potent inhibition of copper corrosion in neutral chloride media by novel non-toxic thiadiazole derivatives, Corros. Sci. 73 (2013) 281–291.

[45] S. Zor, Sulfathiazole as potential corrosion inhibitor for copper in 0.1 M NaCl, Prot. Met. Phys. Chem. Surf. 50 (2014) 530–537.

[46] Y. Qiang, S. Zhang, S. Xu, L. Yin, The effect of 5-nitroindazole as an inhibitor for the corrosion of copper in a 3.0% NaCl solution, RSC Adv. 5 (2015) 63866–63873.

[47] Y. Zhou, S. Xu, L. Guo, S. Zhang, H. Lu, Y. Gong, et al., Evaluating two new Schiff bases synthesized on the inhibition of corrosion of copper in NaCl solutions, RSC Adv. 5 (2015) 14804–14813.

[48] Z. Cheng, S. Mo, J. Jia, J. Feng, H.Q. Luo, N.B. Li, Experimental and theoretical studies of 4,6-diamino-2-mercaptopyrimidine as a copper inhibitor in 3.5 wt% NaCl solution, RSC Adv. 6 (2016) 15210–15219.

[49] K. Xhanari, M. Finsgar, Organic corrosion inhibitors for aluminum and its alloys in chloride and alkaline solutions: a review, Arab. J. Chem. 12 (2019) 4646–4663.

[50] A.C. Balaskas, M. Curioni, G.E. Thompson, Effectiveness of 2-mercaptobenzothiazole, 8-hydroxyquinoline and benzotriazole as corrosion inhibitors on AA 2024-T3 assessed by electrochemical methods, Surf. Interface Anal. 47 (2015) 1029–1039.

[51] S. Zor, S. Sağdinc, Experimental and theoretical study of sulfathiazole as environmentally friendly inhibitor on aluminum corrosion in NaCl, Prot. Met. Phys. Chem. Surf. 50 (2014) 244–253.

[52] E.-S.M. Sherif, Effects of 3-amino-1,2,4-triazole-5-thiol on the inhibition of pure aluminum corrosion in aerated stagnant 3.5 wt. % NaCl solution as a corrosion inhibitor, Int. J. Electrochem. Sci. 7 (2012) 4847–4859.

[53] I. Jevremovic, V.M. Stankovic, The inhibitive effect of ethanolamine on corrosion behavior of aluminium in NaCl solution saturated with CO_2, Metall. Mater. Eng. 18 (2012) 241–257.

[54] R. Banerjee, Ranjana, S.S. Panja, M.M. Nandi, An electrochemical and quantum chemical investigation of some corrosion inhibitors on aluminium alloy in 0.6 M aqueous sodium chloride solution, Indian J. Chem. Technol. 18 (2011) 309–313.

[55] P.A.I. Popoola, S. Omotayo, C.A. Lotto, O.M. Popoola, Inhibitive action of ferrous gluconate on aluminum alloy in saline environment, Adv. Mater. Sci. Eng. 2013 (2013).

[56] N.V. Lakshmi, N. Arivazhagan, S. Karthikeyan, The corrosion inhibition of aluminium in 3.5% NaCl by diisopropyl thiourea, Int. J. Chem. Tech. Res. 5 (2013) 1959–1963.

[57] M. Hakeem, S. Rajendran, A.P.P. Regis, Calcium gluconate as a corrosion inhibitor for aluminium, J. Eng. Comput. Appl. Sci. 3 (2014) 1–11.

[58] B. Zhou, Y. Wang, Y. Zuo, Evolution of the corrosion process of AA 2024-T3 in an alkaline NaCl solution with sodium dodecylbenzenesulfonate and lanthanum chloride inhibitors, Appl. Surf. Sci. 357 (2015) 735–744.

[59] A.C. Balaskas, M. Curioni, G.E. Thompson, Effectiveness of 2-mercaptobenzothiazole, 8-hydroxyquinoline and benzotriazole as corrosion inhibitors on AA 2024-T3 assessed by electrochemical methods, Surf. Interface Anal. 47 (2015).

[60] X. Li, B. Xiang, X. Zuo, Q. Wang, Z.D. Wei, Inhibition of tryptophan on AA 2024 in chloride-containing solutions, J. Mater. Eng. Perform. 20 (2011) 265–270.

[61] A. Balbo, A. Frignani, V. Grassi, F. Zucchi, Corrosion inhibition by anionic surfactants of AA2198 Li-containing aluminium alloy in chloride solutions, Corros. Sci. 73 (2013) 80–88.

[62] D. Wang, D. Yang, D. Zhang, K. Li, L. Gao, T. Lin, Electrochemical and DFT studies of quinolone derivatives on corrosion inhibition of AA5052 aluminium alloy in NaCl solution, Appl. Surf. Sci. 357 (2015) 2176–2183.

[63] J.-A. Hill, T. Markley, T.M. Forsyth, P.C. Howlett, B.R.W. Hintonac, Corrosion inhibition of 7000 series aluminium alloys with cerium diphenyl phosphate, J. Alloy. Compd. 509 (2011) 1683–1690.

[64] W. Liu, A. Singh, Y. Lin, E.E. Ebenso, L. Zho, B. Huang, 8-Hydroxyquinoline as an effective corrosion inhibitor for 7075 aluminium alloy in 3.5% NaCl solution, Int. J. Electrochem. Sci. 9 (2014) 5574–5584.

[65] E.-S.M. Sherif, Electrochemical investigations on the corrosion inhibition of aluminum by 3-amino-1,2,4-triazole-5- thiol in naturally aerated stagnant seawater, J. Ind. Eng. Chem. 19 (2013) 1884–1889.

[66] X. Ren, S. Xu, S. Chen, N. Chen, S. Zhang, Experimental and theoretical studies of triisopropanolamine as an inhibitor for aluminum alloy in 3% NaCl solution, RSC Adv. 5 (2015) 101693–101700.

[67] O.S.I. Fayomi, I.G. Akande, Corrosion mitigation of aluminium in 3.65% NaCl medium using hexamine, J. Bio- Tribo- Corros. 5 (2019) 23.

[68] A.Y. El-Etre, M. Abdallah, Z.E. El-Tantawy, Corrosion inhibition of some metals using lawsonia extract, Corros. Sci. 47 (2005) 385–395.

[69] C. Nada, B. Katarina, P. Sandra, The inhibition effect of pectin and ascorbic acid on the corrosion of tin in sodium chloride solution, in: 47th Annual Meeting of the International Society of Electrochemistry. Abstracts/The International Society of Electrochemistry (ed).—The Electrochemical Society, 1996. L6a-3. 47th Annual Meeting of the International Society of Electrochemistry, Veszprem-Balatonfured, Hungary, 1.-6.09. (1996).

[70] N. Dang, Y.H. Wei, L.F. Hou, Y.G. Li, C.L. Guo, Investigation of the inhibition effect of the environmentally friendly inhibitor sodium alginate on magnesium alloy in sodium chloride solution, Mater. Corros. 66 (2015) 1354–1362.

[71] L. Wang, C. Zhang, H. Xie, W. Sun, X. Chen, X. Wang, et al., Calcium alginate gel capsules loaded with inhibitor for corrosion protection of downhole tube in oilfields, Corros. Sci. 90 (2015) 296–304.

8

Corrosion inhibitors for sweet (CO$_2$ corrosion) and sour (H$_2$S corrosion) oilfield environments

Jeenat Aslam[1], Ruby Aslam[2], Saman Zehra[2], Marziya Rizvi[3]

[1]DEPARTMENT OF CHEMISTRY, COLLEGE OF SCIENCE, YANBU, TAIBAH UNIVERSITY, AL-MADINA, SAUDI ARABIA [2]CORROSION RESEARCH LABORATORY, DEPARTMENT OF APPLIED CHEMISTRY, FACULTY OF ENGINEERING AND TECHNOLOGY, ALIGARH MUSLIM UNIVERSITY, ALIGARH, INDIA [3]CORROSION RESEARCH LABORATORY, DEPARTMENT OF MECHANICAL ENGINEERING, FACULTY OF ENGINEERING, DUZCE UNIVERSITY, DUZCE, TURKEY

Chapter outline

8.1 Introduction

The transportation and production industries of oil and gas are a powerful division of the world economy, currently contributing up to 3% of the world economy, over US$ 2 trillion of total global income [1]. The natural gases and raw unrefined oil as per predictions would dominate most of the fraction of the world economy, that is, 25% and 27%, respectively, by 2040 [2]. This division however faces a dangerous situation of corrosion in transportation,

Environmentally Sustainable Corrosion Inhibitors. DOI: https://doi.org/10.1016/B978-0-323-85405-4.00021-5

165

manufacturing, processing, and containment [3]. According to a current survey, oil and gas firms have wasted over \$1.372 billion, over the years, to suppress this situation [4]. These data reveal that \$589 million were exhausted on facilities for pipelines and surface, \$320 million on capital expenditures, and \$463 million during refining processes related to corrosion [4]. Corrosion problems are not exclusive to the petroleum and natural gas sector but are also prevalent in daily situations, threatening human life and the world's economies equally. A characteristic situation arose during the explosion of the Donghuang II oil pipeline in Qingdao, China on November 22, 2013. One hundred and thirty six people were injured and more than 62 people were killed in this accident [5,6]. The main culprit for this accident proposed by the Chinese Government was the inflammation of oil vapors which spilled from a corroded pipeline. This was named as China's scariest spill since the last benzene oil spill that destroyed the ecology of the Songhua River back in 2005 [7]. An incident of similar fashion was recorded on August 19, 2000, when El Pasco's 30-inch natural gas pipeline burst, causing 12 casualties including infants [8,9]. According to reports [8,9], the explosions were caused by a major decrease in the wall thickness of the pipe due to internal corrosion. Some other notable losses due to corrosion are San Francisco Bay Bridge bolt failure [10], Leo Frigo Memorial Bridge failure [11], Fukushima nuclear plant tank leak [12], and the collapsed bridge of Lowe's Motor Speedway [13]. In reality, 20% of refinery accidents according to the "Major Accident Reporting System" or eMARS are related to the failure caused by corrosion [14].

Numerous techniques have been developed by engineers to mitigate corrosion. The world can save up to 35% (US \$875 billion) by monitoring corrosion and applying preventive techniques wherever possible [15–17].

8.2 Types of corrosion and causing agents in the oilfield industries

Classifying oilfield-related corrosion in an orderly manner is no easy task. Still, corrosion may be categorized based on preventive methods, industry section, mechanism of attack, and appearance of corrosion damage. There are a lot of causes and corrosion categories. The reaction pathway of corrosion detected in a known pipeline differs based on service location, fluid contents, temperature, geometrical aspects, etc. However, in every case, the corrosion is always caused by the surrounding electrolyte. Corrosion can be of many forms, named as oxygen, sour, sweet, galvanic, crevice, stress corrosion cracking, erosion, and microbiologically induced corrosion in the oil and gas field [18,19].

In oilfields, hydrogen sulfide (H_2S), and carbon dioxide (CO_2) are generally present, and corrosion is caused by water. On the reaction of H_2O with H_2S and CO_2, the following reaction happens:

H_2CO_3 reaction:

$$Fe + H_2CO_3 \rightarrow FeCO_3 + H_2 \qquad (8.1)$$

H_2S reaction:

$$Fe + H_2S + H_2O \rightarrow FeS + 2H \tag{8.2}$$

Equations (8.1) and (8.2) might occur individually but also in combination in the presence of CO_2 and H_2S. The resultant entities connect with a cathode or are freely available in electrolyte to continue the degradation.

8.3 Chemistry of sweet and steel corrosion

8.3.1 CO_2 corrosion (sweet corrosion)

Amongst the serious corrosive agents in oil and gas industries is CO_2 gas, especially in a dissolved state [21]. The dry CO_2 gas by itself, like H_2S gas, is not a degrading agent at the temperatures applied in the production of oil and gas. To execute an electrochemical process carbon dioxide gas must be in an aqueous phase and dissolved state in contact with steel [22]. CO_2 is brine soluble and water-soluble. CO_2 will combine with H_2O, to form H_2CO_3 making the fluid acidic. Sweet corrosion is affected via the increase in basicity, temperature, nonaqueous medium, aqueous stream content, metal characteristics, and dynamics of the flow [21,23]. At higher temperatures, the scale of iron carbide produced in the oil and gas pipe is of a protective nature, but still metal corrodes in these circumstances. The sweet corrosion manifests itself in two forms: pitting corrosion [24] and mesa attack [21].

Numerous mechanisms have been suggested [25]. The majority of them are associated with CO_2 corrosion. Therefore, the precise mechanism is still not determined. The most acknowledged CO_2 mechanism is given below:

$$CO_{2(g)} \rightarrow CO_{2(aq)} \tag{8.3}$$

$$CO_{2(aq)} + H_2O_{(l)} \leftrightarrow H_2CO_{3(aq)} \tag{8.4}$$

$$H_2CO_{3(aq)} + e^- \rightarrow H^+_{(aq)} + HCO^-_{3(aq)}; \ (pH \ 4-6) \tag{8.5}$$

$$HCO^-_{3(aq)} + e^- \rightarrow H^+_{(aq)} + CO^{2-}_{3(aq)}; \ (pH \geq 6) \tag{8.6}$$

$$2H^+_{(aq)} + 2e^- \rightarrow H_{2(g)}; \ (pH < 4) \tag{8.7}$$

$$Fe^{2+}_{(aq)} + CO^{2-}_{3(aq)} \rightarrow FeCO_{3(s)} \tag{8.8}$$

Generally sweet corrosion of carbon steel can be depicted by the following reaction:

$$Fe_{(s)} + CO_{2(aq)} + HO_{2(l)} \rightarrow FeCO_{3(s)} + H_{2(g)} \tag{8.9}$$

8.3.2 H₂S corrosion (sour corrosion)

The deterioration of metal because of contact with moisture and H_2S is known as sour corrosion, mostly detectable in the drill pipe. H_2S is a harshly corrosive agent when dissolved in water [23], causing the pipeline to become brittle [20]. H_2S when dissolved in water forms a weak acid, acting as a source of H^+. The by-products formed are hydrogen and iron sulfides ($FeS_{(x)}$). At low temperatures, $FeS_{(x)}$ forms scale-preventing low corrosion [26]. Pitting, uniform, stepwise cracking are the main forms of H_2S corrosion. The common H_2S corrosion reactions can be expressed as follows [27]:

$$H_2S + Fe + H_2O \rightarrow FeS_{(x)} + 2H + H_2O \tag{8.10}$$

Sun [28] proposed the formation of a mackinawite film, that might be expressed as follows (Scheme 8−1).

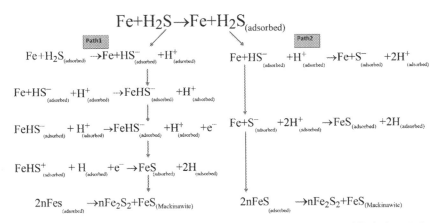

SCHEME 8–1 Mechanism for iron dissolution in aqueous solutions of H₂S. Source: *Modified after W. Sun, Kinetics of iron carbonate and iron sulfide scale formation in CO₂/H₂S corrosion (Ph.D. dissertation), 2006.*

8.4 Factors affecting sweet (CO₂) corrosion

Sweet corrosion might be promoted by metallurgical or ecological factors. Metallurgical factors explain the steel composition itself. Conversely, the ecological factors explain the H_2O reactions of the solution in which the steel is being degraded. Factors which gain attention are dissolved O_2 concentration, fluid flow rate, pH, CO_2 partial pressure, and temperature.

8.5 Factors affecting sour (H₂S) corrosion

Numerous interdependent factors affect the H_2S corrosion like concentration of H_2S, exposure duration and temperature, wax, scales, corrosion products etc. deposited on surface, solution feature and contents, that is, pH, water, organic acids, oil wettability, phase ratios, water cut, and metal chemistry.

8.6 Corrosion inhibitors for sweet and sour environments

The corrosion inhibitors in oil and gas environments are categorized as cathodic, anodic, neutralizing, active, and passivating inhibitors, film-forming, and vapor phase, according to the adsorption of interface processes [29,30]. Commonly, imidazoline (IM) and its derivatives have emerged to become the most studied corrosion inhibitors for sweet and sour corrosion. IM is a special class of nitrogen-containing heterocyclic organic compound. IM has three types of bonding positions which contain an aromatic ring, a pyrrole nitrogen, and a pyridine-like nitrogen atom. However, the IM derivatives have been considered better inhibitors against CO_2 and H_2S-assisted corrosion in steel storage wells and flow lines because of their cationic surface activity and exceptional film-forming capabilities. These compounds form inhibitive layers stopping the flow of electrolytes toward the substrate. Moreover, the interface adsorption paths of IMs are related to precise factors, for example, the type of corrosive media, their chemical structures, the metal surface charge, the presence of dissolved gases, etc. [31].

8.6.1 Imidazolines-based corrosion inhibitors for sweet corrosion

This section will discuss some IMs used to evade CO_2 metal corrosion of various electrolytes (Table 8–1) that are widely used in the transmissions of oil and gas. It is important to note these inhibitors have not been yet applied to industry; a few have been applied but just in fractions.

Zhang et al. [32] reported the synergistic corrosion inhibitive effect among an IM-based dissymmetric bis-quaternary ammonium salt (DBA), which is an ecological surfactant, and thiourea (TU) compound against the CO_2 corrosion of Q235 steel at 80°C. Although, the DBA and TU combination have synergy, notably decreasing the corrosion even at very low concentrations. DBA—TU is a mixed inhibitor and forms a protective layer. The inhibition efficiency (IE) reduced gradually with soaking duration, which assures its sustainability. It might inhibit high-temperature corrosion.

Jevremovic et al. [33] investigated the inhibitive effect of diethylenetriamine IM for steel immersed in CO_2-saturated 3% NaCl solution. From polarization plots, the researchers have inferred a mixed-type corrosion inhibition predominantly anodic. Protection might be governed by inhibited electrode processes owing to the elevated R_{ct}. A maximum corrosion IE (92%) was calculated for 100 ppm at room temperature. This inhibitor derivative form a stable, protective inhibitory layer on steel and reduces corrosive ions attack. The adsorption process of the inhibitor was consistent with Langmuir and Temkin adsorption isotherm.

Villamizar et al. [34] studied the corrosion inhibitory performance of hydroxyethyl, amino ethyl, and amide ethyl IM derivatives for steel exposed to diesel-contaminated 3% NaCl solution saturated with CO_2 gas at 50°C. The diesel phase displayed improved performance owing to

Table 8–1 Imidazoline derivatives used as an efficient corrosion inhibitors for steel in the CO_2 environment.

S.No.	Inhibitor	Metal/corrosive medium	IE	Techniques	Reference
1.	Imidazoline-based quaternary ammonium salt and thiourea	Q235 steel/CO_2-saturated chloride, T 80°C	90.5%	WL, polarization, CV, EIS, and SEM	[32]
2.	Diethylenetriamine imidazoline	Mild steel/CO_2-saturated NaCl	94%	Potentiodynamic sweep, EIS, CV, QCM, and SEM	[33]
3.	Hydroxyethyl, amino ethyl, amide ethyl imidazoline derivatives	Carbon steel/diesel-contaminated 3% NaCl, CO_2, T 50°C	80%	EIS	[34]
6.	Imidazoline derivative	N80 carbon steel/CO_2-saturated NaCl solution containing acetic acid	99%	WL, PDP, EIS, and SEM	[35]
8.	Imidazoline derivative (IM) in combination with l-cysteine	Carbon steel/CO_2-saturated brine, T RT	96.7%	Contact angle, XPS EIS, PDP and weight loss,	[36]
9.	Imidazoline derivative (IM) in combination with sodium dodecyl benzenesulfonate (SDBS)	X52 carbon steel/CO_2-saturated chloride, T 60°C	90.1%	SEM, electrochemical measurements and weight loss	[37]
4.	2-(4-Methoxyphenyl)-4,5-diphenyl-imidazole (M-1),4,5-diphenyl-2-(p-tolyl)-imidazole (M-2), 2-(4-nitrophenyl)-4,5-diphenyl-imidazole	J55 steel/CO_2-saturated 3.5% NaCl solution	93%, 84%, and 81%	Weight loss, EIS, PDP,SECM, contact angle, SEM,XPS, and AFM	[38]
5.	Imidazoline derivative (R1 and R2 undefined)	C−Mn steel/deoxygenated 5% NaCl solution, saturated with CO_2, T 40°C	98%, 93%, and 82%	PDP and SEM	[39]
7.	Imidazoline derivative	Carbon steel/CO_2-saturated 3.5% NaCl, T 60°C	98.6%	PDP, EIS, WL, SKPM	[40]
9.	Imidazoline derivative	Low alloy steel/CO_2-saturated 3% NaCl, T 60°C	96.4%, 81.5%, and 84.8%	PDP, EIS, WL	[41]
10.	N-pendants and electron-rich amidine motifs in 2-(p-alkoxyphenyl)-2-imidazolines	Mild steel/CO_2-saturated 0.5% NaCl, T 40°C	71.8%, 93.1%, 72.5%, and 93.3%	WL, PDP, XPS	[42]
11.	2-Undecyl-1-aminoethylimidazoline (AEI) and2-undecyl-1-aminoethyl-1-hydroxyethyl quaternary imidazoline	N80 mild steel/CO_2-saturated 3% NaCl, T 50°C	90.8% and 99.2%	WL, LP,PDP, EIS, and SEM	[43]

(Continued)

Table 8–1 (Continued)

S.No.	Inhibitor	Metal/corrosive medium	Techniques	IE	Reference
12.	Hydroxyethyl imidazoline derivative	X-80 pipeline steel/CO_2-saturated saltwater 3% NaCl, T 50°C	PDP, LP, EIS, and EN	62%	[44]
13.	Sodium benzoate and oleic-based imidazoline	Mild steel/CO_2-saturated 3.5% NaCl solution at 50°C	WL, PDP, EIS, and XPS	91.45	[45]
14.	Thioureido imidazoline	Q235 steel/CO_2-saturated salt water 2% NaCl, T room temperature	EIS, AFM, anc XPS	93.2%	[46]
15.	1-(2-Thioureidoethyl)-2-alkylimidazoline	Q235 steel/saltwater-CO_2, T 25°C, pH 4.86	PDP, EIS	97.8%	[47]
16.	Heptadecyl-tailed mono and bis-imidazolines	CO_2-saturated 3% NaCl, T 40°C	STM, PDP, EIS, and XPS	82.1%, 86.0%, and 83.1%	[48]
17.	2-Undecyl-1-ethylaminoimidazoline (UEI) + KI	N80 mild steel/CO_2-saturated 3% NaCl	PDP, EIS, and SEM	97.30%	[49]
18.	Imidazoline derivative	API X65 steel/CO_2-saturated salt water 5% NaCl, T RT	PDP and EIS	98.95%	[50]
19.	Carboxyethyl-imidazoline	API X-120pipeline steel/CO_2-saturated 3% NaCl	PDP, EIS, and EN	81.6%	[51]

inhibitive and water-resistant films of IM compounds. Amide ethyl IM is the most efficient derivative. The better hydrophobicity inherent to the alkyl groups within this compound is the key to its success. IM inhibitor [35] was studied as the corrosion inhibitor of N80 carbon steel crevice corrosion in CO_2-saturated NaCl solution with acetic acid (HAc). Performance of the solution with $3 \, g \, L^{-1}$ HAc with IM as an inhibitor displayed mixed-type protection with predominant cathodic effectiveness (Fig. 8–1). In $1 \, g \, L^{-1}$ HAc, IM acted as a mixed-type inhibitor with predominant anodic effectiveness. IM might initially inhibit the crevice corrosion of the steel (Fig. 8–2B), but not at a later stage when added after corrosion for a period of time, that is, 24 hours (Fig. 8–2C).

Zhang et al. [36] reported another IM derivative and l-cysteine (CYS) for preventing carbon steel corrosion in CO_2-saturated brine solution at 60°C. The results explain that the individual IM's action is slight for CO_2-induced corrosion that may be amplified significantly by adding in cysteine. From the Nyquist plots shown in Fig. 8–3A, it is clear that the capacitive loop of IM and CYS combined is much bigger than that of IM or CYS alone, which indicates retardation of the corrosion of L360 steel.

Qian and Cheng [37] studied IM and sodium dodecyl benzenesulfonate (SDBS) inhibitors for corrosion of X52 carbon steel in CO_2-saturated chloride solutions at 60°C. The chemical structures of the studied inhibitors are given in Fig. 8–4A and B. Fig. 8–5 shows corrosion reduction on adding IM/SDBS/IM + SDBS. The corrosion rate is least in the solution containing IM + SDBS at individual concentrations.

The synergism is due to the coadsorption of sulfur from SDBS and nitrogen from IM on the exposed Fe atoms, building up a dense, uniform layer on the steel surface in the presence of either IM or SDBS where the repulsion is present in-between the adsorbed inhibitor molecules. Chemisorption occurs, and Temkin adsorption isotherm is obeyed.

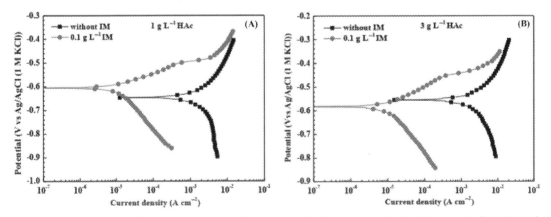

FIGURE 8–1 Potentiodynamic polarization curve of carbon steel (without crevice) in the CO_2-saturated 1.65% NaCl solution containing various concentrations of HAc in the absence and presence $0.1 \, g \, L^{-1}$ IM: (A) $1 \, g \, L^{-1}$ HAc, (B) $3 \, g \, L^{-1}$ HAc. *IM*, Imidazoline. Source: *Y.Z. Li, N. Xu, X.P. Guo, G.A. Zhang, Inhibition effect of imidazoline inhibitor on the crevice corrosion of N80 carbon steel in the CO2-saturated NaCl solution containing acetic acid. Corros. Sci. 126 (2017) 127–141.*

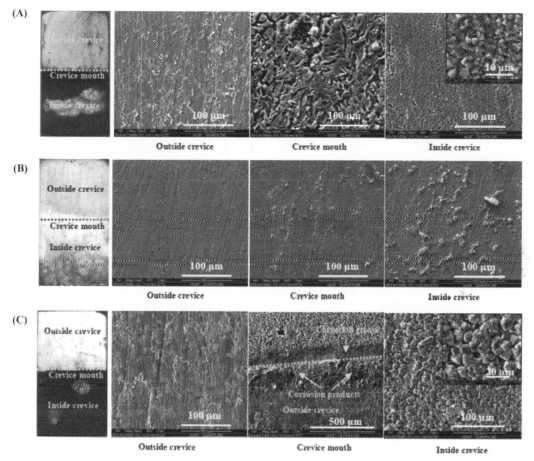

FIGURE 8–2 Macro and micromorphologies of carbon steel after corrosion in the CO_2-saturated 1.65% NaCl solution containing 1 g L^{-1} HAc in the absence or presence of IM for 72 h: (A) absence of IM, (B) presence of IM added at the beginning, (C) presence of IM added after corrosion for 24 h. *Source: Y.Z. Li, N. Xu, X.P. Guo, G.A. Zhang, Inhibition effect of imidazoline inhibitor on the crevice corrosion of N80 carbon steel in the CO2-saturated NaCl solution containing acetic acid. Corros. Sci. 126 (2017) 127–141.*

8.6.2 Imidazolines-based corrosion inhibitors for H$_2$S corrosion of steel

Since steels, broadly used in oil and gas transportation, are plagued by sour corrosion (H$_2$S), scholars worldwide have come up with diverse formulations of compounds and different tools to overcome this issue. This part provides an outline of the corrosion inhibitors used so far in the H$_2$S environment. IMs and their derivatives are also the most efficient corrosion inhibitors for H$_2$S corrosion in the oil and gas industry.

Edwards et al. [52] reported corrosion inhibition by the oleic IM via the bubble test and flow loop tests. According to the researchers, the long hydrocarbon chain is playing a significant role in the inhibition mechanism, and varying the chemistry of the pendant side chain

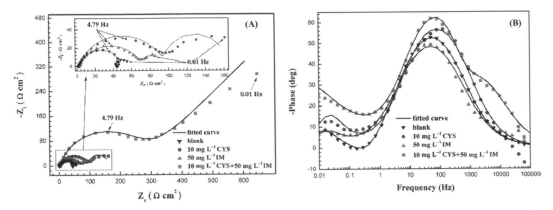

FIGURE 8–3 (A) Nyquist and (B) Bode plots of L360 carbon steel measured in CO$_2$-saturated 3% NaCl solution with different inhibitors at 60°C. Source: *C. Zhang, H. Zhang, J. Zhao, Synergistic inhibition effect of imidazoline derivative and l-cysteine on carbon steel corrosion in a CO$_2$-saturated brine solution. Corros. Sci. 112 (2016) 160–169.*

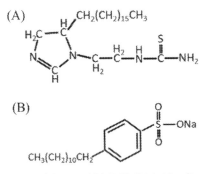

FIGURE 8–4 Molecular structures of inhibitors: (A) IM and (B) SDBS. *IM*, Imidazoline; *SDBS*, sodium dodecyl benzenesulfonate. Source: *Y. Shan Qian, F. Cheng, Synergism of imidazoline and sodium dodecylbenzenesulphonate inhibitors on corrosion inhibition of X52 carbon steel in CO$_2$-saturated chloride solutions, J. Mol. Liq. 294 (2019) 111674.*

does not affect the molecule performance to a major extent. Gusmano et al. [53] agreed with Edwards et al. on the fact that strong interaction occurs among the IM ring and metal, though on the role of the pendant group they do not agree. One of the authors reported that the pendant group might enhance the inhibition. Zhao et al. [54] confirmed that IM derivatives have anticorrosive effects on carbon steel and are generally used in acid-corroded pipelines for conveying oil and gas. The corrosion inhibitor has a better corrosion inhibition effect in the H$_2$S/CO$_2$ environment. When the dosage is only 50 mg L^{-1}, the corrosion IE can reach 94%. SEM results explain that fewer pits and corrosion products were found on the surface of the sample after adding the inhibitor. The adsorption performance of IM derivatives on the iron surface was obtained via molecular dynamics (MD)

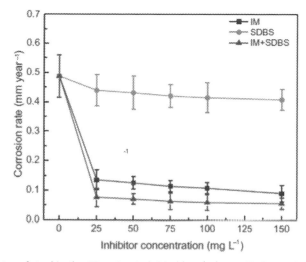

FIGURE 8–5 Corrosion rates of steel in the CO_2-saturated chloride solution at 60°C containing IM, SDBS or IM + SDBS at different concentrations determined by the weight loss measurement. *IM*, Imidazoline; *SDBS*, sodium dodecyl benzenesulfonate. Source: *Y. Shan Qian, F. Cheng, Synergism of imidazoline and sodium dodecylbenzenesulphonate inhibitors on corrosion inhibition of X52 carbon steel in CO_2-saturated chloride solutions, J. Mol. Liq. 294 (2019) 111674.*

simulation. The results of MD simulation showed that IM derivatives had better adsorption on water than iron so that the corrosion inhibitor will first be adsorbed on the iron surface to play a role in the corrosion inhibition, the frontier orbital adsorption sites of the inhibitor molecules are located around the phenyl group, thus forming a more stable adsorption layer on the surface of the metal. A topic of argument was the corrosion inhibition mechanism of IM, which might be a mutual effect of activation and blocking, and the energy-related factors [55]. Szyprowski [56] synthesized seven homogeneous IMs, derivatives of cyclopentyl- and cyclohexyl naphthenic acids, and studied their corrosion inhibition performance on 1H18N9T austenitic stainless steel and St3S carbon steel in 2% NaCl containing hydrocarbon fraction and H_2S gas saturation. The inhibitors offered high protective effectiveness at a low concentration of 25 ppm, attaining 99% IE for mild steel and 80% IE for austenitic steel at 40°C. The IE is enhanced by raising the substituent chain length. Therefore, the least efficient compounds were short-chained IM rings. The inhibitors investigated were chemisorbed on the electrode surface and it obeyed the Frumkin adsorption isotherm. Ramachandran et al. [57] reported the atomistic simulations (MD and quantum mechanics to clarify the mechanism of corrosion inhibition via oleic IM and reported the result that a self-assembled single layer is developed on the initial oxide surface of the iron. This film carried out the pathway of inhibition by hydrophobic barrier formation against the ingression of mini ions. Fig. 8–6 displays topography for the optimum structure of OI-n-CCN with n equal to 7, 11, and 12 Cs. With *n* less than 12 or more carbons the surface is fully covered.

This suggestion verified the explanation povided by Blair et al. [58]. The authors explained in a US Patent that the most efficient IM inhibitors bear long hydrocarbon chains attached to a

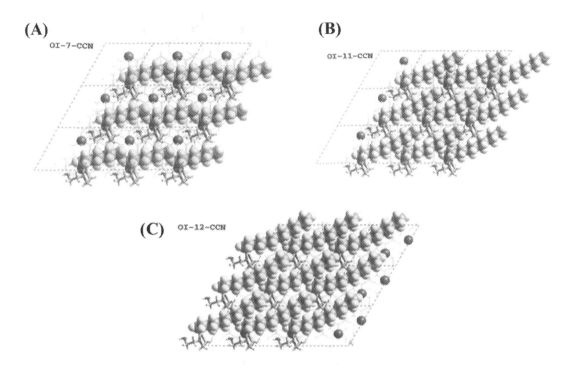

(A)   OI-7-CCN

(B)   OI-11-CCN

(C)   OI-12-CCN

FIGURE 8–6 Top view of surface showing nine unit cells with one OI per unit cell: (A) OI-7-CCN (17% observed corrosion inhibition) shows considerable uncovered surface; (B) OI-11-CCN (77% observed corrosion inhibition) shows that the head group is not covered; and (C) OI-12-CCN (95% observed corrosion inhibition) shows complete coverage. Source: *S. Ramachandran, B.-L. Tsai, M. Blanco, H. Chen, Y. Tang, W.A. Goddard, Self-assembled monolayer mechanism for corrosion inhibition of iron by imidazolines, Langmuir 12 (1996) 6419.*

nitrogen atom of the IM ring or on a relatively small organic radical attached to a nitrogen atom of the ring.

The corrosion IE of IMs might be improvised by its corrosion reactions with appropriate contents such as organic acids that are able to increase its separation in the aqueous medium. Meyer [59] patented a distinctive instance of this reaction and Martins et al. [60] and Benitez Aguilar et al. [61] have patented IM-based inhibitors with excellent performance. The IM compounds affinity to bioaccumulate is a limiting factor to their usage. Considerable investigations have been made in order to graft IM green compounds. Rivera-Grau et al. [62] reported IM mutated by coconut oil, namely aminoethyl-amine IM, a green inhibitor, to explore the H_2S corrosion inhibition performance of 1018 carbon steel. It was found that the coconut oil-modified IM was a better inhibitor compared with the commercial one.

Diaz et al. [63] evaluated carboxyethyl-IM for ultra-high strength API X120 pipeline steel corrosion in H_2S + 3% NaCl in the presence and absence of 10 vol.% diesel. The studies were performed in varying concentrations of inhibitor (0, 5, 10, 25, 50, and 100 ppm) at a

Table 8–2   Imidazoline derivatives used as an efficient corrosion inhibitors for steel in H_2S environment.

S.No.	Inhibitor	Metal/corrosive medium	Techniques	IE	Reference
1.	Aminoethyl-amine imidazoline	1018 carbon steel/H_2S-containing 3% NaCl solutions, T 50°C	PDP and EIS	84.6	[62]
2.	Hydroxyethyl-imidazoline	1018 carbon steel/H_2S-containing 3% NaCl solutions, T 50°C	PDP and EIS	84.6	[62]
3.	Imidazoline (derivatives of cyclopentyl- and cyclohexylonaphthenic acids)	Austenitic steel 1H18N9T & carbon steel St_3S/H_2S containing 2% NaCl solution, T 20, 40, and 60°C	Potentiodynamic method	99	[56]
4.	Carboxyethylimidazoline	API X-120 pipeline steel/3% NaCl + H_2S without and with 10% diesel, T 50°C	PDP and EIS	98	[63]
5.	Hydroxyethyl imidazoline	Oilfield pipeline/H_2S containing H_2S containing 3% NaCl, T 50°C	Surface interaction	95	[64]
6.	2-Phenylbenzimidazole	Carbon steel/H_2S-containing 3% NaCl solutions, T 25°C	Electrochemical measurements, UV, SEM, EDS	87	[65]
7.	Imidazoline inhibitors derived from vegetable oil from canola, sun flower, safflower, soybean, corn, and mixtures thereof reacted with polyalkylated polyamines.	Steel/90% brine, 10% kerosene (500–600 mg L^{-1}) H_2S, T 60°C	EIS, PDP, UV, FTIR, SEM and AFM	95.33	[66]
8	10%–20% sodium benzoate, 20%–30% heptadecylethoxyl imidazoline, 5%–10% sodium molybdate, 5%–10% zinc vitriol, and 30%–60% water	H_2S conc. (200 mg L^{-1}), T 80°C	Dynamic indoor weight loss method	89.3	[67]

temperature of 50°C. The concentration of 50 ppm afforded 98% protection. Similarly, Lucio-Garcia et al. [64] have reported H_2S corrosion inhibition of steel pipeline by hydroxyethyl IM. The 5, 10, 25, 50, and 100 ppm inhibitor in an H_2S-containing 3% NaCl solution at 50°C was used. The results explained that just 5 and 10 ppm concentration of the inhibitor protected API X120 steel by almost 95% at 50°C. Table 8–2 shows the some of IM-based inhibitors studied in H_2S corrosion inhibition of steel.

8.7 Conclusions

Recently, the problem of sweet and sour corrosion has been a most important hurdle in oil and gas production. This chapter has summarized the factors, mechanisms, and corrosion inhibitors studied in the scientific literature describing both sweet and sour corrosion of steel. IM derivatives are extensively studied inhibitors for sweet and sour corrosion of steel in the oilfield industry. Therefore understanding the corrosion inhibitor functioning is a prerequisite for better production and good quality output.

References

[1] IBISWorld, Global oil & gas exploration & production US industry market research report, 2018. < https://www.ibisworld.com/industry-trends/global-industry-reports/mining/oil-gas-exploration-production.html) > . (Accessed 12 January 2019).

[2] Organization of Petroleum Exporting Countries, World oil outlook 2040, 2019. < https://www.opec.org/opec_web/flipbook/WOO2017/WOO2017/assets/common/downloads/WOO%202017 > . (Accessed 12 January 2019).

[3] C.M. Menendez, B. Hughes, J. Jardine, W.Y. Mok, S. Ramachandran, V. Jovancicevic, et al., New sour gas corrosion inhibitor compatible with kinetic hydrate inhibitor. Paper No. IPTC 17440, International Petroleum Technology Conference (2014).

[4] T.E. Perez, Corrosion in the oil and gas industry: an increasing challenge for materials, JOM 65 (8) (2013) 1033. Available from: https://doi.org/10.1007/s11837-013-0675-3.

[5] A. Chen, Sinopec oil pipeline blast kills 35 in Eastern China. REUTERS, China Daily, 2013. < http://www.reuters.com/article/2013/11/22/ussinopec-blastidUSBRE9AL08E20131122 > . (Accessed 06 February 2019).

[6] M. Wayne, C. Tejada, China cites lapses in Sinopec pipeline blasts. The Wall Street Journal (January 2014), 2014. < https://www.wsj.com/articles/china-cites-lapses-in-sinopec-pipeline-blasts-1389254357 > . (Accessed 06 February 2019).

[7] United Press International Inc, Sinopec reveals causes of fatal November pipeline blast, 2014. < http://www.upi.com/Business_News/EnergyResources/2014/01/14/Sinopec-reveals-causes-of-fatal-November-pipelineblast/UPI-95061389699897/#ixzz2qQEs1w14 > . (Accessed 06 February 2019).

[8] F. Tom, El Paso natural gas to pay $15.5 million penalty and perform comprehensive reforms to pipeline system, TDD (202) 514-1888. Washington D. C.: Department of Justice, 2007. < https://www.justice.gov/archive/opa/pr/2007/July/07_enrd_548.html > . (Accessed 06 February 2019).

[9] Infographic, The El Paso natural gas company pipeline explosion, 2019. < https://www.corrosionpedia.com/infographic-the-el-paso-natural-gas-companypipeline-explosion/2/6977 > . (Accessed 10 March 2019).

[10] C. Yun, L.K. Thomas, High strength steel anchor rod problems on the new bay bridge, Revision 1, November 12, 2013, Prepared (unsolicited) for Senator Mark DeSaulnier, Committee on Transportation and Housing, California State Senate, 2013. < http://media.sacbee.com/smedia/2013/12/07/21/47/Djfhs.So.4.pdf > .

[11] M. Lydia, K. Crowe, Corrosion suspected in Green Bay bridge sag; packers traffic affected, 2013. < http://archive.jsonline.com/news/traffic/no-newsagging-on-leo-frigo-bridge-on-i-43-in-green-bay-dot-says-b99112046z1-226307681.html/ > . (Accessed 06 February 2019).

[12] A. Jacob, Y. Okada, Fukushima workers exposed to radiation in filter system leak, 2013. < http://www.bloomberg.com/news/2013-10-09/fukushimatank-leaks-probably-caused-by-corrosion-around-sealant.html > . (Accessed 06 February 2019).

[13] L. Amanda, Lowe's walkway collapse raises concerns about pedestrian bridge safety, inspections. WRAL (May 2000), 2000. < https://www.wral.com/news/local/story/141306/ > . (Accessed 06 February 2019).

[14] Major Accident Reporting System (eMARS), 2019. < https://ec.europa.eu/knowledge4policy/projects-activities/major-accident-reporting-systememars_en > . (Accessed 06 February 2019).

[15] M. Goyal, S. Kumar, I. Bahadur, C. Verma, E.E. Ebenso, Organic corrosion inhibitors for industrial cleaning of ferrous and non-ferrous metals in acidic solutions: a review, J. Mol. Liq. 256 (2018) 565−573.

[16] C. Verma, L. Olasunkanmi, E.E. Ebenso, M. Quraishi, Substituents effect on corrosion inhibition performance of organic compounds in aggressive ionic solutions: a review, J. Mol. Liq. 251 (2018) 100−118.

[17] C. Verma, E.E. Ebenso, I. Bahadur, M. Quraishi, An overview on plant extracts as environmental sustainable and green corrosion inhibitors for metals and alloys in aggressive corrosive media, J. Mol. Liq. 266 (2018) 577–590.

[18] W.F. Oxford, R.E. Foss, Corrosion of Oil and Gas Well Equipment, 87th edition, Division of Production, American Petroleum Institute, Dallas, 1958.

[19] D. Brondel, R. Edwards, A. Hayman, D. Hill, S. Mehta, T. Semerad, Corrosion in the oil industry, Oilfield Rev. 6 (1994) 4–18.

[20] F. Dean, S. Powell, Hydrogen flux and high temperature acid corrosion, 06436th edition, in: NACExpo 2006 Conference (2006).

[21] K. Nalli, Corrosion and its mitigation in the oil and gas industry. An overview. PM-Pipeliner Report, 2010.

[22] A. Dugstad, The importance of $FeCO_3$ supersaturation on the CO_2 corrosion of carbon steels, corrosion ˝92, paper 14. NACE, Houston, 1992.

[23] L.K. Gatzky, R.H. Hausler, A novel correlation of tubing corrosion rates and gas production rates, Adv. CO_2 Corros. 1 (1984) 87.

[24] M.B. Kermani, D. Harrop, The impact of corrosion on the oil and gas industry, SPE Prod. Facilities 11 (1996) 186–190.

[25] C. de Waard, U. Lotz, EFC publication number 13, in: Prediction of CO_2 Corrosion of Carbon Steel, The Institute of Materials, London, 1994.

[26] D. Corbin, E. Willson, New technology for real-time corrosion detection, in: Tri-service Corrosion Conference, USA, 2007.

[27] G.V. Chilingar, C.M. Beeson, Surface Operations in Petroleum Production, American Elsevier, New York, 1969, p. 397.

[28] W. Sun, Kinetics of iron carbonate and iron sulfide scale formation in CO_2/H_2S corrosion (Ph.D. dissertation), 2006.

[29] B.J. Usman, S.A. Ali, Carbon dioxide corrosion inhibitors: a review, Arab. J. Sci. Eng. 43 (2018) 1–22.

[30] P. Rajeev, Corrosion mitigation of the oil well steels using organic inhibitors—a review, J. Mater. Environ. Sci. 3 (2012) 856–869.

[31] M. Dudukcu, The inhibitive effect of 5-amino-indole on the corrosion of mild steel in acidic media, Mater. Corros. 62 (2011) 264–268.

[32] J. Zhang, X. Sun, Y. Ren, M. Du, The synergistic effect between imidazoline-based dissymmetric bisquaternary ammonium salts and thiourea against CO_2 corrosion at high temperature, J. Surfactants Deterg. 18 (2015) 981–987.

[33] I. Jevremovic, M. Singer, S. Nesic, V. Miskovic-Stankovi, Electrochemistry of carbon dioxide corrosion mitigation using tall oil diethylenetriamineimidazoline as corrosion inhibitor for mild steel, Mater. Corros. 67 (2016) 756–768.

[34] W. Villamizar, M. Casales, J.G. Gonzales-Rodriguez, L. Martinez, An EIS study of the effect of the pedant group in imidazolines as corrosion inhibitors for carbon steel in CO_2 environments, Mater. Corros. 57 (2006) 696–704.

[35] Y.Z. Li, N. Xu, X.P. Guo, G.A. Zhang, Inhibition effect of imidazoline inhibitor on the crevice corrosion of N80 carbon steel in the CO_2-saturated NaCl solution containing acetic acid, Corros. Sci. 126 (2017) 127–141.

[36] C. Zhang, H. Duana, J. Zhao, Synergistic inhibition effect of imidazoline derivative and l-cysteine on carbon steel corrosion in a CO_2-saturated brine solution, Corros. Sci. 112 (2016) 160–169.

[37] Y. Shan Qian, F. Cheng, Synergism of imidazoline and sodium dodecylbenzenesulphonate inhibitors on corrosion inhibition of X52 carbon steel in CO_2-saturated chloride solutions, J. Mol. Liq. 294 (2019) 111674.

[38] A. Singh, K.R. Ansari, A. Kumar, et al., Electrochemical, surface and quantum chemical studies of novel imidazole derivatives as corrosion inhibitors for J55 steel in sweet corrosive environment, J. Alloys Compd. 712 (2017) 121−133.

[39] L.D. Paolinelli, T. Pérez, S.N. Simison, The incidence of chromium-rich corrosion products on the efficiency of an imidazoline-based inhibitor used for CO_2 corrosion prevention, Mater. Chem. Phys. 126 (2011) 938−947.

[40] H.H. Zhang, X. Pang, K. Gao, Localized CO_2 corrosion of carbon steel with different microstructures in brine solutions with an imidazoline-based inhibitor, Appl. Surf. Sci. 442 (2018) 446−460.

[41] H.H. Zhang, X. Pang, M. Zhou, et al., The behavior of pre-corrosion effect on the performance of imidazoline-based inhibitor in 3 wt.%NaCl solution saturated with CO_2, Appl. Surf. Sci. 356 (2015) 63−72.

[42] M.A.J. Mazumder, H.A. Al-Muallem, S.A. Ali, The effects of N-pendants and electron-rich amidine motifs in 2-(p-alkoxyphenyl)-2 imidazolines on mild steel corrosion in CO_2-saturated 0.5 M NaCl, Corros. Sci. 90 (2015) 54−68.

[43] X. Liu, P.C. Okafor, Y.G. Zheng, The inhibition of CO_2 corrosion of N80 mild steel in single liquid phase and liquid/particle two-phase flow by aminoethylimidazoline derivatives, Corros. Sci. 51 (2009) 744−751.

[44] D.M. Ortega-Toledo, J.G. Gonzalez-Rodriguez, M. Casales, et al., The CO_2 corrosion inhibition of a high strength pipeline steel by hydroxyethyl imidazoline, Mater. Chem. Phys. 122 (2010) 485−490.

[45] J. Zhao, G. Chen, The synergistic inhibition effect of oleic-based imidazoline and sodium benzoate on mild steel corrosion in a CO_2-saturated brine solution, Electrochim. Acta 69 (2012) 247−255.

[46] B. Wang, M. Du, J. Zhang, C.J. Gao, Electrochemical and surface analysis studies on corrosion inhibition of Q235 steel by imidazoline derivative against CO_2 corrosion, Corros. Sci. 53 (2011) 353−361.

[47] F.G. Liu, M. Du, J. Zhang, M. Qiu, Electrochemical behavior of Q235 steel in saltwater saturated with carbon dioxide based on new imidazoline derivative inhibitor, Corros. Sci. 51 (2009) 102−109.

[48] M.W.S. Jawich, G.A. Oweimreen, S.A. Ali, Heptadecyl-tailed mono- and bis-imidazolines: a study of the newly synthesized compounds on the inhibition of mild steel corrosion in a carbon dioxide-saturated saline medium, Corros. Sci. 65 (2012) 104−112.

[49] P.C. Okafor, X. Liu, Y.G. Zheng, Corrosion inhibition of mild steel by ethylaminoimidazoline derivative in CO_2-saturated solution, Corros. Sci. 51 (2009) 761−768.

[50] G. Zhang, C. Chen, M. Lu, et al., Evaluation of inhibition efficiency of an imidazoline derivative in CO_2-containing aqueous solution, Mater. Chem. Phys. 105 (2007) 331−340.

[51] D.M. Ortega-Toledo, J.G. Gonzalez-Rodriguez, M. Casales, et al., CO_2 corrosion inhibition of X-120 pipeline steel by a modified imidazoline under flow conditions, Corros. Sci. 53 (2011) 3780−3787.

[52] A. Edwards, C. Osborne, S. Webster, D. Klenerman, M. Joseph, P. Ostovar, et al., Mechanistic studies of the corrosion inhibitor oleic imidazoline, Corros. Sci. 36 (2) (1994) 315.

[53] G. Gusmano, P. Labella, G. Montesperelli, A. Privitera, S. Tassinari, Study of the Inhibition Mechanism of Imidazolines by Electrochemical Impedance Spectroscopy, Corrosion, NACE International, 2006, p. 576.

[54] X. Zhao, C. Chen, H. Yu, Q. Chen, Evaluation and mechanism of corrosion inhibition performance of new corrosion inhibitor. Paper No. 11029, in: NACE International Corrosion Conference & Expo (2018).

[55] Y.X. Dai, F.N. Lv, B. Wang, Y. Chen, Thermoresponsive phenolic formaldehyde amines with strong intrinsic photoluminescence: preparation, characterization and application as hardeners in waterborne epoxy resin formulations, Polymer 145 (2018) 454. Available from: https://doi.org/10.1016/j.polymer.2018.05.007.

[56] A.J. Szyprowski, Relationship between chemical structure of imidazoline inhibitors and their effectiveness against hydrogen sulphide corrosion of steels, Br. Corros. J. 35 (2) (2000) 155.

[57] S. Ramachandran, B.-L. Tsai, M. Blanco, H. Chen, Y. Tang, W.A. Goddard, Self-assembled monolayer mechanism for corrosion inhibition of iron by imidazolines, Langmuir 12 (1996) 6419.

[58] C.M. Blair, Jr., W. Groves, W.F. Gross. Processes for preventing corrosion and corrosion inhibitors. US Patent No. 2,466,517, April, 1949.

[59] G.R. Meyer. Corrosion inhibition compositions. US Patent No. US 6448411 B1, 2001.

[60] R.L. Martins, J.A. McMahon, B.A. Oude Alink. Biodegradable corrosion inhibitors of low toxicity. European Patent No. EP 0 651 074 B1, 1998.

[61] J.L.R. Benitez Aguilar, A. Tobo Cervantes, A. Estrada Martinez, N. Navarro Ordonez. Corrosion inhibitors derived from vegetable oils and it's process of obtaining. US Patent No. US 2017/0029960 A1, 2017.

[62] L.M. Rivera-Grau, M. Casales, I. Regla, D.M. Ortega-Toledo, J.A. AscencioGutierrez, J.G. Gonzalez-Rodriguez, et al., H2S corrosion inhibition of carbon steel by a coconut-modified imidazoline, Int. J. Electrochem. Sci. 7 (12) (2012) 12391.

[63] E.F. Diaz, J.G. Gonzalez-Rodriguez, A. Martinez-Villafañe, C. Gaona-Tiburcio, H_2S corrosion inhibition of an ultra high strength pipeline by carboxyethyl-imidazoline, J. Appl. Electrochem. 40 (2010) 1633. Available from: https://doi.org/10.1007/s10800-010-0149-z.

[64] M.A. Lucio-Garcia, J.G. Gonzalez-Rodriguez, A. Martinez-Villafañe, G. Dominguez-Patiño, M.A. Neri-Flores, J.G. Chacon-Nava, A study of hydroxyethyl imidazoline as H_2S corrosion inhibitor using electrochemical noise and electrochemical impedance spectroscopy, J. Appl. Electrochem. 40 (2010) 393−399.

[65] S. Pournazari, M.H. Moayed, M. Rahimizadeh, In situ synthesis of 2-phenylbenzimidazole as an hydrogen sulfide corrosion inhibitor of carbon steel, Corrosion 69 (12) (2013) 1195.

[66] J.L.R. Benitez Aguilar, A. Tobo Cervantes, A. Estrada Martinez, N. Navarro Ordonez. Corrosion inhibitors derived from vegetable oils and it's process of obtaining. US Patent No. US 2017/0029960 A1, 2017.

[67] X. Zhang, L. Guo, S. Li, J. Xu, B. Wu, B. Xue, et al. Hydrogen sulfide with a water-soluble anti-corrosion inhibitors. Chinese Patent No. CN103409122B, 2015.

Sustainable corrosion inhibitors for environmental industry

9

Carbohydrate polymers protecting metals in aggressive *environment*

Marziya Rizvi

CORROSION RESEARCH LABORATORY, DEPARTMENT OF MECHANICAL ENGINEERING, FACULTY OF ENGINEERING, DUZCE UNIVERSITY, DUZCE, TURKEY

Chapter outline

Environmentally Sustainable Corrosion Inhibitors. DOI: https://doi.org/10.1016/B978-0-323-85405-4.00014-8

9.1 A revisit to carbohydrate polymers

When we imagine polymers generally what comes to mind are the synthetic substances commonly used in daily routines, such as plastics. You may be surprised to know that nature is itself a huge reservoir of polymers which are harmless and biodegradable (Fig. 9−1).

With a little modification in their chemical structure or when they are applied in crude form appropriately, their immense potential can be noted. Let us talk about the most abundant class of natural polymers, that are better known as carbohydrate polymers. Carbohydrates are a basic and most important component of a healthy diet. They are made up of carbon, hydrogen, and oxygen. In biochemistry, the term "carbohydrate" is synonymous with "saccharide." Based on the number of building units, the saccharides are subcategorized to "mono-," "di-," "oligo-," and "polysaccharides." Amongst these four subcategories of the saccharides it has been observed that the polysaccharides, also most commonly known as carbohydrate polymers, have made their mark in various scientific and industrial processes, ranging from being vessels for drug delivery, forming corrosion-resistant films on metals, starting materials for fuel cells, forming active surfaces for heterogeneous catalysis, etc.

The lengthy chains of cyclic sugars connected via -O bridges are found in the living bodies of organisms as well as parts of other biological systems. As stated they have a long chain of cyclic sugars, and their molecular lengths and molecular weights are very high. Another way the biochemists have categorized these interesting substances is based on the role they perform in the natural systems, these carbohydrate polymers are either "structural carbohydrate polymers" or "protective carbohydrate polymers." Some very commonly known carbohydrate polymers are starch, cellulose, dextrins, chitin/chitosan, and gums. Some carbohydrate polymers like chitin can be derived from exoskeletons of arthropods, shells of

FIGURE 9–1 Sources of naturally occurring carbohydrate polymers.

mollusks and crustaceans, sometimes even the scales of the fishes. Another major type of carbohydrate which is abundant in nature is obtained from plants as plant polysaccharides like pectin, arabinogalactans, arabinoxylans, glucomannans, xyloglucans, etc. There is yet another types of carbohydrate polymer which is obtained as exudate from bark of the trees like acacia gum, garcinia gum, etc. Sometimes some microorganisms like *Xanthomonas* work upon plant cellulose and convert it into useful substance like xanthan gum. The list of the potentially useful carbohydrate polymers is almost endless. The carbohydrate polymers never cease to amaze us, as these eco-friendly substances can be crafted into reliable and superior material with a little chemical modification. They are quite inexpensive substances, pose no harm, and are easily accessible. The unmatched properties of this class of compounds has attracted the attention of scientists all around the world from synthetic compounds to their eco-friendly alternatives.

When the rest of the scientific world was busy exploring the list of applications of carbohydrate polymers, some corrosion scientists in the past decades have successfully applied these renewable resources to prevent the metals from corroding in a much more natural and humane way by using them as "inhibitors." Inhibitors are the substances which can be introduced to the environment which is surrounding the metal and causing it to degrade. The role of the corrosion inhibitors is to keep the metals safe from the ongoing changes in pH, temperature, composition, and dynamics of the environment. The inhibitors which occupied the corrosion prevention studies a couple of decades ago were toxic substances like chromates, arsenate sugars, etc. Not only were these compounds toxic for the workers who handle the industrial equipment where inhibitors are used, they are expensive and because of their toxic nature there are laws and regulations to their usage too. To the contrary the carbohydrate polymers are natural, nontoxic, biodegradable, inexhaustible/renewable, and easily accessible sets of chemical compounds which nature has to offer us. These fascinating substances inhibit the oxidation of pure metal surfaces by offering specific active adsorption sites for the functional molecules which may cause the metal surfaces to otherwise corrode if left to react with the aggressive ions in the corrosion cell. The heteroatom "O" is the basic building block of all the carbohydrates, which coordinates with the metals to stop their oxidation in corrosive solution. Sometimes the cyclic rings in the long chains trap the corrosion precursors to prevent the metallic oxidation. The researchers have carried out lengthy studies on natural polymers in the past two decades. Let us explore some notable carbohydrate polymers with the potential to prevent the corrosion of metals, looking at their basics, mechanisms, and efficiencies.

9.2 Mucilage and gums from plants

The viscous exudates and mucilage from bark, stems, and sometimes leaves and seeds of some plants have properties which make them important components of adhesives, binders, thickeners, stabilizers, microencapsulating components of drugs in medicine technology and most importantly corrosion inhibitors in oil and gas industries. They are generally water-soluble compounds, which increases their importance. Depending on their source, they bear faint to highly pungent odors.

9.2.1 Guar gum

Some beans, like guar, bear polysaccharides in their seed endosperm, which is commonly known as guar gum. This gum was evaluated as an inhibitor in 1 M H_2SO_4 to prevent the rusting of carbon steel. It was noted that heterocyclic pyran groups in the molecule trap the corrosion-inducing molecules and comprehensively inhibit the corrosion by bonding chemically and physically with the surface moieties responsible for oxidizing the steel surface. The adsorption analysis reveals the adsorption of a monolayer, Langmuir isotherm being followed by the inhibition process. 1500 ppm of guar gum in 1 M H_2SO_4 may inhibit 93.8% of corrosion [1].

9.2.2 Acacia gum

The acacia trees exude a water-soluble complex mixture of oligosaccharide, polysaccharides, and glycoproteins, which is generally called gum Arabic or Acacia gum. It is amongst the oldest gums which have been used for the prevention of corrosion. The researchers have evaluated its efficiency for multiple substrates like aluminum and mild steel in unit molar of sulfuric acid environment. It effectively protects both the metals but acts more efficiently for preventing the oxidation of aluminum where it interestingly acts as a physiosorbed inhibitor compared to mild steel where it is chemisorbed on the surface. At the highest temperature of evaluation which was 60°C in the case of mild steel, 500 ppm of gum Arabic prevented 37.88% of corrosion. On aluminum substrate 500 ppm of gum Arabic evaded almost 80% of corrosion at 30°C [2]. Gum Arabic has also been evaluated for mild steel immersed in HCl by gravimetry, hydrogen evolution analysis, electrochemical analysis Fourier transform infrared spectroscopy (FTIR), scanning electron microscopy (SEM), and X-ray photoelectron spectroscopic (XPS) techniques [3]. It was observed that gum Arabic displayed more synergism in hydrochloric acid compared to sulfuric acid when an external magnetic field was applied. This occurred probably due to changes in interaction mode and oxide layer formed by the assistance of a field which bears superior inhibition toward pitting on repassivation. Apart from the modification of inhibition by applying a field, the researchers have attempted to synergize the reactions by adding chemicals like surfactants and halides. Thus SDBS and CTAB were added along with gum Arabic to a less aggressive solution of 0.1 M H_2SO_4 in a temperature range of 30°C−60°C. The response of the metal to the addition of surfactants to its environment in this way was inferred using gravimetric analysis, solution analysis, SEM, and atomic force microscopy (AFM) [4]. The surface morphologies displayed a clear increase in the protection offered by gum Arabic after the addition of surfactants. The adsorption studies inferred intramolecular interaction upon adsorption as Freundlich adsorption isotherm was followed. The researchers reported a 83.36% efficiency for 1000 ppm gum Arabic which further improved to 90.74% on adding just 1 ppm CTAB at 30°C. These studies were carried out with mineral acids in acidic environments. Let us talk about the behavior of the same polymer in an alkaline environment. The researchers have also conducted the study using gum Arabic as a corrosion inhibitor in NAOH at 303K and 313K using gravimetric studies and hydrogen evolution. Iodide ion was introduced as a synergizing agent to the aggressive solution. The maximum efficiency of 75% was obtained and the adsorption isotherm followed was Temkin's [5].

9.2.3 Xanthan gum

A very interesting carbohydrate polymer is obtained when the bacteria *Xanthomonas campestris* acts on plant cellulose and changes it to yet another polysaccharide, xanthan gum. This carbohydrate polymer is well-known to bakers and is already used as a thickener in food industries. In recent years it has caught the attention of the corrosion researchers because of its unique molecular structure (Fig. 9−2), water solubility, inexpensiveness, and easy availability. It was observed that this compound inhibits the corrosion of aluminum in 1 M HCl [6]. At 40°C xanthan gum was 69.05% efficient in protecting aluminum from 1 M HCl. The adsorption process followed the Temkin and El-Awady adsorption isotherms. Potentiodynamic polarization suggested mixed-type inhibition and cathodic partial inhibitive reaction and the anodic dissolution reaction determined the corrosion rate of the system.

9.2.4 *Ficus* gum or fig gum

Gum exudates obtained from various species of fig tree were commonly named as *Ficus* gum. The gum obtained from African rock fig was used as an inhibitor for mild steel corrosion in H_2SO_4 medium [7]. Various corrosion tests revealed that this gum was 65% efficient at 333K. The inhibitor causes the protective effect by chemically adsorbing to the surface of the mild steel. Tannins, glucoproteins, and polysaccharides constitute this gum, which ensures the effective inhibition of corrosion of the surface of mild steel. With so many components constituting a single gum there is a chance of formation of multiple layers of adsorption. The researchers have reported the Langmuir mode of adsorption where the adsorption is not only exothermic but spontaneous in nature as well as.

Xanthan gum

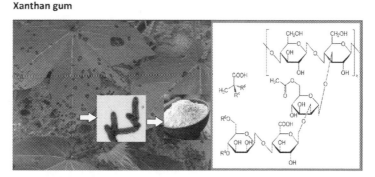

FIGURE 9–2 Molecular structure of xanthan gum.

Eddy et al. [8] obtained *Ficus* gum from a different variant of this plant known as *Ficus benjamina* and used this as an inhibitor for aluminum corrosion in 0.1 M H_2SO_4 through gravimetric study. Like the previous study this gum had multicomponents too and hence the formation of multiple layers of adsorption was proposed by the researchers. With multilayered adsorption it was reported that the process followed the Frumkin and Dubinin–Radushkevich adsorption models. A mere 0.5 g L^{-1} of this gum successfully inhibited 87% degradation of aluminum at 333K.

9.2.5 Hutch Gum

Gum is derived from the African copaiba balsam tree, *Daniella oliverri*. Collected by tapping the bark of this plant, this gum is commonly called hutch gum. This polysaccharide was used to inhibit the corrosion of mild steel in hydrochloric acid environment by the weight loss method and FTIR [9]. It was studied that this gum can inhibit 72.36% mild steel corrosion at a concentration of 0.5 g L^{-1} at 303K. Endothermy and a spontaneous process was recorded which supported the mechanism of physical adsorption. It was found that this inhibitor formed a monolayer of molecules on the metal surface following Langmuir adsorption.

9.2.6 Okra mucilage

Sometimes a minor modification in the chemistry of the natural substance enhances the characteristics as well as the properties, making the substance more useful. A research team modified polyacrylamide by grafting it with Okra mucilage, a natural vegetable polysaccharide. This modified mucilage was tested as an inhibitor for mild steel corrosion in 0.5 M H_2SO_4 environment using weight loss and electrochemical investigation. A film of polymer was found to adsorb physically on metal substrate obeying the Langmuir adsorption isotherm. The inhibition kept on improving on adding this copolymer to the corrosion cell and a maximum inhibition was obtained at 100 ppm at 25°C when this copolymer inhibited almost 97% of corrosion. Thus we may say that a little modification in the natural structures may render beneficial outcomes when we discuss the natural polymers [10].

9.2.7 Corn polysaccharide

The polysaccharide from kernels of corn was modified and used as an inhibitor of mild steel in 1 M HCl through gravimetric, surface, and electrochemical evaluations [11]. A comprehensive and mixed adsorption was observed following Langmuir adsorption isotherm with maximum efficiency of 91.26% at 313K. The researchers prefer conducting the binding studies on the adsorbed polymer through XPS, UV–Visible spectroscopy or FTIR. In this study the XPS spectra detected the bonded N and C atoms of the heterocycle and the O atom of the hydroxyl group on the metal surface.

9.2.8 Tannins

Mangroves are small shrubs growing in tropical coastal areas. These interesting plants contain polyphenols called tannins which impart characteristic colors and odors to these plants. These substances are antifungal in nature as well as act as natural pesticides against small insects and pests which may invade this plants. The researchers have incorporated these polyphenols as corrosion inhibitors. Along with phosphoric acid, the action of these tannins was tested on prerusted steel in 3.5% sodium chloride solution. The efficiencies thus obtained were compared with the tannins from *Mimosa* [12]. At pH 0.5 and pH 2.0, inhibition increased but solely when mangrove and mimosa tannins were added, while at pH 5.5 the addition of phosphoric acid gave even higher efficiency of 79% at 30°C.

9.2.9 *Raphia hookeri* gum

Gum exudated from *Raphia hookeri*, commonly called ivory coast Raphia palms, was also considered to be a potential corrosion inhibitor by some researchers. These gum exudates were used as corrosion inhibitors of aluminum in HCl solution. This gum yielded a moderate corrosion inhibition efficiency of 56.3% at a concentration of 500 ppm at 30°C [13]. The constituent phytochemicals in the exudates adsorbed on the surface of the aluminum metal obeys the Temkin adsorption isotherm and the kinetic thermodynamic model of El-Awady et al.

9.2.10 Gum from the variants of butter-fruit trees

Bush pear is native to the African subcontinent and the scientific name of this fruit is *Dacryodes edulis*. The phytochemical content of this plant inspired the researchers to use it as a corrosion inhibitor of aluminum in 2 M hydrochloric acid environment [14]. It was observed that the inhibitor physiosorbed on the metals surface and yielded an efficiency of 42%. *Pachylobus edulis* is very closely related to *Dacryodes edulis*, or maybe it is the same plant. The literature does not have much to say about the similarity of these plants but both of them are commonly called bush pears or butter fruits. This bush pear gum was also tested for mild steel in 2 M H_2SO_4 synergized by potassium halides using hydrogen evolution and thermometric methods in the range of 30°C−60°C [15]. Synergistic effects increased the inhibition efficiency of the exudates in the presence of potassium halides in the order $KI > KBr > KCl$. The adsorption of the exudates gum alone and in combination with the potassium halides was approximated by the Temkin adsorption isotherm.

9.2.11 Tragacanth gum

Astragalus and its various species, which are a set of Middle Eastern legumes, naturally render a ribbon-like dried sap which is commonly called "goat's thorn" (Greek: tragacanth −goat thorn). The gum refined from this dried sap is called Shiraz gum, elect gum, or dragon gum. Currently Iran is the largest producer of this gum which is used for medicines and herbal remedies. It is an odorless, viscous, tasteless, yet water-soluble compound which may have single or multiple polysaccharides. The major part is water soluble, called tragacanthin, and the minor part is a gel, which swells on proximity to water, called bassorin. This substance is generally obtained from roots of the plants and dried for use. The major water-soluble part may be precipitated out and is mostly comprised of the polysaccharide arabinogalactans. When tested as a corrosion inhibitor of low-carbon steel in 1 M HCl it gives a corrosion inhibition efficiency of 96.35% at 500 ppm concentration [16]. The researchers have studied it extensively using computational methods to judge the exact mechanism of its adsorption. The gravimetric, electrochemical, and the binding studies in collaboration with the computational studies suggest that this polymer with its long chain covers the metal substrate as a monolayer obeying the Langmuir adsorption isotherm to form a protective film on the surface of steel. The macromolecular long-chain structure of the polysaccharide along with the structure comprising abundant heteroatom O and functional groups are responsible for efficiently protecting the metallic surface against acidic attack.

9.2.12 Plantago gum

This group of flowering plants are called plantains or flea worts and are found distributed all over the world with over 200 species. The species of *Plantago* which produce mucilaginous seed coat are called psyllium. This psyllium is a commonly used dietary fiber in many parts of the world. The mucilage extracted from the seed coat comprises repeating unit of arabinose and xylose which are together called as arabixylans. These arabinoxylans are separated from the plant and used as corrosion inhibitors by corrosion researchers. The abundance of this natural polymer, its inexpensiveness, and ease of availability prompted the researchers to use this compound as a corrosion inhibitor of low-carbon steel in 1 M HCl. At a concentration of 1 g L^{-1} in 1 M HCl it can efficiently inhibit 94% of low-carbon steel corrosion [17]. The adsorption of this polysaccharide obeyed the Langmuir adsorption isotherm with a comprehensive mode of adsorption involving both physiosorption as well as chemisorption. But chemisorption was found to play a predominant role in the adsorption process.

9.2.13 Cellulose and its modified variants

A molecule of cellulose consists of innumerable, often thousands, of C, H, and O atoms (Fig. 9–3). It is responsible for imparting the stiffness to the plants. Nondigestible by

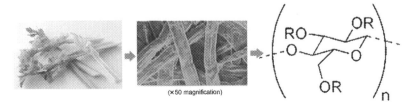

(×50 magnification)

FIGURE 9–3 Basic molecular structure of cellulose.

humans, it still forms an important dietary fiber for us. It is found in many forms and has many variants. For example the most commonly used is carboxymethyl cellulose or CMC.

9.2.13.1 Carboxymethyl cellulose

Researchers have studied the behavior of CMC on mild steel corrosion in 2 M sulfuric acid environment using gravimetric study, hydrogen evolution, and thermometric methods. The adsorption process of CMC obeyed the Langmuir and Dubinin–Radushkevich isotherm models. The adsorption studies strongly suggest that CMC physiosorbed on the mild steel surface. 500 ppm in corrosive sulfuric acid medium yielded an efficiency of 64.8% at 30°C [18].

9.2.13.2 Sodium carboxymethyl cellulose

In yet another study, with a little modification sodium carboxymethyl cellulose or Na-CMC was used as a corrosion inhibitor of mild steel in 1 M HCl solution [19]. The Na-CMC adsorbed on the metal by bridging via the hydroxyl groups and obeyed the Langmuir adsorption isotherm. At a weight concentration of 0.04%, an efficiency of 72% was achieved at 298K.

9.2.13.3 Hydroxyethyl cellulose

Another important cellulose derivative is HEC. Apart from general use as a thickening, binding, and gelling/stabilizing agent hydroxyethyl cellulose or HEC is also employed medicinally for dissolving the drugs in gastrointestinal fluids. Structurally it is similar to CMC except for the presence of the hydroxyethyl group in the place of carboxymethyl in the same position. The successful utility of this particular cellulose as a corrosion inhibitor in numerous media is drawn from its uniquely placed functional moieties (-OH, -COOH) on its cellulose backbone as well as a macro-sized chain molecule which ensures greater coverage of the metal surface, barring the degrading ions. This interesting variant of cellulose was not only used to protect the metal in neutral saline environment [20] but it was also efficiently utilized as a corrosion inhibitor of zinc carbon batteries [21]. In both cases HEC was more than 90% efficient as an inhibitor.

9.2.13.4 Hydroxypropyl cellulose

Yet another variant of cellulose, hydroxypropyl cellulose is very well-known to corrosion researchers studying the natural polymers. It was used for the acidic corrosion prevention of cast iron. The investigations were performed by gravimetric and electrochemical testing [22]. Another noteworthy addition to the research work was the addition of potassium iodide to the system which acted antagonistically and synergistically in a simultaneous fashion. Electrochemistry suggested a mixed type of adsorption and the adsorption analysis suggested it was obeying the Langmuir isotherm. The thermochemical parameters suggested a physisorbed film of inhibitors on the metal surface. Cast iron was 89.5% saved by 500 ppm hydroxypropyl cellulose at 298K.

9.2.13.5 Hydroxypropyl methyl cellulose

Another derivative of cellulose which could be effectively used as a corrosion inhibitor of mild steel in sulfuric acid environment was hydroxypropyl methyl cellulose or HPMC. Investigation techniques comprised weight analysis, impedance, and polarization calculations [23]. Potassium iodide was added to enhance the corrosion inhibition efficiency. The electrochemical results suggested that HPMS inhibited both cathodic and anodic partial reactions. Quantum chemical descriptors indicate effective adsorption of molecule on the metal surface. The adsorption process follows the Freundlich adsorption isotherm.

9.2.13.6 Ethyl hydroxyethyl cellulose or EHEC

This cellulose derivative was effectively applied for acid corrosion inhibition of mild steel corrosion in 1 M H_2SO_4 solution using weight loss calculations, EIS, PDP, and quantum chemical calculation techniques [24]. The IE increased with EHEC concentration and further on addition of KI. The effect of EHEC on corrosion of mild steel was attributed to the general adsorption of both protonated and molecular species of the additive on the cathodic and anodic sites. Formation of a chemisorbed film on the mild steel surface was observed. Electrochemical tests showed that EHEC and EHEC + KI were a mixed-type inhibitor with a predominant cathodic effect.

9.3 Starch and its derivatives

Starch is a carbohydrate macromolecule with multiple glucose units which are connected by glycoside bonds. Normally, starch consists of varying percentages by weight of amylase. When linear and helical the chain is amylose, and it is amylopectin when branched (Fig. 9—4A and B). Starch is the major source of energy for higher animals as it is derived from a plant source. They are sometimes processed as simple sugars, thickeners, and glues for general use. Starches have electron-rich hydroxyl groups which coordinate and complete the voids in the orbitals of metal substrates. Many research have used starch as corrosion inhibitors. Amongst them a notable study was conducted by Mobin et al. [25], using starch for corrosion inhibition of mild steel in H_2SO_4 by using gravimetric and electrochemical measurements. It was observed to inhibit 67% corrosion in 0.1 M HCl at 200 ppm concentration. With the synergistic influence of surfactants its efficiency increased. Similar extensive studies were conducted by other researchers too where modified starches, tapioca starch, and cassava starch were used as corrosion inhibitors [26—29]. In almost all the cases starch was found to be a cathodic inhibitor suppressing the cathodic reactions and yielding a high efficiency of protection.

9.3.1 Pectin

Pectins are heteropolysaccharides which are abundant in the cell walls of terrestrial non-woody vegetation. They are easily available commercially as powder or granules as they are already an integral part of the food industry. They may be primarily obtained from citrus

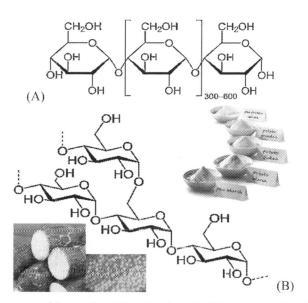

(A)

(B)

FIGURE 9–4 Molecular structures of the amylose (A) and amylopectin (B) molecules of starch.

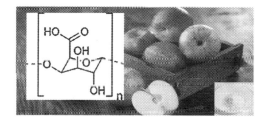

FIGURE 9–5 Molecular structure of pectate.

fruits and apples. The carboxyl and the carboxymethyl groups on its backbone increase its functionality not only as a corrosion inhibitor but also as a scale remover (Fig. 9–5).

Umoren et al. [30] used pectin to protect the X60 pipeline steel from hydrochloride attack. It was 98% efficient at 60°C at a concentration of 1000 ppm. Fares et al. [31] used pectin from citrus fruits as a corrosion inhibitor of aluminum in HCl. But the concentration reported for highest obtained IE was very high, that is, at $8 \, g \, L^{-1}$ it might inhibit 91% corrosion. Some researchers have also used lemon peel pectin for mild steel in 1 M HCl [32]. It proved itself to be a mixed type inhibitor, geometrically blocking the attack of hydrochloride by chemisorbing on the surface. The inhibition efficiency of 2000 ppm pectin at 298K was observed to be 90.3%, which further increased to 94.2% at 318K. Grassino et al. studied the effect of tomato pectin [33], while others studied pectin from *Opuntia* [34] as a corrosion inhibitor yielding considerably good efficiencies. Pectates obtained solely from very ripe fruits are already in industrial applications as emulsifying and foaming agents for food and medicine. In 2012 a researcher reported corrosion inhibition of pectates for aluminum in 4 M NaOH using gasometric and weight loss methods [35]. 88% of corrosion was evaded when 1.6% pectate was added to the corrosion system. The same researcher reconducted the work this time using sodium pectate for pure aluminum substrate in the same 4 M NaOH. The computed results showed very negligible difference in the results [36].

9.3.2 Chitosan

Chitosan is obtained either from the exoskeletons of marine crustaceans and mollusks or by *N*-deacetylation of fungal cell-wall chitin. Possessing antibacterial and antifungal properties it is widely used in cosmetics and skin therapeutics. Chitosan's anticorrosion ability could be drawn from its molecular structure (Fig. 9—6), which bears the electron-rich hydroxyl and amino groups in the structure of its molecules, which are capable of bonding to steel surface via coordinate bonds.

Umoren et al. [37] have reported the application of chitosan as a corrosion inhibitor for mild steel in 0.1 M HCl. The chemisorption of this polymer was found to accord with Langmuir adsorption isotherm. Applying this polymer to the corrosion system could protect up to 96% of the metal at 60°C and then drops further increasing the temperature. Increasing the chitosan concentration to 4 μM slightly elevated the IE again. At the same time El Haddad [55] studied the application of chitosan to protect Cu in 0.5 M HCl acid. The weight loss and electrochemical measurement resulted in the establishment of chitosan as a mixed kind of inhibitor with the maximum efficiency of 93% at 25°C when 8 μM of inhibitor was added. The quantum computations show that the N and O atoms in the chitosan molecule are the two main active sites that cause it to adsorb on the Cu surface. Cheng et al. [38] have modified this simple polymer to a carboxymethyl chitosan-Cu^{2+} ($CMCT-Cu^{2+}$) mixture for inhibiting mild steel corrosion in 1 M HCl and studied the process using gravimetric and electrochemical methods. On addition of CMCT and the mixture of Cu^{2+} + CMCT the corrosion effect was controlled. A complex formation occurred between Cu^{2+} ion and CMCT, which as an inhibitor was much more effective (91.9% IE) when compared to its building constituents. β cyclodextrin modified natural chitosan was used by Liu et al. [39] to inhibit carbon steel corrosion in 0.5 M HCl solution. β-CD-chitosan (β-cyclodextrin modified chitosan) acted as a mixed inhibitor with a maximum inhibition of 96.02% at a small concentration of 230 ppm. Sangeetha et al. [40] synthesized

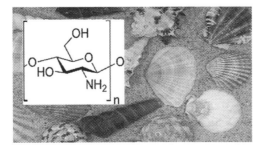

FIGURE 9–6 Molecular structure of chitosan.

O-fumaryl-chitosan for the inhibition of mild steel corrosion in 1 M HCl. A corrosion inhibition efficiency of 93.2% was observed at room temperature on adding 500 ppm of inhibitor. There are many other studies where chitosan derivatives were effectively employed as corrosion inhibitors [41].

9.3.3 Carrageenan

Carrageenan are a group of gelatinous linear polysaccharides having sulfated β-D-galactose and anhydrous-α-D-galactose backbone (Fig. 9−7).

The most common source of carrageenan is seaweeds. Bearing a flexible molecular symmetry and building up unstable helixes they are gels at room temperature. This gelling capacity makes them suitable as food thickeners and stabilizers. Fares et al. [42] reported the usage of i-carrageenan for inhibition of aluminum sheets in HCl medium at different concentrations. When pefloxacin mesylate was added as a mediator, it caused an improvement in the magnitude of IE, increasing it from 66.7% to 91.8%. SEM helped in the detection of the inhibitor mediator film on the surface. Zaafarany [43] also studied all the three available variants of carrageenan, i-, k-, and λ-carrageenan, as corrosion inhibitors of low-carbon steel in 1 M HCl. All these variants were established as being anodic-type inhibitors for steel in the acidic medium studied. The 500 ppm concentration of carrageenan displayed efficiencies of 76%, 72%, and 80%, respectively.

9.3.4 Dextrin

Dextrin is a carbohydrate having glucose (D) units linked by glycosidic bonds [α-(1→4) or α-(1→6)], as shown in Fig. 9−8.

Dextrins are present in the human digestive system as the hydrolysis product of starch upon action of amylases. An alternate way to produce them is synthesis by heat treatment in acidic solution. Numerous dextrins are naturally present, like α,β-dextrin, maltodextrin, amylodextrincyclic, and highly branched cyclic dextrin compounds. Researchers have reported dextrin to be an inhibitor of zinc-plated mild steel in HCl using weight loss and surface morphology tests [44]. In combination with thiourea additive, it has demonstrated improved protection, as examined by SEM/EDX. Other research has been performed to evaluate the use

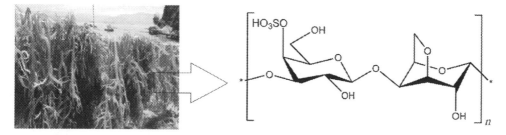

FIGURE 9–7 Molecular structure of carrageenan.

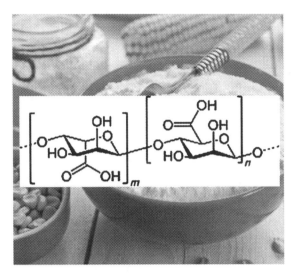

FIGURE 9–8 Molecular structure of dextrin.

of cyclodextrins for different metals and medium. The cyclodextrins have imparted high efficiencies in all cases, thus proving how successfully they inhibit the degradation of metals due to their unique structures and properties [45–48].

9.3.5 Alginates

Alginate are the sugars that have a carboxylic acid functional group attached to their molecular structure. Also known as "aligns" or "alginic acid" these are linear copolymers that have covalently bonded (1−4)-linked β-D-mannuronate and C-5 epimer β-L-guluronate homopolymeric blocks shown in Fig. 9−9.

This anionic polysaccharide is the main constituent of the hydrocolloids algal cell walls and seaweeds where it attaches with molecular water. Conventional corrosion testing along with surface analysis evaluated the potential of these alginate derivatives as inhibitors for carbon steel degradation in acidic solutions [49]. The results suggested an improvement in efficiency of these inhibitors on elevation of the solution temperature or their own concentration in the test solution. The electrochemical polarization suggested a mixed-type inhibition with predominant cathodic control. Their adsorption was approximated by Langmuir adsorption isotherm model. It is worth mentioning one of its recently studied derivatives, hydroxyl propyl alginate, as an inhibitor of mild steel corrosion in 1 M HCl at room temperature using chemical and electrochemical techniques. Corrosion inhibition efficiency was observed to improve with the increasing concentration of this compound. A physisorbed layer of molecules was observed on the metal surface which was further assured and confirmed by the results of SEM, AFM, and FTIR [50].

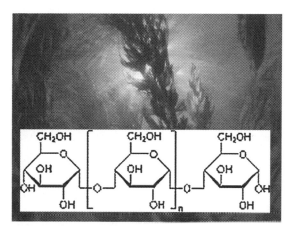

FIGURE 9–9 Molecular structure of alginates.

9.4 Conclusive remarks for the future and application of carbohydrate polymers in corrosion inhibition studies

It has already been noted by analyzing the previous research that a little modification in the structures of the carbohydrate polymers may lead to obtaining very high inhibition performances in very aggressive environments of high molarity mineral acid. The only minor shortcoming associated with these carbohydrate polymers as corrosion inhibitors is that few of them are partially soluble and have a lesser stability when kept for days, that is, after a set period of time they may not remain functional in a highly aggressive solution of 1 or 2 molar mineral acids. The chemical modifications should be aimed to increase the solubility of the carbohydrate polymers and increase their stability at high temperatures to reduce further the amount and concentration of inhibitor which is required. Considering the industrial and practical applications of corrosion inhibitors, the addition of polymers at concentrations ranging from 500 to 1000 ppm, repeatedly, may make the solutions turbid or they may even pose a risk to the efficiency of equipment. So the research that is trending currently is the application of modified carbohydrate polymers resulting in efficient protection of the metal surface at extremely low concentrations.

Another noteworthy aspect is the application of computational analysis to study properly the reaction pathways and adsorption mechanisms of these carbohydrate polymers when added as inhibitors. Sometimes a single molecule has more than one type of functionality attributed to it. Theoretical calculations are necessary to deduce all the possible points of bonding and the energies involved in the adsorption process. Not only the bonding sites and bonding energies are relevant to the corrosion studies, another very important aspect is the spatial orientation of the carbohydrate polymer molecules on the metal surfaces. It is worth mentioning that these polymers have lengthy molecular chains and high molecular weights, which researchers have often considered as one aspect behind the successful application of carbohydrate polymers as corrosion inhibitors. It is very important to study how these molecules align themselves on the metal substrate while being adsorbed on it. Some of the extensively studied carbohydrate polymers used as corrosion inhibitors are enlisted in Table 9−1.

The conventional gravimetric, electrochemical, and surface studies provide a clear view of binding occurring on the surface of the metal substrate, however, some of the inhibition may be attributed to the passivation of the metal. Sometimes a researcher keeps on adding the inhibitor to the solution and achieves high values of resistance from the electrochemical instruments, but when the sample surface is studied for the bonded inhibitor, the surface is devoid of any inhibitor molecules, only the passive layer constituents are detected. Referring to Pourbaix diagrams does help in setting up a demarcation between the passivation of the surface and inhibition by polymer but it would be best if all the changes are measured in real time by some precise and accurate electrochemical techniques. Darowicki was the first researcher to successfully apply a real-time corrosion monitoring method in electrochemical analysis [53]. Later many researchers started to apply it in various electroanalytical studies. Gerengi and his research group are some of the earliest researchers to work on using a technique that can monitor the corrosion processes occurring in the electrochemical cell in real time [54]. This accurate and precise technique, which is called dynamic electrochemical impedance spectroscopy (DEIS), can detect the changes occurring on the metal sample continuously in real time for a set period of time. With multifunctional structures and lengthy chain, carbohydrate polymers can be well studied and evaluated by combined DEIS and theoretical investigations to judge the exact process of inhibition occurring on the metal surfaces.

Table 9–1 Comparison of reported inhibition efficiency of some other corrosion inhibitors originated from natural products with present inhibitor.

Carbohydrate polymers	Substrate	Medium	Inhibitors conc. (ppm)	Temp. (°C)	Inhibition efficiency IE (%)	Reference
Guar gum	Carbon steel	1 M H_2SO_4 containing NaCl	1500	25	93.88	[1]
Mangrove tannin	Mild steel	3.5% NaCl	3000	25	90	[12]
Tapioca starch	AA6061 Alloy	Seawater	1000	25	96	[28]
Iota carrageenan	Aluminum	1 M, 1.5 M, 2 M HCl	1600	40	74.2	[42]
Gum acacia	Mild steel	1 M H_2SO_4	1500	30	91.71	[4]
Pectin	Carbon steel	1 M HCl	2000	45	94.2	[15]
Hydroxypropyl methylcellulose	Aluminum	0.5 M H_2SO_4	2000	30	63.5	[23]
Xanthan gum	A1020 carbon steel	1 M HCl	1000	30	83.17	[51]
Schinopsis lorentzii extract	Low-carbon steel	1 M HCl	2000	29	66	[52]

References

[1] M. Abdallah, Guar gum as corrosion inhibitor for carbon steel in sulfuric acid solutions, Port. Electrochim. Acta 22 (2004) 161–175.

[2] S.A. Umoren, Inhibition of aluminum and mild steel corrosion in acidic medium using Gum Arabic, Cellulose 15 (2008) 751–761.

[3] M.A. Abu-Dalo, A.A. Othman, N.A.F. Al-Rawashdeh, Exudate gum from acacia trees as green corrosion inhibitor for mild steel in acidic media, Int. J. Electrochem. Sci. 7 (2012) 9303–9324.

[4] M. Mobin, M.A. Khan, Investigation on the adsorption and corrosion inhibition behaviour of gum acacia and synergistic surfactant additives on mild steel in 1 M H_2SO_4, J. Disper. Sci. Technol. 34 (2013) 1496–1506.

[5] S.A. Umoren, Synergistic influence of gum arabic and iodide ion on the corrosion inhibition of aluminum in alkaline medium, Port. Electrochim. Acta 27 (2009) 565–577.

[6] I.O. Arukalam, C.O. Alaohuru, C.O. Ugbo, K.N. Jideofor, P.N. Ehirim, I.C. Madufor, Effect of xanthan gum on the corrosion protection of aluminum in HCl medium, Int. J. Adv. Res. Technol. 3 (2014) 5–15.

[7] P.O. Ameh, L. Magaji, T. Salihu, Corrosion inhibition and adsorption behaviour for mild steel by *Ficus glumosa* gum in H_2SO_4 solution, Afr. J. Pure Appl. Chem. 6 (2012) 100–106.

[8] N.O. Eddy, P.O. Ameh, A.O. Odiongenyi, Physicochemical characterization and corrosion inhibition potential of *Ficus benjamina* (FB) gum for aluminum in 0.1 M H_2SO_4, Port. Electrochim. Acta 32 (2014) 183–197.

[9] N.O. Eddy, A.O. Odiongenyi, P.O. Ameh, E.E. Ebenso, Corrosion inhibition potential of *Daniella oliverri* gum exudate for mild steel in acidic medium, Int. J. Electrochem. Sci. 7 (2012) 7425–7439.

[10] S. Banerjee, V. Srivastava, M.M. Singh, Chemically modified natural polysaccharide as green corrosion inhibitor for mild steel in acidic medium, Corros. Sci. 59 (2012) 35–41.

[11] H. Zhang, D. Wang, F. Wang, X. Jin, T. Yang, Z. Cai, et al., Corrosion inhibition of mild steel in hydrochloric acid solution by quaternary ammonium salt derivatives of corn stalk polysaccharide, Desalination 372 (2015) 57–66.

[12] A.A. Rahim, E. Rocca, E.J. Steinmetz, M.J. Kassim, Inhibitive action of mangrove tannins and phosphoric acid on pre-rusted steel via electrochemical methods, Corros. Sci. 50 (2008) 1546–1550.

[13] S.A. Umoren, I.B. Obot, E.E. Ebenso, N.O. Obi-Egbedi, The inhibition of aluminum corrosion in hydrochloric acid solution by exudate gum from *Raphia hookeri*, Desalination 247 (2009) 561–572.

[14] S.A. Umoren, I.B. Obot, E.E. Ebenso, N. Obi-Egbedi, Studies on the inhibitive effect of exudate gum from *Dacroydes edulis* on the acid corrosion of aluminum, Port. Electrochim. Acta 26 (2008) 199–209.

[15] S.A. Umoren, U.F. Ekanem, Inhibition of mild steel corrosion in H_2SO_4 using exudate gum from *Pachylobus edulis* and synergistic potassium halide additives, Chem. Eng. Commun. 197 (2010) 1339–1356.

[16] M. Mobin, M. Rizvi, L.O. Olasunkanmi, E.E. Ebenso, Biopolymer from tragacanth gum as a green corrosion inhibitor for carbon steel in 1 M HCl solution, ACS Omega 2 (2017) 3997–4008.

[17] M. Mobin, M. Rizvi, Polysaccharide from Plantago as a green corrosion inhibitor for carbon steel in 1 M HCl solution, Carbohydr. Polym. 160 (2017) 172–193.

[18] M.M. Solomon, S.A. Umoren, I.I. Udosoro, A.P. Udoh, Inhibitive and adsorption behaviour of carboxymethyl cellulose on mild steel corrosion in sulphuric acid solution, Corros. Sci. 52 (2010) 1317–1325.

[19] E. Bayol, A.A. Gürten, M. Dursun, K. Kayakırılmaz, Adsorption behavior and inhibition corrosion effect of sodium carboxymethyl cellulose on mild steel in acidic medium, Acta Phys. Chim. Sin. 24 (2008) 2236–2242.

[20] M.N. El-Haddad, Hydroxyethyl cellulose used as an eco-friendly inhibitor for 1018 c-steel corrosion in 3.5% NaCl solution, Carbohydr. Polym. 112 (2014) 595–602.

[21] M.A. Deyab, Hydroxyethyl cellulose as efficient organic inhibitor of zinc carbon battery corrosion in ammonium chloride solution: electrochemical and surface morphology studies, J. Power Sources 280 (2015) 190–194.

[22] V. Rajeswari, D. Kesavan, M. Gopiraman, P. Viswanathamurthi, Physicochemical studies of glucose, gellan gum, and hydroxypropyl celluloseInhibition of cast iron corrosion, Carbohydr. Polym. 95 (2013) 288–294.

[23] I.O. Arukalam, Durability and synergistic effects of KI on the acid corrosion inhibition of mild steel by hydroxypropyl methylcellulose, Carbohydr. Polym. 112 (2014) 291–299.

[24] I.O. Arukalam, I.O. Madu, N.T. Ijomah, C.M. Ewulonu, G.N. Onyeagoro, Acid corrosion inhibition and adsorption behaviour of ethyl hydroxyethyl cellulose on mild steel corrosion, J. Mater. 1 (2014) 1–11.

[25] M. Mobin, M.A. Khan, M. Parveen, Inhibition of mild steel corrosion in acidic medium using starch and surfactants additives, Appl. Polym. Sci. 121 (2011) 1558–1565.

[26] T. Brindha, J. Mallika, V.S. Moorthy, Synergistic effect between starch and substituted piperidin-4-one on the corrosion inhibition of mild steel in acidic medium, Mater. Environ. Sci. 6 (2015) 191–200.

[27] M. Bello, N. Ochoa, V. Balsamo, F. López-Carrasquero, S. Coll, A. Monsalve, et al., Modified cassava starches as corrosion inhibitors of carbon steel: an electrochemical and morphological approach, Carbohydr. Polym. 82 (2010) 561–568.

[28] R. Rosliza, W.B. Nik, Improvement of corrosion resistance of AA6061 alloy by tapioca starch in seawater, Curr. Appl. Phys. 10 (2010) 221–229.

[29] X. Li, S. Deng, Cassava starch graft copolymer as an eco-friendly corrosion inhibitor for steel in H_2SO_4 solution, Korean J. Chem. Eng. 32 (2015) 2347–2354.

[30] S.A. Umoren, I.B. Obot, A. Madhankumar, Z.M. Gasem, Performance evaluation of pectin as eco-friendly corrosion inhibitor for X60 pipeline steel in acid medium: experimental and theoretical approaches, Carbohydr. Polym. 124 (2015) 280–291.

[31] M.M. Fares, A.K. Maayta, M.M. Al-Qudah, Pectin as promising green corrosion inhibitor of aluminum in hydrochloric acid solution, Corros. Sci. 60 (2012) 112–117.

[32] M.V. Fiori-Bimbi, P.E. Alvarez, H. Vaca, C.A. Gervasi, Corrosion inhibition of mild steel in HCl solution by pectin, Corros. Sci. 92 (2015) 192–199.

[33] A.N. Grassino, J. Halambek, S. Djakovi, S.R. Brncic, M. Dent, Z. Grabari, Utilization of tomato peel waste from canning factory as a potential source for pectin production and application as tin corrosion inhibitor, Food Hydrocoll. 52 (2016) 265–274.

[34] N. Saidi, H. Elmsellem, M. Ramdani, A. Chetouani, K. Azzaoui, F. Yousfi, Using pectin extract as eco-friendly inhibitor for steel corrosion in 1 M HCl media, Der Pharm. Chem. 7 (2015) 87–94.

[35] I. Zaafarany, Corrosion inhibition of aluminum in aqueous alkaline solutions by alginate and pectate water-soluble natural polymer anionic polyelectrolytes, Port. Electrochim. Acta 30 (2012) 419–426.

[36] R. Hassan, I. Zaafarany, A. Gobouri, H. Takagi, A revisit to the corrosion inhibition of aluminum in aqueous alkaline solutions by water-soluble alginates and pectates as anionic polyelectrolyte inhibitors, Int. J. Corros. 30 (2013) 419–426.

[37] S.A. Umoren, M.J. Banera, T.A. Garcia, C.A. Gervasi, M.V. Mirıfico, Inhibition of mild steel corrosion in HCl solution using chitosan, Cellulose 20 (2013) 2529–2545.

[38] S. Cheng, S. Chen, T. Liu, X. Chang, Y. Yin, Carboxymethyl chitosan–Cu^{2+} mixture as an inhibitor used for mild steel in 1.0 M HCl, Electrochim. Acta 52 (2007) 5932–5938.

[39] Y. Liu, C. Zou, X. Yan, R. Xiao, T. Wang, M. Li, β-Cyclodextrin modified natural chitosan as a green inhibitor for carbon steel in acid solutions, Ind. Eng. Chem. Res. 54 (2015) 5664–5672.

[40] Y. Sangeetha, S. Meenakshi, C. Sundaram, Interactions at the mild steel acid solution interface in the presence of O-fumaryl-chitosan: electrochemical and surface studies, Carbohydr. Polym. 136 (2016) 38—45.

[41] R.R. Mohamed, A.M. Fekry, Antimicrobial and anticorrosive activity of adsorbents based on chitosan Schiff's base, Int. J. Electrochem. Sci. 6 (2011) 2488—2508.

[42] M.M. Fares, A.K. Maayta, J.A. Al-Mustafa, Corrosion inhibition of iotacarrageenan natural polymer on aluminum in presence of zwitterions mediator in HCl media, Corros. Sci. 65 (2012) 223—230.

[43] I. Zaafarany, Inhibition of acidic corrosion of iron by some carrageenan compounds, Curr. World Environ. 1 (2006) 101—108.

[44] C.A. Loto, R.T. Loto, Effect of dextrin and thiourea additives on the zinc electroplated mild steel in acid chloride solution, Int. J. Electrochem. Sci. 8 (2013) 12434—12450.

[45] C. Zou, X. Yan, Y. Qin, M. Wang, Y. Liu, Inhibiting evaluation of βCyclodextrin-modified acrylamide polymer on alloy steel in sulfuric acid solution, Corros. Sci. 85 (2014) 445—454.

[46] X. Yan, C. Zou, Y. Qin, A new sight of water soluble polyacrylamide modified by β-cyclodextrin as corrosion inhibitor for X70 steel, Starch/Stärke 66 (2014) 968—975.

[47] B. Fan, G. Wei, Z. Zhang, N. Qiao, Preparation of supramolecular corrosion inhibitor based on hydroxypropyl-bcyclodextrin/octadecylamine and its anticorrosion properties in the simulated condensate water, Anti-Corros. Methods Mater. 61 (2014) 104—111.

[48] Y. Liu, C. Zou, C. Li, L. Lin, W. Chen, Evaluation of β-cyclodextrin polyethylene glycol as green scale inhibitors for produced-water in shale gas well, Desalination 377 (2016) 28—33.

[49] S.M. Tawfik, Alginate surfactant derivatives as eco friendly corrosion inhibitor for carbon steel in acidic environment, RSC Adv. 5 (2015) 104535—104550.

[50] Y. Sangeetha, S. Meenakshi, C.S. Sundaram, Investigation of corrosion inhibitory effect of hydroxyl propyl alginate on mild steel in acidic media, J. Appl. Polym. Sci. 133 (2016) 43004—43010.

[51] M. Mobin, M. Rizvi, Inhibitory effect of xanthan gum and synergistic surfactant additives for mild steel corrosion in 1 M HCl, Carbohydr. Polym. 136 (2016) 384—393.

[52] H. Gerengi, H.I. Sahin, Schinopsis lorentzii extract as a green corrosion inhibitor for low carbon steel in 1 M HCl solution, Ind. Eng. Chem. Res. 51 (2012) 780—787.

[53] K. Darowicki, Theoretical description of the measuring method of instantaneous impedance spectra, J. Electroanal. Chem. 486 (2000) 101—105.

[54] H. Gerengi, P. Slepski, E. Ozgan, M. Kurtay, Investigation of corrosion behavior of 6060 and 6082 aluminum alloys under simulated acid rain conditions, Mater. Corros. 66 (2015) 233—240.

[55] M.N. El-Haddad, Chitosan as a green inhibitor for copper corrosion in acidic medium, Int. J. Biol. Macromol. 55 (2013) 142—149.

10

Chitosan and its derivatives as environmental benign corrosion inhibitors: Recent advancements

Chandrabhan Verma[1], Chaudhery Mustansar Hussain[2]

[1]INTERDISCIPLINARY RESEARCH CENTER FOR ADVANCED MATERIALS, KING FAHD UNIVERSITY OF PETROLEUM AND MINERALS, DHAHRAN, SAUDI ARABIA [2]DEPARTMENT OF CHEMISTRY AND ENVIRONMENTAL SCIENCE, NEW JERSEY INSTITUTE OF TECHNOLOGY, NEWARK, NJ, UNITED STATES

Chapter outline

Abbreviations

WL	Weight loss
HE	Hydrogen evolution
GDP	Gross domestic product
%IE	Percentage inhibition efficiency
C_R	Corrosion rate
EIS	Electrochemical impedance spectroscopy
PDP	Potentiodynamic polarization
LPR	Linear polarization resistance

Environmentally Sustainable Corrosion Inhibitors. DOI: https://doi.org/10.1016/B978-0-323-85405-4.00016-1

θ	Surface coverage
SEM	Scanning electron microscopy
EDX	Energy dispersive X-ray
XRD	X-ray powder diffraction
XPS	X-ray photoelectron spectroscopy
AFM	Atomic force microscopy
DFT	Density functional theory
MDS	Molecular dynamics simulation
MCS	Monte Carlo simulation
REACH	Registration, Evaluation, Authorization, and Restriction of the chemical
OSPAR	Oslo Paris Commission
LC$_{50}$	Lethal concentration
EC$_{50}$	Effective concentration
log K$_{OW}$ or D^{OW}	Partition coefficient
MCRs	Multicomponent reactions
MW	Microwave
US	Ultrasound
CH	TS-Cht: Chitosan

10.1 Introduction

10.1.1 Corrosion and its mitigation

Corrosion is defined as decomposition of metallic materials by electrochemical (or chemical) reactions with the environment. Corrosion has become a worldwide problem as it causes about 3%–5% loss of global GDP [1,2]. Apart from economic losses, several accidents have also been reported worldwide that have caused numerous safety and fatality losses [3]. The problem of corrosion becomes more pronounced during metallurgical acid cleaning of the metallic structures and ores where a large amount of metal is dissolved by the concentrated acidic solutions. In view of big losses associated with corrosion, numerous methods have been developed depending on the nature of the metal and environment [4,5]. The use of synthetic compounds is one of the most effective and economic methods of corrosion mitigation [6,7]. Several classes of compounds that can be broadly categorized into inorganic and organic compounds are used as effective corrosion inhibitors [8,9]. However, the implementation of inorganic compounds as corrosion inhibitors is restricted because of their toxic nature that adversely affect the aquatic and soil life [10]. Nevertheless, organic compounds are established as the most effective and economic alternatives for corrosion mitigation. They show high efficiency against metallic corrosion and can be synthesized effectively and economically at a high yield. Their high efficiency against metallic corrosion is attributed to their ability to form stable surface protective complexes through coordination bonding. They contain several electron-rich centers including homo-atomic such as $-N{=}N-$, $>C{=}C<$, $-C{\equiv}C-$, etc. and hetero-atomic $>C{=}O$, $>C{=}N-$, $>C{=}S$, $-N{=}O$ and $-C{\equiv}N$, etc. multiple bonds, through which they interact and adsorb on the metallic surface. The presence of polar functional moieties, such as $-OH$, $-OMe$, $-NH_2$, $-SH$, $-NO_2$, $-COOH$, and $-CN$, also enhance the inhibition efficiency of organic compounds [11,12]. The heteroatoms of these functional groups transfer their nonbonding electrons into the d-orbitals of surface metallic atoms and form coordination bonding [13,14]. The presence of these moieties also enhances the solubility of organic compounds in polar electrolytes. Several classes of organic compounds such as heterocyclics [7,15], polymers [12,16], macromolecules [17], and oligomers [18] are used as corrosion inhibitors.

10.1.2 Adsorption mechanism of corrosion mitigation

It is well established that organic compounds retards metal corrosion by adsorbing on the metallic surface. In fact, organic inhibitors adsorb at the interface of metal and electrolytes to avoid the contact of metal with its environment. The adsorption of organic compounds over a metallic surface results due to the opposite potential developed over the metallic surface. Adsorption of the organic inhibitors depends upon numerous factors [19,20]. Generally, a rise in temperature causes a reduction in adsorption efficiency as at elevated temperatures increased kinetic energy decreases the interactions with metallic surface [21,22]. Furthermore, depending upon the nature of the electrolytes they may undergo a catalytic decomposition and/or degradation that can adversely affect the inhibition efficiency of the organic inhibitors [11]. It is estimated that an increase in temperature by $10°C$ can cause a rise in corrosion rate (CR) by two times [23,24]. The nature and effectiveness of adsorption of organic inhibitors on the metallic surface can be explained on the basis of the adsorption isotherm model [25,26]. Among the different commonly used isotherm models the Langmuir isotherm model is best fitted and it can be presented as [25,26]:

$$\frac{C}{\theta} = \frac{1}{K_{ads}} + C \tag{10.1}$$

$$\log\left(\frac{\theta}{1-\theta}\right) = \log C + \log K_{ads} \tag{10.2}$$

where C, K_{ads}, and θ represent the inhibitor concentration, adsorption constant, and surface coverage, respectively. Using K_{ads} values, the magnitude of Gibb's free energy for adsorption (ΔG_{ads}) can be derived for inhibitors adsorption. Obviously, the value of ΔG_{ads} indicates the nature of adsorption as its negative value is consistent with the spontaneous nature of adsorption and vice versa. Furthermore, ΔG_{ads} value also gives information about the mode (mechanism) of adsorption. Generally, its value of -20 kJ mol^{-1} or less negative (or more positive) is consistent with physisorption mechanism and its value of -40 kJ mol^{-1} or more negative is consistent with chemisorption mechanism [27]. A literature study reveals that the adsorption of most of the organic compounds obeyed the mixed (physiochemisorption) mode of adsorption [25,27].

10.1.3 Techniques of corrosion testing

A literature survey shows that several experimental and computational methods have been used to investigate the corrosion inhibition efficiency of inhibitors [28,29]. Every method has its own advantages and limitations. The weight loss (WL), gravimetric, or mass loss method is the most frequently used method for testing the inhibition effect of inhibitors. The extensive use of this method is based its connection with several advantages including simplicity, high reproducibility, ease to perform, and high accuracy. Furthermore, the WL method can be considered as one of the most cost-effective methods of as it does not require the use of any expensive instruments or computer-based software [30,31]. The WL technique is a nondestructive and qualitative method that gives information about the CR. WL measurements give useful information in the form of several parameters, including CR (C_R), inhibition efficiency (%IE), and surface coverage (θ). Obviously, an increase in %IE and decrease in C_R is consistent with high inhibition efficiency. Apart from WL, electrochemical studies including electrochemical impedance spectroscopy (EIS) and potentiodynamic polarization (PDP) give information about the electrochemical nature of the inhibitors [32,33]. EIS study shows that most of the previously investigated compounds acted as interface-type corrosion inhibitors, that is, they adsorb and increase the barrier for the charge transfer process. PDP study reveals that most of the previously studied compounds behaved as mixed-type corrosion inhibitors. The surface morphological studies, such as SEM, EDX, XRD, XPS, and AFM, are widely used to demonstrate the inhibition effectiveness of corrosion inhibitors [34,35].

Recently, the implementation of computational modeling, specially using DFT, Monte Carlo (MC), and molecular dynamics (MD) simulations, has been developed as an effective tool for the measurement of the anticorrosive effect of corrosion inhibitors [36−38]. These techniques provide some vital information through which corrosion inhibition effectiveness of a compound can be described. One of the greatest advantages of computational simulations is that the implementation of these techniques is a cost-effective and eco-friendly approach as these techniques do not involve the use of any expensive instruments or toxic chemicals. These techniques also provide information about the nature of interactions taking place between the metal surface and inhibitor molecule. Computational techniques, especially MD and MC simulations, provide pictorial orientation of inhibitor molecules over the metallic surface. Generally, a molecule with a flat orientation would be a better corrosion inhibitor compared with compounds having a vertical orientation. Common experimental and computational methods of corrosion monitoring are diagrammatically presented in Fig. 10−1.

10.1.4 Environmental sustainable corrosion inhibitors

Green or sustainable chemistry or process is defined as the nonpolluting and safe manufacturing in high yield or preventing pollution through superior synthetic design [39−41]. Nowadays, greenness is an essential requirement for chemicals to be used for various industrial and household applications including corrosion inhibition. According to REACH (Registration, Evaluation, Authorization and Restriction of the chemical) and the Oslo Paris Commission (OSPAR), biodegradable and nonbioaccumulative chemicals that have very low or zero environmental toxicity are considered as green or environment-friendly [42,43]. Toxicity of any chemicals can be measured using EC_{50} and LC_{50} values [44]. EC reflects the effective concentration and EC_{50} is the concentration of chemical that adversely affects the growth of 50% of a living population, whereas LC reflects the lethal

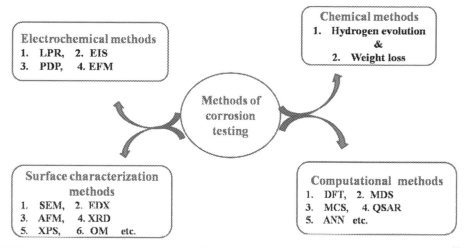

FIGURE 10–1 Experimental and computational methods of corrosion monitoring.

concentration and LC_{50} is the concentration of chemical that would kill 50% of the living population. It is estimated that for a chemical to be green, EC_{50}/LC_{50} value should be greater than 10 mg L^{-1} [45]. According to OSPAR, an acceptable biodegradation level for any chemical is 60% in 28 days [46,47]. Bioaccumulation is deliberate in partition coefficient (log K_{OW} or D^{OW}) which is a gage of distribution of the chemical in water and octanol mixture. For a chemical to be green, log K^{OW} should be less than 3 [48].

Inorganic inhibitors (mostly chromates) were used alone as the earliest, versatile, and most effective corrosion inhibitors for ferrous and nonferrous alloys because of their cost-effectiveness and versatility. Mostly, inorganic inhibitors act as an anodic or passivator type of inhibitor and for better effectiveness a critical concentration is highly required, as below the critical concentration chromate causes pitting corrosion [49,50]. Inorganic inhibitors are recognized as the "gold standard" in corrosion inhibition, however now their use has been phased out because of their toxicity and expensive discharge [51]. The increasing ecological sensitivity and awareness restrict the consumption of traditional toxic inorganic and organic corrosion inhibitors [52]. Therefore nowadays, corrosion scientists and engineers are frequently paying attention to the expansion of eco-friendly and sustainable alternatives of either artificial or natural origins. In this course, one-step multicomponent reactions (MCRs), during which three or more reactants combined in a single step, have been seen as one of the greenest protocols for the development of environmentally benign alternative corrosion inhibitors [53]. Obviously, MCRs are interrelated by means of quite a lot of profitable features, including high synthetic yield, ease of operation, lower number of purification and workup steps, particularly in catalyst and solvent-free situations. A literature study suggested that numerous green inhibitors derived using MCRs are successfully employed as effective corrosion inhibitors for metals and alloys. A detailed description of compounds synthesized using MCRs as green corrosion inhibitors are published elsewhere [53]. Nowadays, chemicals synthesized using microwave (MW) and ultrasound (US) irradiations, especially in the

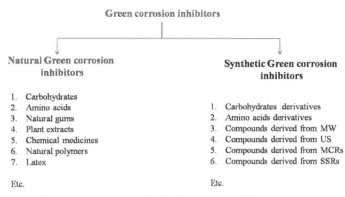

FIGURE 10–2 Classification of green or environment-friendly corrosion inhibitors.

association with MCRs, are considered as another green protocol as these nonconventional heating techniques are allied by means of plentiful advantageous uniqueness together with instantaneous heating of reaction mixture, generation of high temperature, and high chemical selectivity [3]. Owing to the association of MW and US irradiations with environmental friendliness, several classes of heterocyclic compounds are synthesized and evaluated as effective corrosion inhibitors.

Chemical syntheses using biomass and biologically originated chemicals such as amino acids and carbohydrates and their derivatives can also be regarded as green corrosion inhibitors [12]. Apart from these, water, supercritical carbon dioxide and ionic liquids can be considered as environment-friendly solvents, therefore chemicals synthesized using the solvents are regarded as green chemicals and can be used as nontoxic alternatives for traditional corrosion inhibitors. Even, a lot of ionic liquids, especially imidazolium based are extensively used as environmentally benign corrosion inhibitors for metals and alloys. A detail documentation on ILs as green alternatives for traditional toxic corrosion inhibitors can be found elsewhere [54]. The high thermal and chemical stability, high solubility in polar medium, low vapor pressure, and low volatility make the ILs a suitable alternative source of corrosion inhibitors. Chemical medicines (drugs) and plant extracts can also be thought as environmentally benign alternative sources for metallic corrosion inhibition because of their natural and/or biological origin [55,56]. The environment-friendly or green corrosion inhibitors can be classified as natural and synthetic types (Fig. 10−2). There are numerous examples of each kind of environmental alternatives.

10.2 Chitosan and its derivatives as corrosion inhibitors: Literature survey

Chitosan (CH) is a linear polysaccharide which consists of a random distribution of D-glucosamine and N-acetyl-D-glucosamine joined together with 1−4-glycosidic linkage. It is mainly distributed in crustacean shells. It is chemically synthesized using deacetylation of

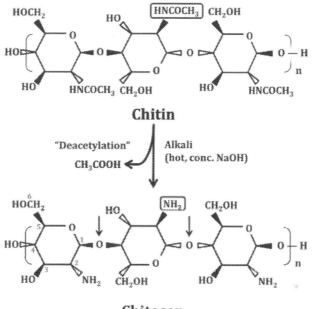

Chitin

"Deacetylation"

CH$_3$COOH ⟵   Alkali (hot, conc. NaOH)

Chitosan

FIGURE 10–3 Conversion of chitin into chitosan.

chitin (Fig. 10−3). Owing to the environment-friendly behavior of CH and its derivatives several studies dealing with the anticorrosive effect of CH and its derivatives based on organic compounds are reported.

Recently, Mouaden and coworkers [57] demonstrated the inhibition effect of CH on copper corrosion in synthetic seawater containing 20 ppm sulfide using chemical, electrochemical, and surface analyses methods. Results showed that CH showed optimum inhibition effectiveness of 89% at 800 ppm concentration. Polarization measurements showed that CH behaved as a mixed-type inhibitor. CH inhibits corrosion by adsorbing on the metallic surface following the physisorption mechanism. EIS studies revealed that CH adsorbed at the metal−electrolyte interface and increased the resistance for the charge transfer process, thereby it acts as an interface-type corrosion inhibitor. Gupta et al. [58] studied the anticorrosive effect of CH for mild steel corrosion in 1 M sulfamic acid using several methods. The synergistic effect of KI at 5 ppm concentration was investigated on the inhibition property of CH. Results showed that at 200 ppm concentration CH showed 73.5% inhibition efficiency, whereas at the same concentration in the presence of 5 ppm KI CH showed inhibition efficiency of more than 90%. CH with and without KI behaves as a mixed and interface-type corrosion inhibitor. SEM and EDX analyses were undertaken to describe the adsorption behavior of CH on metallic surface in sulfamic acid medium.

Our research team [17] synthesized Schiff's base, designated as SCSB, from salicylaldehyde and CH and evaluated it as an anticorrosive material for carbon steel in 3.5% NaCl solution at 65°C using several methods. Analyses showed that SCSB acted as a mixed-type inhibitor. The SCSB showed the highest protection power of 95.2% at 150 mg L^{-1}

concentration. Adsorption mechanism of SCSB followed the Langmuir adsorption isotherm. The adsorption mechanism of SCSB was supported by several surface study methods including SEM, EDX, XPD, and SKP techniques. Corrosion inhibition properties of PEG cross-linked CH for mild steel corrosion in 1 M HCl is reported elsewhere [59]. The corrosion inhibition properties of CH cross-linked with thiocarbohydrazide (TS-Cht) for stainless steel in 3.5% NaCl medium was reported recently [60]. Polarization and EIS studies showed that TS-Cht acted as a mixed- and interface-type corrosion inhibitor, respectively. The TS-Cht inhibits metallic dissolution by an adsorption mechanism that followed the Langmuir isotherm model. The TS-Cht showed more than 94% inhibition effectiveness at 500 mg L^{-1} concentration. Adsorption of TS-Cht on metallic surface was supported by SEM and EDX analyses. EIS and PDP studies were performed after establishing the open circuit potential.

Polyaspartic acid/chitosan complex (PASP/CS) was synthesized, characterized, and tested as a corrosion inhibitor for A3 carbon steel in 3.5% NaCl medium [61]. At 8 mg L^{-1} concentration PASP showed the optimum efficiency of only 58.8% that increased to 83.5% in the presence of chitosan (PASP/CH). A study showed that PASP/CH acted as an anodic type corrosion inhibitor and it formed an inhibitive film through its adsorption at the interface of metal and electrolyte. SEM analysis confirms the adsorption mechanism of corrosion inhibition. Chauhan et al. [62] modified CH using 4-amino-5-methyl-1,2,4-triazole-3-thio to form a compound designated as CH-AMT. CH-AMT was evaluated as a corrosion inhibitor for MS in 1 M HCl using chemical (WL), electrochemical, and surface analyses methods. CH-AMT showed more than 95% efficiency at 200 mg L^{-1} concentration. The experimental results were reinforced with computational studies carried out using DFT and MD simulations methods. Polarization results showed that CH-AMT acts as a mixed-type inhibitor with slight cathodic supremacy. CH-AMT inhibits metallic corrosive dissolution by an adsorption mechanism that obeyed the Langmuir adsorption isotherm model. DFT studies showed that CH-AMT has a very small energy bandgap, therefore it showed high chemical reactivity and thus inhibition effectiveness. MD simulations studies revealed that CH-AMT showed strong bonding with a metallic surface. A polyaniline/chitosan composite (PANI/CH) was developed and tested as a corrosion inhibitor for mild steel in 0.5 M HCl [63]. Polarization analysis revealed that PANI/CH acted as a mixed-type corrosion inhibitor. Adsorption of PANI/CH on the metallic surface was supported by an SEM study. DFT studies showed that PANI/CH interacts using donor—acceptor interactions.

Our research group developed three CH-based environment-friendly Schiff's bases (CSBs) using different aromatic aldehydes and evaluated them as corrosion inhibitors for mild steel in acidic medium [64]. The environment-friendly nature of the CSBs is based on the fact that they were synthesized using energy-efficient MW irradiation. Results showed that the inhibition efficiency of the CSBs varies depending upon the nature of the aldehyde. Among the tested CSBs, CBS-3 showed the highest protection power of 90.65% at 50 ppm concentration. All evaluated CSBs behaved as mixed-type corrosion inhibitors. Adsorption of the CSBs over a metallic surface obeyed the Langmuir adsorption isotherm model. The adsorption mechanism of corrosion inhibition was supported by SEM-EDX analyses. DFT and MD simulations analyses were undertaken to demonstrate the nature of adsorption and mechanism of interactions between

CSBs and metallic surface. In another study, our research group demonstrated the anticorrosive effect of modified CH using thiosemicarbazide and thiocarbohydrazide for mild steel in 1 M HCl [65]. The inhibitors were designated as CH-TS and CH-TCH, respectively. The synthesized CH-based compounds were characterized using FT-IR and ^{1}H NMR methods. Both tested compounds showed more than 90% efficiency at 50 ppm concentration. Adsorption of CH-TS and CH-TCH on metallic surface obeyed the Langmuir adsorption isotherm model. CH-TS and CH-TCH acted as predominantly cathodic type inhibitors. AFM analyses were undertaken to demonstrate the adsorption of CH-TS and CH-TCH on metallic surface. DFT studies well supported the experimental results. Inhibition effectiveness of the CH derivatives has also been reported upon extensively in other reports [66–71].

Useful links

https://linkinghub.elsevier.com/retrieve/pii/S0144861715012163.
https://www.degruyter.com/view/journals/zna/68/8-9/article-p581.xml.
https://pubs.rsc.org/ko/content/articlehtml/2016/ra/c6ra24026g.

Conflict of interest

The authors declare that there is no conflict of interest in any way.

Author's contribution

The authors collectively designed and wrote the chapter.

References

[1] C. Verma, M. Quraishi, H. Lgaz, L. Olasunkanmi, E.-S.M. Sherif, R. Salghi, et al., Adsorption and anticorrosion behaviour of mild steel treated with 2-((1H-indol-2-yl) thio)-6-amino-4-phenylpyridine-3, 5-dicarbonitriles in a hydrochloric acid solution: experimental and computational studies, J. Mol. Liq. 283 (2019) 491–506.

[2] V. Srivastava, J. Haque, C. Verma, P. Singh, H. Lgaz, R. Salghi, et al., Amino acid based imidazolium zwitterions as novel and green corrosion inhibitors for mild steel: experimental, DFT and MD studies, J. Mol. Liq. 244 (2017) 340–352.

[3] C. Verma, M. Quraishi, E.E. Ebenso, Microwave and ultrasound irradiations for the synthesis of environmentally sustainable corrosion inhibitors: an overview, Sustain. Chem. Pharm. 10 (2018) 134–147.

[4] Z. Foroulis, Corrosion and corrosion inhibition in the petroleum industry, Mater. Corros./Werkstoffe Korros. 33 (1982) 121–131.

[5] G.L. Mendonça, S.N. Costa, V.N. Freire, P.N. Casciano, A.N. Correia, P. de Lima-Neto, Understanding the corrosion inhibition of carbon steel and copper in sulphuric acid medium by amino acids using electrochemical techniques allied to molecular modelling methods, Corros. Sci. 115 (2017) 41–55.

[6] L.M. Rodríguez-Valdez, A. Martínez-Villafañe, D. Glossman-Mitnik, Computational simulation of the molecular structure and properties of heterocyclic organic compounds with possible corrosion inhibition properties, J. Mol. Struct. THEOCHEM 713 (2005) 65−70.

[7] A. Singh, M. Talha, X. Xu, Z. Sun, Y. Lin, Heterocyclic corrosion inhibitors for J55 steel in a sweet corrosive medium, ACS Omega 2 (2017) 8177−8186.

[8] P. Marcus, Corrosion Mechanisms in Theory and Practice, CRC Press, 2011.

[9] M.G. Fontana, Corrosion Engineering, Tata McGraw-Hill Education, 2005.

[10] A. Peter, I. Obot, S.K. Sharma, Use of natural gums as green corrosion inhibitors: an overview, Int. J. Ind. Chem. 6 (2015) 153−164.

[11] C. Verma, L.O. Olasunkanmi, E.E. Ebenso, M.A. Quraishi, I.B. Obot, Adsorption behavior of glucosamine-based, pyrimidine-fused heterocycles as green corrosion inhibitors for mild steel: experimental and theoretical studies, J. Phys. Chem. C 120 (2016) 11598−11611.

[12] S.A. Umoren, U.M. Eduok, Application of carbohydrate polymers as corrosion inhibitors for metal substrates in different media: a review, Carbohydr. Polym. 140 (2016) 314−341.

[13] G. Bahlakeh, M. Ramezanzadeh, B. Ramezanzadeh, Experimental and theoretical studies of the synergistic inhibition effects between the plant leaves extract (PLE) and zinc salt (ZS) in corrosion control of carbon steel in chloride solution, J. Mol. Liq. 248 (2017) 854−870.

[14] O. Abiola, N. Oforka, E. Ebenso, N. Nwinuka, Eco-friendly corrosion inhibitors: the inhibitive action of Delonix Regia extract for the corrosion of aluminium in acidic media, Anti-Corros. Methods Mater. 54 (4) (2007) 219−224.

[15] M.A. Quraishi, D.S. Chauhan, V.S. Saji, Heterocyclic Organic Corrosion Inhibitors: Principles and Applications, Elsevier, 2020.

[16] S. Umoren, Polymers as corrosion inhibitors for metals in different media—a review, Open Corros. J. 2 (2009).

[17] K. Ansari, D.S. Chauhan, M. Quraishi, M.A. Mazumder, A. Singh, Chitosan Schiff base: an environmentally benign biological macromolecule as a new corrosion inhibitor for oil & gas industries, Int. J. Biol. Macromol. 144 (2020) 305−315.

[18] P.P. Deshpande, N.G. Jadhav, V.J. Gelling, D. Sazou, Conducting polymers for corrosion protection: a review, J. Coat. Technol. Res. 11 (2014) 473−494.

[19] F. Bentiss, M. Traisnel, M. Lagrenee, Influence of 2,5-bis(4-dimethylaminophenyl)-1,3,4-thiadiazole on corrosion inhibition of mild steel in acidic media, J. Appl. Electrochem. 31 (2001) 41−48.

[20] T. Zhao, G. Mu, The adsorption and corrosion inhibition of anion surfactants on aluminium surface in hydrochloric acid, Corros. Sci. 41 (1999) 1937−1944.

[21] I. Obot, N. Obi-Egbedi, S. Umoren, Antifungal drugs as corrosion inhibitors for aluminium in 0.1 M HCl, Corros. Sci. 51 (2009) 1868−1875.

[22] S. Umoren, E. Ebenso, The synergistic effect of polyacrylamide and iodide ions on the corrosion inhibition of mild steel in H_2SO_4, Mater. Chem. Phys. 106 (2007) 387−393.

[23] G. Burstein, C. Liu, R. Souto, The effect of temperature on the nucleation of corrosion pits on titanium in Ringer's physiological solution, Biomaterials 26 (2005) 245−256.

[24] G. Glass, C. Page, N. Short, Factors affecting the corrosion rate of steel in carbonated mortars, Corros. Sci. 32 (1991) 1283−1294.

[25] I. Obot, N. Obi-Egbedi, S. Umoren, Adsorption characteristics and corrosion inhibitive properties of clotrimazole for aluminium corrosion in hydrochloric acid, Int. J. Electrochem. Sci. 4 (2009) 863−877.

[26] E. Ituen, O. Akaranta, A. James, Evaluation of performance of corrosion inhibitors using adsorption isotherm models: an overview, Chem. Sci. Int. J. (2017) 1−34.

[27] L. Nnanna, I. Anozie, A. Avoaja, C. Akoma, E. Eti, Comparative study of corrosion inhibition of aluminium alloy of type AA3003 in acidic and alkaline media by Euphorbia hirta extract, Afr. J. Pure Appl. Chem. 5 (2011) 265–271.

[28] J. Bhawsar, P. Jain, P. Jain, Experimental and computational studies of *Nicotiana tabacum* leaves extract as green corrosion inhibitor for mild steel in acidic medium, Alex. Eng. J. 54 (2015) 769–775.

[29] D. Daoud, T. Douadi, H. Hamani, S. Chafaa, M. Al-Noaimi, Corrosion inhibition of mild steel by two new S-heterocyclic compounds in 1 M HCl: experimental and computational study, Corros. Sci. 94 (2015) 21–37.

[30] M. Abdallah, Rhodanine azosulpha drugs as corrosion inhibitors for corrosion of 304 stainless steel in hydrochloric acid solution, Corros. Sci. 44 (2002) 717–728.

[31] A. Al-Sarawy, A. Fouda, W.S. El-Dein, Some thiazole derivatives as corrosion inhibitors for carbon steel in acidic medium, Desalination 229 (2008) 279–293.

[32] F. Ijsseling, Electrochemical methods in crevice corrosion testing: report prepared for the European Federation of Corrosion Working Party 'Physico-chemical testing methods of corrosion: fundamentals and applications', Br. Corros. J. 15 (1980) 51–69.

[33] J. Vega, H. Scheerer, G. Andersohn, M. Oechsner, Experimental studies of the effect of Ti interlayers on the corrosion resistance of TiN PVD coatings by using electrochemical methods, Corros. Sci. 133 (2018) 240–250.

[34] A.S. Hamdy, D. Butt, A. Ismail, Electrochemical impedance studies of sol–gel based ceramic coatings systems in 3.5% NaCl solution, Electrochim. acta 52 (2007) 3310–3316.

[35] A.S. Hamdy, Advanced nano-particles anti-corrosion ceria based sol gel coatings for aluminum alloys, Mater. Lett. 60 (2006) 2633–2637.

[36] I. Obot, D. Macdonald, Z. Gasem, Density functional theory (DFT) as a powerful tool for designing new organic corrosion inhibitors. Part 1: an overview, Corros. Sci. 99 (2015) 1–30.

[37] D.K. Verma, Density functional theory (DFT) as a powerful tool for designing corrosion inhibitors in aqueous phase, Adv. Eng. Test. (2018) 87.

[38] C. Verma, H. Lgaz, D. Verma, E.E. Ebenso, I. Bahadur, M. Quraishi, Molecular dynamics and Monte Carlo simulations as powerful tools for study of interfacial adsorption behavior of corrosion inhibitors in aqueous phase: a review, J. Mol. Liq. 260 (2018) 99–120.

[39] J.L. Tucker, Green chemistry, a pharmaceutical perspective, Org. Process. Res. Dev. 10 (2006) 315–319.

[40] P. Glavič, R. Lukman, Review of sustainability terms and their definitions, J. Clean. Prod. 15 (2007) 1875–1885.

[41] D.J. Constable, A.D. Curzons, L.M.F. dos Santos, G.R. Geen, R.E. Hannah, J.D. Hayler, et al., Green chemistry measures for process research and development, Green Chem. 3 (2001) 7–9.

[42] E. Donchevskaya, REACH, Registration, Evaluation, Authorization and Restriction of Chemicals, Видавництво СумДу, 2009.

[43] J. Killaars, R. Tholens, Effects of the harmonised mandatory control system (OSPAR Decision 2000/2) on service companies, IN: SPE International Conference on Health, Safety and Environment in Oil and Gas Exploration and Production, Society of Petroleum Engineers, 2002.

[44] A. Castano, M. Cantarino, P. Castillo, J. Tarazona, Correlations between the RTG-2 cytotoxicity test EC50 and in vivo LC50 rainbow trout bioassay, Chemosphere 32 (1996) 2141–2157.

[45] M.H. Li, Toxicity of perfluorooctane sulfonate and perfluorooctanoic acid to plants and aquatic invertebrates, Environ. Toxicol. 24 (2009) 95–101.

[46] N.S. Battersby, The biodegradability and microbial toxicity testing of lubricants—some recommendations, Chemosphere 41 (2000) 1011–1027.

[47] H.C. Fisher, A.F. Miles, S.H. Bodnar, S.D. Fidoe, C.D. Sitz, Progress towards biodegradable phosphonate scale inhibitors, IN: Proceedings of the RSC Chemistry in the Oil Industry XI, (2009) 2–4.

[48] A. Finizio, M. Vighi, D. Sandroni, Determination of n-octanol/water partition coefficient (Kow) of pesticide critical review and comparison of methods, Chemosphere 34 (1997) 131–161.

[49] D. Winkler, M. Breedon, P. White, A. Hughes, E. Sapper, I. Cole, Using high throughput experimental data and in silico models to discover alternatives to toxic chromate corrosion inhibitors, Corros. Sci. 106 (2016) 229–235.

[50] D.A. Winkler, M. Breedon, A.E. Hughes, F.R. Burden, A.S. Barnard, T.G. Harvey, et al., Towards chromate-free corrosion inhibitors: structure–property models for organic alternatives, Green Chem. 16 (2014) 3349–3357.

[51] J. Sinko, Challenges of chromate inhibitor pigments replacement in organic coatings, Prog. Org. Coat. 42 (2001) 267–282.

[52] M. Bethencourt, F. Botana, J. Calvino, M. Marcos, M. Rodriguez-Chacon, Lanthanide compounds as environmentally-friendly corrosion inhibitors of aluminium alloys: a review, Corros. Sci. 40 (1998) 1803–1819.

[53] C. Verma, J. Haque, M. Quraishi, E.E. Ebenso, Aqueous phase environmental friendly organic corrosion inhibitors derived from one step multicomponent reactions: a review, J. Mol. Liq. 275 (2019) 18–40.

[54] C. Verma, E.E. Ebenso, M. Quraishi, Ionic liquids as green and sustainable corrosion inhibitors for metals and alloys: an overview, J. Mol. Liq. 233 (2017) 403–414.

[55] G. Gece, Drugs: A review of promising novel corrosion inhibitors, Corros. Sci. 53 (2011) 3873–3898.

[56] C. Verma, D. Chauhan, M. Quraishi, Drugs as environmentally benign corrosion inhibitors for ferrous and nonferrous materials in acid environment: an overview, J. Mater. Env. Sci. 8 (2017) 4040–4051.

[57] K. El Mouaden, B. El Ibrahimi, R. Oukhrib, L. Bazzi, B. Hammouti, O. Jbara, et al., Chitosan polymer as a green corrosion inhibitor for copper in sulfide-containing synthetic seawater, Int. J. Biol. Macromol. 119 (2018) 1311–1323.

[58] N.K. Gupta, P. Joshi, V. Srivastava, M. Quraishi, Chitosan: a macromolecule as green corrosion inhibitor for mild steel in sulfamic acid useful for sugar industry, Int. J. Biol. Macromol. 106 (2018) 704–711.

[59] D. Chauhan, V. Srivastava, P. Joshi, M. Quraishi, PEG cross-linked chitosan: a biomacromolecule as corrosion inhibitor for sugar industry, Int. J. Ind. Chem. 9 (2018) 363–377.

[60] K.E. Mouaden, D. Chauhan, M. Quraishi, L. Bazzi, Thiocarbohydrazide-crosslinked chitosan as a bioinspired corrosion inhibitor for protection of stainless steel in 3.5% NaCl, Sustain. Chem. Pharm. 15 (2020) 100213.

[61] T. Chen, D. Zeng, S. Zhou, Study of polyaspartic acid and chitosan complex corrosion inhibition and mechanisms, Pol. J. Environ. Stud. 27 (2018) 1441.

[62] D.S. Chauhan, M. Quraishi, A. Sorour, S.K. Saha, P. Banerjee, Triazole-modified chitosan: a biomacromolecule as a new environmentally benign corrosion inhibitor for carbon steel in a hydrochloric acid solution, RSC Adv. 9 (2019) 14990–15003.

[63] P. Kong, H. Feng, N. Chen, Y. Lu, S. Li, P. Wang, Polyaniline/chitosan as a corrosion inhibitor for mild steel in acidic medium, RSC Adv. 9 (2019) 9211–9217.

[64] J. Haque, V. Srivastava, D.S. Chauhan, H. Lgaz, M.A. Quraishi, Microwave-induced synthesis of chitosan Schiff bases and their application as novel and green corrosion inhibitors: experimental and theoretical approach, ACS Omega 3 (2018) 5654–5668.

[65] D.S. Chauhan, K. Ansari, A. Sorour, M. Quraishi, H. Lgaz, R. Salghi, Thiosemicarbazide and thiocarbohydrazide functionalized chitosan as ecofriendly corrosion inhibitors for carbon steel in hydrochloric acid solution, Int. J. Biol. Macromol. 107 (2018) 1747–1757.

[66] G. Cui, J. Guo, Y. Zhang, Q. Zhao, S. Fu, T. Han, et al., Chitosan oligosaccharide derivatives as green corrosion inhibitors for P110 steel in a carbon-dioxide-saturated chloride solution, Carbohydr. Polym. 203 (2019) 386–395.

[67] D.S. Chauhan, K.E. Mouaden, M. Quraishi, L. Bazzi, Aminotriazolethiol-functionalized chitosan as a macromolecule-based bioinspired corrosion inhibitor for surface protection of stainless steel in 3.5% NaCl, Int. J. Biol. Macromol. (2020).

[68] A. Farhadian, M.A. Varfolomeev, A. Shaabani, S. Nasiri, I. Vakhitov, Y.F. Zaripova, et al., Sulfonated chitosan as green and high cloud point kinetic methane hydrate and corrosion inhibitor: experimental and theoretical studies, Carbohydr. Polym. 236 (2020) 116035.

[69] S. John, A. Salam, A.M. Baby, A. Joseph, Corrosion inhibition of mild steel using chitosan/TiO$_2$ nanocomposite coatings, Prog. Org. Coat. 129 (2019) 254–259.

[70] A. Keerthana, P.M. Ashraf, Carbon nanodots synthesized from chitosan and its application as a corrosion inhibitor in boat-building carbon steel BIS2062, Appl. Nanosci. 10 (2020) 1061–1071.

[71] M. Rbaa, M. Fardioui, C. Verma, A.S. Abousalem, M. Galai, E. Ebenso, et al., 8-Hydroxyquinoline based chitosan derived carbohydrate polymer as biodegradable and sustainable acid corrosion inhibitor for mild steel: experimental and computational analyses, Int. J. Biol. Macromol. 155 (2020) 645–655.

11

Ionic liquids as green corrosion inhibitors

Yeestdev Dewangan[1], Amit Kumar Dewangan[1], Fahmida Khan[2], Perla Akhil Kumar[3], Vivek Mishra[3], Dakeshwar Kumar Verma[1]

[1]DEPARTMENT OF CHEMISTRY, GOVERNMENT DIGVIJAY AUTONOMOUS POST GRADUATE COLLEGE, RAJNANDGAON, INDIA [2]DEPARTMENT OF CHEMISTRY, NATIONAL INSTITUTE OF TECHNOLOGY RAIPUR, RAIPUR, INDIA [3]VISHWAVIDYALAYA ENGINEERING COLLEGE, LAKHANPUR, AMBIKAPUR, CC OF CSVTU, BHILAI, INDIA

Chapter outline

Abbreviations

WL	Weight loss
EIS	Electrochemical impedance spectroscopy
PDP	Potentiodynamic polarization
OCP	Open circuit potential
SEM	Scanning electron microscopy
EDS	Electron dispersion X-ray spectroscopy
QCC	Quantum chemical calculation
DFT	Density function theory

Environmentally Sustainable Corrosion Inhibitors. DOI: https://doi.org/10.1016/B978-0-323-85405-4.00011-2

XPES	X-ray photoelectron spectroscopy
MDS	Molecular dynamic simulation
XRD	X-ray diffraction
FTIR	Fourier transform infrared
AFM	Atomic force microscopy
EDX	Energy dispersive X-ray
CM	Confocal microscopy
OI	Optical interferometry
RS	Raman spectroscopy
CV	Cyclic voltammetry
ENA	Electrochemical noise analysis
GA	Gravimetric analysis
EMIM	1-Ethyl-3-methylimidazolium
BMIM	1-Butyl-3-methylimidazolium
AOIM	1-Allyl-3-octylimidazolium
AEIM	1-Allyl-3-ethylimidazolium
AHIM	1-Allyl-3-hexylimidazolium
[OTP][NTf$_2$]	Octadecyl triphenyl phosphonium bis(trifluoromethylsulfonyl)amide

11.1 Introduction

Metals, such as iron, copper, aluminum, zinc, and their alloys, are applied extensively as industrial materials, due to their availability, cost-effective and most important physical properties [1]. Commonly metals are used in various industries such as metalworking, chemical processing, construction, marine applications, refining, power plants, electronic equipment, and aeronautics. These metals are susceptible to corrosion, due to which they have a corrosion environment, which is used during various processes like descaling and acid pickling. Hydrochloric acid, sulfuric acid, nitric acid, and phosphoric acid are used prominently for these functions. During these processes, there is loss of rust as well as bare metal above the metal surface [2,3]. In order to save this loss, a very small amount of inhibitor molecules are used in these acidic solutions, which cover the metal surface and protect the base metal from the attack of the corrosive anion. Thus the rate of corrosion can be reduced or completely prevented by using inhibitors. These corrosion inhibitors are usually deposited above the metal surface as a protective sheet/blanket sheet. These are mainly deposited above the metal surface via chemisorption or physisorption, which is primarily the interaction between lone pair electrons, pi electrons, and inhibitors, and the charged surface of metal present in heteroatoms of inhibitor molecules. Synthetic organic molecules are mainly used as inhibitors, in which both heteroatoms (O, S, N, P) and pi electrons are present in their structure. Although they protect the metal well from corrosion, the methods used in their production are environmentally harmful and toxic [4–7]. For this reason, it has become important to search for corrosion inhibitors which are environment-friendly, nontoxic, low-cost, biodegradable, and present heteroatoms and unsaturated pi bonds in their structure and also can potentially protect against metal corrosion like synthetic inhibitors [8,9]. In this context, ionic liquids (ILs) have been used successfully as corrosion inhibitors in various corrosion environments for the past decade. The low volatility, less toxicity, inflammability, high thermal stability, and wide range property of ILs make them useful for various applications [10−12]. They are usually composed of organic cations with inorganic and organic anions and consist of liquid organic salts at normal temperature with complete ions. They are mainly used in analytical chemistry, synthetic chemistry, extraction, and electrochemistry. Being green and sustainable material, they are eco-friendly for organisms and the environment [13,14]. Commonly used environmental procedures such as mass loss and electrochemical techniques [potentiodynamic polarization (PDP)] and electrochemical impedance spectroscopy (EIS) are used to determine the potential inhibition properties of ILs. For the characterization of surface morphology, X-ray diffraction (XRD), scanning electron microscopy (SEM), electron dispersion X-ray spectroscopy (EDS), and atomic force microscopy (AFM) are used prominently. Theoretical-based calculations such as quantum chemical calculation (QCC) and molecular simulation have been used prominently for decades to determine the molecular and adsorption property of inhibitor molecules. Fig. 11−1 shows the development of ILs as green corrosion inhibitors and other applications.

11.2 General characteristics, classifications, and applications of ionic liquids

ILs are usually a type of salt that is stable in the liquid state. Moderate conductance, inflammability, mild vapor pressure, higher molecular stability, solvating nature, and wide liquid region are the key characteristics of ILs [15−17]. Also they have the tendency to dissolve and solubilize the materials (inorganic and organic compounds) and gases like CO_2, CH_4, H_2, O_2, etc. Due to these characteristics of ILs, they are widely used as green and sustainable materials, for example, for CO_2 removal, catalyst in the organic reaction, nonmaterial synthesis, separation methods, and in electrochemistry, for example, in batteries and supercapacitors (Fig. 11−2) [18−21]. Also, ILs are used in solar cells, as a catalyst in biochemical transformations, sensitizer, electrodeposition, nanoconfined materials, etc. (Fig. 11−3) [22−24].

11.3 Literature survey on experimental analysis of ionic liquids toward corrosion mitigation

Experimental techniques are important tools for the determination of corrosion mitigation on metal and alloys into aggressive media. Various experimental techniques, such as gravimetric analysis (GA) and electrochemical analysis are applied extensively for these purpose. These techniques are also useful for the primary determination of the effective nature of inhibitors toward corrosion inhibition. Hanzaa et al. reported that an inhibitor molecule shows maximum IE (86%) at a concentration of 600 mg L^{-1} as per electrochemical analysis [25]. Ha and coworkers applied IL $Py_{1,2}$ Cl as a corrosion inhibitor on stainless steel in a chloride-based electrolyte for rechargeable Mg batteries [26]. The inhibition efficiency of

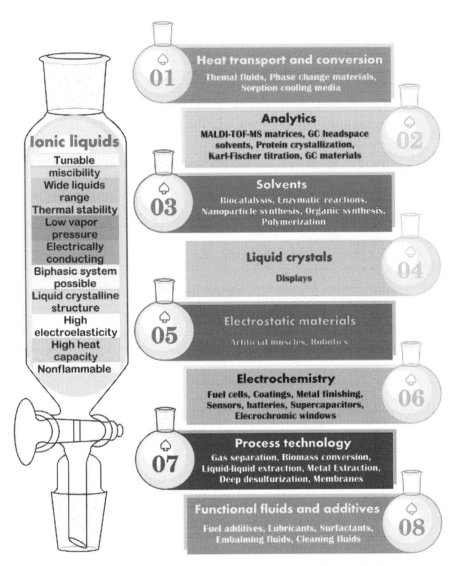

FIGURE 11–1 Development of ionic liquids as green corrosion inhibitors and other applications.

three inhibitors followed the order of IL1, IL2, and IL3 at 94% > 92.4% > 90.5%, respectively, according to electrochemical analysis [27]. Similarly MBT-functionalized ILs are used as corrosion inhibitors. EIS and PDP curve measurements revealed the inhibition of MBT functionalized inhibitors was much more than for conventional ILs [BMIM][BF$_4$] toward bronze [28]. All of the synthesized ILs (IL1, IL2, IL3, and IL4) acted as corrosion inhibitors for carbon steel in 1.0 M HCl with the order of the inhibition efficiency of given ILs as follows IL$_4$ > IL$_3$ > IL$_2$ > IL$_1$ [29]. The two novel synthesized green ILs [VAIM][PF$_6$] and [VAIM][BF$_4$] exhibited good corrosion inhibition for the carbon steels in 1.0 M HCl solution in which

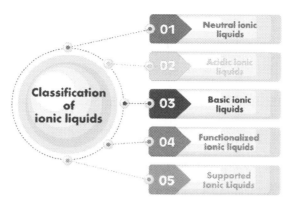

FIGURE 11–2 Classification of ionic liquids.

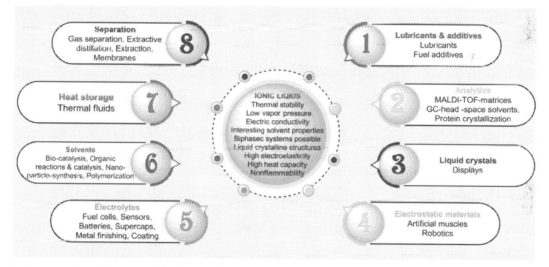

FIGURE 11–3 General characteristics of ionic liquids.

[VAIM][PF$_6$] had a better efficiency compared to [VAIM] [BF$_4$] [30]. MOBB's corrosion efficiency in HCl and H$_2$SO$_4$ media were 98.7% and 98.8%, respectively, at 50°C [31]. The two novel synthesized green ILs [VAIM][PF$_6$] and [VAIM][BF$_4$] exhibited good corrosion inhibition for the carbon steels in 1.0 M HCl solution. As temperature increased, [VAIM][PF$_6$] had better efficiency compared [VAIM] [BF$_4$] [32]. [(HOC$_2$)MIm]PF$_6$ showed better corrosion inhibition as compared to [(HOC$_2$)MIm] NTF2 and the microscopic adsorption mechanism was described by molecular dynamic simulations (MDSs) [33]. Two ammonium derivative ILs, *N*-trioctyl-*N*-methyl ammonium (TMA) methyl sulfate and *N*-tetradecyl-*N*-trimethyl ammonium (TTA) methylsulfate were applied as a corrosion inhibitor for API-X52 steel (1HCl); TMA showed 85% inhibition efficiency and TTA showed 68% inhibition efficiency according to electrochemical analysis at their optimum concentration of 0.209 and 0.272 mM concentration, respectively [34]. Three pyrazine derivative ILs (IL$_1$, IL$_2$, and IL$_3$) proved to be efficient corrosion inhibitors, showing 90.7%, 97.3%, and 82.6% inhibition efficiency, respectively, in the concentration of 100 mg L^{-1} at 303K as per EIS analysis [35]. Three pyrazine derivative, namely HMIMI, BMIMI, and MPIMI, were applied as corrosion inhibitors for mild steel (1HCl) and showed 93.1%, 87.8%, and 80.4% inhibition efficiency, respectively, according to EIS measurements at their optimum concentration of 1×10^{-3} M concentration [36]. At the low temp [BsMIM][HSO$_4$] and [BsMIM][BF$_4$] showed good inhibition properties in aggressive media 1 M H$_2$SO$_4$ solution for the corrosion of 304SS. The inhibiting efficiency shown by [BsMIM][BF$_4$] and [BsMIM][HSO$_4$] is quite similar for the inhibitor of 304SS [37]. IL 1-hexylpyridinium bromide (NR3) was used as a corrosion inhibitor for the corrosion of carbon steel in aggressive media of 1.0 M HCl using both electrochemical (EIS and linear polarization) and chemical (i.e., gravimetric) techniques [38]. Immersion tests were conducted for mild steel in the immersed mixture of 0.01 M NaCl and 4 mM 2-MeHImn 4-OHCin at pH 8, 2, and neutral conditions. Better corrosion inhibition efficiency was found at lower pH [39]. As the concentration of the inhibitor and alkyl chain length increases, the corrosion efficiency of the inhibitor also increases. The 500 ppm concentration of [C4MIM] Br, [C6MIM] Brand, [C8MIM] Br achieved the efficiency of 53.7%, 85.5%, and 93.3%, respectively [40]. Corrosion of Q235 steel has been investigated carefully by electrochemical tests, SEM-EDX, and XPS depth profiling methods in the presence of [(CH$_2$)$_4$SO$_3$HMIm][HSO$_4$] inhibitor at the condition of water-free and different weight percentages of IL [41]. The PDP studies showed the corrosion inhibition reaction (both anodic and cathodic reaction) and the inhibition efficiency obtained for BAIL1 and BAIL2 were 88.0% and 91.1%, respectively [42]. CPOMPB acts as a good corrosion inhibitor in aggressive media of 1 M HCl. EIS analysis exhibited good corrosion inhibition efficiency at the concentration of 10^{-3} M of CPOMPB [43]. Two systems of IL, ChCl-oxalic acid (1:0.5 molar ratio) and ChCl-Malonic acid (1:2 molar ratio), were taken for different steel samples such as OL44, OL52, and SS1.4571 for polarization and electrochemical tests. SS1.4571 showed good corrosion resistance due to the formation of the passive layer on its surface being more stable and resistant [44]. Dicationic imidazolium IL inhibitor was investigated as a corrosion protection layer on a titanium metal surface and the inhibitor showed corrosion efficiency in the following order: IL1, IL2, IL3, and IL4 [45]. Table 11−1 summarizes the molecular structure, metals, media, techniques applied, and outcomes of experimental investigations.

11.4 Literature survey on computational calculations of ionic liquids towards corrosion mitigation

Recently computational calculation-based techniques have been employed extensively in order to determine the optimized structure, adsorption behavior, and most preferred attacking sites of inhibitors. Density functional theory and molecular simulations have been used. Imidazole-based IL was applied as a corrosion inhibitor and the results revealed that BMIM-Cl and GO-BMIM enhance corrosion resistance of the epoxy coating/graphene oxide composite [46]. According to

Table 11–1 The molecular structure, metals, media, techniques applied, and outcomes of experimental analysis.

S. No.	Molecular structure	Metal(s) and media	Techniques	Nature of inhibitor(s)	Outcome	Reference
1.		Mild steel/ 1.0 M H_2SO_4	SEM, EDX, AFM AND EIS	Mixed type/Langmuir's adsorption isotherm	Inhibitor molecule shows maximum IE (86%) at concentration of 600 mg L^{-1}	[25]
2.		Mild steel/ 1.0 M HCl	EIS, EDS, SEM		Ionic liquid $Py_{1,2}$ Cl used as a corrosion inhibitor on stainless steel in chloride-based electrolyte for rechargeable Mg batteries	[26]
3.	R = 12 : N^1,N^1,N^1,N^2,N^2,N^2,-1,2-diaminium bromide R = 12= tridodecylamin	Mild steel/ 1.0 M HCl	EIS, PDC, WL	Mixed type/Langmuir's adsorption isotherm	The inhibition efficiency of three inhibitors followed the order of IL1, IL2, and IL3, respectively, 94% > 92.4% > 90.5%	[27]
	R=12: N^1,N^1,N^1,N^2,N^2,N^2,-1,2-diaminium bromide R = 12= tridodecylamin					
	R=12: N^1,N^1,N^1,N^2,N^2,N^2,-1,2-diaminium bromide R = 12= tridodecylamin					
4.			EIS, SEM, AFM, DFT	Mixed type/Langmuir's adsorption isotherm	M3T functionalized Ls used as corrosion inhibitors. EIS and potentiodynamic polarization curves measurements	[28]

(Continued)

Table 11–1 (Continued)

S. No.	Molecular structure	Metal(s) and media	Techniques	Nature of inhibitor(s)	Outcome	Reference
		Mild steel/ 1.0 M H$_2$SO$_4$			revealed the inhibition of MBT functionalized inhibitors much than conventional ILs[BMIM][BF$_4$] towards bronze	
5		Carbon steel/ 1.0 M HCl	EIS, SEM, FTIR		All of the synthesized ionic liquids (IL1, IL2, IL3 and IL4) applied corrosion inhibitors for carbon steel in 1.0 M HCl and The order of the inhibition efficiency of given ionic liquids in following order IL$_4$ > IL$_3$ > IL$_2$ > IL$_1$	[29]
6		Cold Rolled steel/	EIS, WL, PDP	Langmuir's adsorption isotherm	The two novel synthesized green ILs [VAIM][PF$_6$] and [VAIM][BF$_4$] exhibited good corrosion inhibition for the carbon steels in 1.0 M HCl solution. As temperature	[30]

(Continued)

Table 11–1 (Continued)

S. No.	Molecular structure	Metal(s) and media	Techniques	Nature of inhibitor(s)	Outcome	Reference
		1.0 M HCl			increased [VAIM][PF$_6$] behaved a better efficiency compared [VAIM] [BF$_4$]	
7		Mild steel/ 0.1 M HCl and 0.1 M H$_2$SO$_4$	EIS, OCP, EDX, FTIR, NMR, LC-MS, XRD	Langmuir's adsorption isotherm	MOBB's corrosion efficiency is displayed in HCl and H$_2$SO$_4$ media by 98.7% and 98.8% respectively, at 50°C	[31]
8		Cold rolled steel/ 1.0 M HCl	EIS, WL, PDP, MD	Langmuir's adsorption isotherm	The two novel synthesized green ILs [VAIM][PF$_6$] and [VAIM][BF$_4$] exhibited good corrosion inhibition for the carbon steels in 1.0 M HCl solution. As temperature increased, [VAIM][PF$_6$] behaved a better efficiency compared to [VAIM] [BF$_4$]	[32]
9		Mild steel/ 1.0 M HCl	EIS, PDP, SEM		[(HOC2)MIm]PF$_6$ showed better corrosion inhibition as compared to [(HOC2)MIm] NTF2 and the microscopic adsorption mechanism were described by Molecular dynamic simulations	[33]

(Continued)

Table 11−1 (Continued)

S. No.	Molecular structure	Metal(s) and media	Techniques	Nature of inhibitor(s)	Outcome	Reference
9		Mild steel/ 1 M HCl	EIS, PDP, XPS, AFM, SEM, WL. TMA and TTA	Mixed type/Langmuir adsorption isotherm	Two ammonium derivative ionic liquid N-trioctyl-N-methyl ammonium (TMA) methyl sulfate and N-tetradecyl-N-trimethyl ammonium (TTA) methylsulfate applied for corrosion inhibitor for API-X52 steel (1HCl), in which TMA shows 85% inhibition efficiency and TTA shows 68% inhibition efficiency according to electrochemical analysis at their optimum concentration of 0.209 and 0.272 mM concentration, respectively	[34]
10	 R = (-C8H17), 3,3'-(1,4-phenylenebis(methylene)) bis(1-octyl-1Himidazol-3-ium)bromide. (IL1). R = (-C10H21), 3,3'-(1,4-phenylenebis(methylene) bis(1-decyl-1Himidazol-3-ium)bromide. (IL2). R = (-C12H25), 3,3'-(1,4-phenylenebis(methylene) bis(1-dodecyl-1Himidazol-3-ium)bromide. (IL3).	Stainless steel/ 0.5 M H₂SO₄ (304 SS)	EIS, ¹H NMR, FTIR	Mixed type/Langmuir adsorption Model	Three pyrazine derivative gemini ionic liquids (IL₁, IL₂, and IL₃) proved to be efficient corrosion inhibitors showing 90.7%, 97.3%, and 82.6% inhibition efficiency, respectively, in the concentration of 100 mg L⁻¹ at 303K as per EIS analysis	[35]
11		C38 Steel/ 1 M HCl	PDP, EFM, EIS, AFM, SEM, FTIR	Mixed type/Langmuir adsorption isotherm	Three pyrazine derivative, namely, HMIMI, BMIMI, and MPIMI were applied as corrosion inhibitors for mild steel (1HCl) and showed 93.1%, 87.8%, and 80.4% inhibition efficiency, respectively according to EIS measurements at their optimum concentration of 1 × 10⁻³ M concentration	[36]
12			EIS			[37]

(Continued)

Table 11–1 (Continued)

S. No.	Molecular structure	Metal(s) and media	Techniques	Nature of inhibitor(s)	Outcome	Reference
13		Mild steel/ 1.0 M H_2SO_4		Mixed-type/Langmuir adsorption isotherm	At low temp [BsMIM] [HSO$_4$] and [BsMIM][BF$_4$] showed good inhibitor properties in aggressive media 1 M H_2SO_4 solution for the corrosion of 304SS. The inhibiting efficiency shown by [BsMIM][BF$_4$] and [BsMIM][HSO$_4$] is quite similar to the inhibitor of 304SS	
14		Carbon steel/ 1.0 M HCl	EIS, LPR, SEM	Langmuir adsorption isotherm Freundlich, Temkin, Flory–Huggins, Frumkin, and Langmuir	Ionic liquid 1-hexylpyridinium bromide (NR3) used as a corrosion inhibitor for the corrosion of carbon steel in aggressive media 1.0 M HCl using both electrochemical (electrochemical impedance spectroscopy and linear polarization and chemical (i.e., gravimetric) techniques	[38]
15		Mild steel/ 0.01 M NaCl	SEM, FTIR		Immersion tests were conducted for mild steel in the immersed mixture of 0.01 M NaCl and 4 mM 2-MeH mn 4-OHCin at pH 8, 2, and neutral conditions. Better corrosion inhibition efficiency was found at lower pH	[39]
16		Mild steel/ 1.0 M HCl	EIS, WL, Uv-Visible	Mixed type/Langmuir adsorption isotherm, Flory–Huggins isotherm	As the concentration of the inhibitor and alkyl chain length increases, the corrosion efficiency of the inhibitor also increases. The 500 ppm concentration of [C4MIM] Br, [C6MIM] Brand [C8MIM] Br achieves the efficiency of 53.7%, 85.5%, and 93.3%, respectively	[40]
17		Mild steel/ 1.0 M H_2SO_4 Q235 steel	EIS, PDP, XPS		Corrosion of Q235 steel has been investigated carefully by electrochemical tests, SEM-EDX, and XPS depth profiling methods in presence of [(CH$_2$)$_4$SO$_3$HMIm] [HSO$_4$] inhibitor at the condition of water-free and different weight percentage of IL	[41]
						[42]

(Continued)

Table 11–1 (Continued)

S. No.	Molecular structure	Metal(s) and media	Techniques	Nature of inhibitor(s)	Outcome	Reference
		Carbon steel/ 0.5 M HCl	EIS, PDP, XPS, WL	Mixed type/Langmuir adsorption isotherm	The PDP studies showed the corrosion inhibition reaction (both anodic and cathodic reaction) and the inhibition efficiency has been obtained for BAIL1 and BAIL2 of 88.0% and 91.1%, respectively	
18		Carbon steel/ 1.0 M HCl	WL, EIS, PDP		CPOMPB acts as a good corrosion inhibitor in aggressive media 1 M HCl. EIS analysis exhibited good corrosion inhibition efficiency at the concentration of 10^{-3} M of CPOMPB	[43]
19		Carbon steel/ 1.0 M HCl	PDP, XRD, EIS		Two system ionic liquid ChCl-Oxalic acid (1:0.5 molar ratio) and ChCl-Malonic acid (1:2 molar ratio) at 353 taken for different steel samples such as OL44, OL52, and SS1.4571 for the polarization test and electrochemical test. Among these steel samples, SS1.4571 shows good corrosion resistance due to the formation of a passive layer on its surface that is more stable and resistant	[44]

(Continued)

Table 11–1 (Continued)

S. No.	Molecular structure	Metal(s) and media	Techniques	Nature of inhibitor(s)	Outcome	Reference
20		Ti rod	SEM, XPES, tribological testing		Dicationic imidazolium ionic liquid inhibitor investigated for corrosion protection layer on titanium metal surface and the inhibitor showed corrosion efficiency in the following order: IL1, IL2, IL3, and IL4	[45]

Yousefi and coworkers a computational study suggested the good inhibition properties of EMIm Cl, BMIm Cl, BMIm PF_6, BMIm BF_4, BMIm Br, and HMIm Cl for mild steel in 1 M HCl solution [47]. A theoretical study of imidazolium-based IL showed the good adsorptive nature of the inhibitor [48]. Molecular simulation revealed the highest adsorption value of [bmim][Ac] of the studied inhibitors [49]. Chauhan et al.'s study molecular simulation revealed the highest adsorption value to be from [bmim][Ac] among the studied inhibitors [50]. Ardakani investigated imidazolium-derived polymeric IL applied as a green corrosion inhibitor for mild steel in 1 M HCl and found it exhibited a high HOMO value [51]. Both DFT and MD simulation showed good correlation with other experimental techniques for all heteroatoms-containing inhibitors [52]. MDS showed the optimum adsorption value for [OTP][NTf_2] [53]. DFT calculations have been applied for inhibition analysis which show the greater inhibition ability of DI than TT [54]. According to Luna et al. MDS showed good agreement with other applied techniques [55]. According Luna et al. DFT and MD simulation revealed the good HOMO and adsorption energy for ILs [56]. Likhanova et al.'s applied techniques showed good agreement with theoretical calculations for both inhibitors [57]. MD simulation shows the relatively similar adsorption energies for both ILs [58]. WL techniques revealed maximum IE of 88.0% and 91.1% for BAIL1 and BAIL2, respectively, showing the good agreements with computational study [59]. MD simulation reveals a higher adsorption value for C10-IMIC4-S inhibitor due to extra alkyl chain length [60]. Cao et al. showed higher HOMO and adsorption values of the inhibitor showing good correlation with other experimental techniques [61]. Schmitzhaus reported that B-HEAOL oleate ILs showed higher anticorrosion properties among M-2HEAOL and B-HEAOL oleate ILs for mild steel corrosion in 0.01 M NaCl solution [62]. Analysis such as EIS, SEM, and GA was applied for 304 Stainless Steel in 1 M H_2SO_4 and found efficient corrosion inhibition [37]. Ma and coworkers applied imidazolium ILs as potential corrosion inhibitors for stainless steel [63]. According to Kowsari et al. [TBA][L-Met] showed good HOMO and LUMO values, indicating the adsorptive nature of ILs [64]. According to recent research DFT analysis shown that 12-2-12, 14-2-14, and 16-2-16 exhibited good inhibition properties toward metal corrosion protection [65]. Allylic groups containing imidazolium ILs showed inhibition efficiency in the following order: [AEIM]Br (93.4%) < [ABIM]Br (96.4%) < [AHIM]Br (97.2%) < [AOIM]Br (97.5%). As the alkyl chain length increases in the imidazolium IL, then the inhibition efficiency also increases [66]. Study reveals [Chl][Cl], [Chl][I], and [Chl][Ac] to be corrosion inhibitors at the concentration of 17.91×10^{-4} for mild steel in aggressive acidic solution according to a study of MD simulation and DFT calculation [67]. Pyrido[2,b]pyrazine derivative synthesis was used as a prominent corrosion inhibitor in aggressive solution (1 M HCl) for steel corrosion and revealed 93% inhibition efficiency at 10^{-3} M concentration of inhibition [68]. According to Vastag et al. DFT calculations indicate the interaction between metal surface and bromide ion, in which bromide ion donate electron to metal surface [69]. As per Feng et al. [VBIM]I shows the best protective corrosion efficiency among [VMIM], [VPIM], and [VBIM]. These inhibition results were investigated for X70 steel in 0.5 M H_2SO_4 solution at 298K by QCC [70]. Yesudass and coworkers studied five derivatives, [EMIM] [$EtSO_4$]$^-$, [EMIM][Ac]$^-$, [BMIM] [SCN]$^-$, [BMIM][Ac]$^-$, and [BMIM] [DCA]$^-$, for the protection of corrosion. [EMIM][$EtSO_4$]$^-$ was considered as the best inhibitor among the studied inhibitors and IE was found to be 92.76% at

Table 11–2   The molecular structure, metals, media, techniques applied, and outcomes of DFT, MD/MC simulations.

S. No.	Structure of inhibitor	Metals and electrolytes	Methods applied	Outcome	Reference
1		MS/3.5 wt.% NaCl	FTIR, XRD, UV–Vis, FE-SEM, TGA, EDS	Imidazole-based ionic liquid was applied as corrosion inhibitor and results revealed that BMIM-Cl and GO-BMIM enhance corrosion resistance of the epoxy coating/graphene oxide composite	[46]
2		MS/1 M HCl	EIS, PDP, RDE, AFM, ATR-FTIR, DLS, tensiometry, DFT, QSAR	Computational study suggests the good inhibition properties of EMIm Cl, BMIm Cl, BMIm PF$_6$, BMIm BF$_4$, BMIm Br, and HMIm Cl for mild steel in 1 M HCl solution	[47]
3		MS/0.5 M H$_2$SO$_4$	^{1}H NMR, ^{13}C NMR, FTIR, SEM, AFM, DFT, MD, EIS	Theoretical study of imidazolium-based ionic liquid exhibited the good adsorptive nature of the inhibitor	[48]
4		MS/1 M HCl	SEM, AFM, MD, GA, DFT	Molecular simulation reveals the highest adsorption value of [bmim][Ac] among the studied inhibitors	[49]
5		Cu/3.5% NaCl	EIS, PDP, DFT, CV, ToF-SIMS, AFM, FTIR-ATR	DFT analysis shows the high HOMO and low LUMO values of all studied inhibitors	[50]
6		MS/1 M HCl	EIS, SEM, QCC, DFT	Imidazolium-derived polymeric ionic liquid applied as green corrosion inhibitor for mild steel in 1 M HCl and exhibited high HOMO value	[51]

(Continued)

Table 11−2   (Continued)

S. No.	Structure of inhibitor	Metals and electrolytes	Methods applied	Outcome	Reference
7		MS/1 M HCl	PDP, EIS, DFT, MD, RDF	Both DFT and MD simulation showing good correlation with other experimental techniques for all heteroatoms-containing inhibitors	[52]
8		Mg alloy/ 0.05 wt.% NaCl	SEM, AFM, XPS, MD, EIS	Molecular dynamic simulation showing the optimum adsorption value for [OTP][NTf$_2$]	[53]
9		Cu/0.1 M HCl	WL, SEM, XPS, MD, DFT, OCP, EIS, PDP	DFT calculations have been applied for inhibition analysis which showed greater inhibition ability of DI than TT	[54]
10		API 5 L X52 steel/0.5 M H$_2$SO$_4$	EIS, EDS, XPS, MD, OCP, SEM, AFM, WL, PDP	Molecular dynamic simulation showing good agreement with other applied techniques	[55]
11		API 5 L X52 steel/0.5 M H$_2$SO$_4$	SEM, XPS, WL, PDP, EIS, DFT, MD, UV−Vis, AFM, QCC	DFT and MD simulation reveals the good HOMO and adsorption energy, respectively, for studied ionic liquids	[56]

(Continued)

Table 11–2 (Continued)

S. No.	Structure of inhibitor	Metals and electrolytes	Methods applied	Outcome	Reference
12		API-X60 steel/ 1 M H$_2$SO$_4$	^{1}H & ^{13}C NMR, WL, XPS, DRIFT, SEM/ EDS, PDP, SEM, EIS, PDP	All applied techniques showing good agreements with theoretical calculations for both inhibitors	[57]
13		Carbon steel/ 2 N HCl	ATR-FTIR, ^{1}H NMR, UV–Vis, SEM/EDAX, QCC, MD, MC, DFT, EIS, ENA	MD simulation shows the relatively similar adsorption energies for both the ionic liquids	[58]
14		Carbon steel/ 0.5 M HCl	EIS, PDP, WL, UV–Vis, SEM, XPS, OCP, XPS	WL techniques revealed maximum IE of 88.0% and 91.1% for BAIL1 and BAIL2, respectively, showing the good agreement with computational study	[59]
15		Carbon steel/ 0.5 M HCl	SEM, UV–Vis, XPS, MD, EIS, DFT, MOT, FMOT	MD simulation reveals the higher adsorption value for C10-IMIC4-S inhibitor	[60]
16		Carbon steel/ 0.5 M HCl	EIS, OCP, SEM, XPS, CM, UV–Vis, DFT, MD, QCC	Higher HOMO and adsorption values of inhibitor, showing good correlation with other experimental techniques	[61]
17		MS/ 0.01 mol L^{-1} NaCl	SEM, OI, RS, FTIR-ATR, OCP, PDP, CV	B-HEAOL oleate ionic liquids show high anticorrosion properties among M-2HEAOL and	[62]

(Continued)

Table 11–2 (Continued)

S. No.	Structure of inhibitor	Metals and electrolytes	Methods applied	Outcome	Reference
				B-HEAOL oleate ionic liquids for mild steel corrosion in 0.01 M NaCl solution	
18		304 Stainless steel/1 M H_2SO_4	EIS, SEM, GA	Analysis such as EIS, SEM, and gravimetric analysis was applied for 304 stainless steel in 1 M H_2SO_4 and found efficient corrosion inhibition	[37]
19		Fe, Ni, 304 stainless steel	EIS, SEM, PDP	All studied inhibitors applied as potential corrosion inhibitors for stainless steel	[63]
20		Mild steel/ 1.0 M HCl	EIS, FTIR, PDP, UV–Vis, Quantum chemical study	[TBA][L-Met] showed good HOMO and LUMO values showing the adsorptive nature of ionic liquids	[64]
21		Mild steel/ 0.5 M HCl	EDS, DFT, PDP, EIS, WL	DFT analysis showed that 12-2-12, 14-2-14, and 16-2-16 exhibited good inhibition properties toward metal corrosion protection	[65]

(Continued)

Table 11–2   (Continued)

S. No.	Structure of inhibitor	Metals and electrolytes	Methods applied	Outcome	Reference
22		Mild steel/ 0.5 M H_2SO_4	DFT, EIS, MD, PDC	Allylic group-containing imidazolium ILs show inhibition efficiency in the following order: [AEIM]Br (93.4%) < [ABIM]Br (96.4%) < [AHIM]Br (97.2%) < [AOIM]Br (97.5%). As the alkyl chain length increases in the imidazolium IL then the inhibition efficiency also increases	[66]
23		Mild steel/ 1.0 M HCl	DFT, MC, FIS	Study reveals [Chl][Cl], [Chl][I] and [Chl][Ac] as corrosion inhibitor at the concentration of 17.91×10^{-4} for mild steel in aggressive acidic solution according to MD simulation and DFT calculation	[67]
24		Mild steel/1 M HCl	DFT, EIS	Pyrido[2,b]pyrazine derivative synthesized and used as prominent corrosion inhibitor in aggressive solution (1 M HCl) for steel corrosion and revealed 93% inhibition efficiency at 10^{-3} M concentration of inhibition	[68]
25		Copper/0.1 M Na_2SO_4	QCM, SEM, DFT, PDP. (QCM), (SEM) and (DFT)	DFT calculation indicates the interaction between metal surface and bromide ion, in which the bromide ion donates electron to the metal surface	[69]
26		X70 steel/Mild steel/0.5 M H_2SO_4	WL, PDP, SEM, AFM, FTIR, MD, DFT	[VBIM]I shows the best protective corrosion efficiency among [VMIM], [VPIM], and [VBIM]. These inhibition results investigated for X70 steel in 0.5 M H_2SO_4 solution at 298K by quantum chemical calculation	[70]

(Continued)

Table 11−2   (Continued)

S. No.	Structure of inhibitor	Metals and electrolytes	Methods applied	Outcome	Reference
27		Mild steel/ 1.0 M HCl	SEM, PDS, EIS.	Five derivative [EMIM] [EtSO₄]⁻, [EMIM][Ac]⁻, [BMIM] [SCN]⁻, [BMIM] [Ac]⁻, and [BMIM] [DCA]⁻ for the protection of corrosion in which [EMIM][EtSO₄]⁻ is considered to be a good inhibitor among the studied inhibitors. IE is found to be 92.76% at 500 ppm	[71]

500 ppm [71]. Additionally Table 11−2 reveals the molecular structure, metals, media, techniques applied, and outcomes of DFT and MD/MC simulations.

11.5 Adsorption mechanism

ILs play an important role to prevent corrosion in metals. The metal cations present in the metals behave like a Lewis acid and have the ability to accept electrons. ILs are corrosion inhibitors that contain heteroatoms (N, O, S, P), a lone pair of electrons, and pi bonds, etc., so that they behave like a Lewis base. These organic molecules provide their pi electrons to the metal cation, resulting in the formation of a strong bond as a result of the interaction between the HOMO of the donor atom of ILs and the LUMO of the metal (Fig. 11−4). The bond thus formed deposits the IL and forms a protective layer like a blanket sheet on the metal surface and prevents further corrosion on the metal surface. In the present work ILs such as Ipyr-C_2H_5 and Ipyr-C_4H_9 have been used as a corrosion inhibitor. Both organic molecules have heteroatoms and a lone pair of electrons as well as pi bonds, that is, these organic molecules form a complex by providing a pi electron to a metal cation. In the current work both organic molecules have been used as IL and both these organic molecules fill the property of good inhibitor and efficiently protect the metal from corrosion.

11.6 Conclusion

Based on traditional corrosion monitoring techniques, such as mass loss, electrochemical analysis proves that ILs potentially retard corrosion occurring in metals and alloys and are also green, sustainable, and do not harm the environment and living organisms. Their various physical properties, such as high thermal stability, melting point below than 100°C, solvent nature, wide liquid range, electric conductivity, and solubility in vapor solvents, make

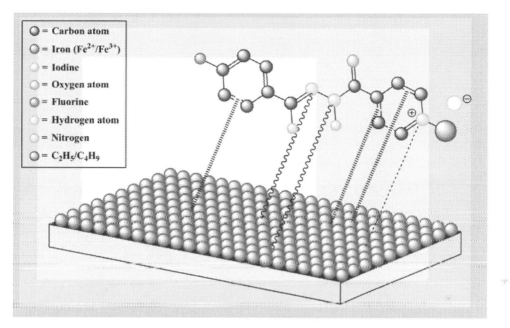

FIGURE 11 4 Proposed chemisorption mechanism of ionic liquids.

them very important. Having active sites (heteroatoms, S, N, O, P, and pi electrons) present in them, they easily coordinate and form strong bonds and protect the metal surface from further corrosion [11,12]. Previous research revealed that the majority of ILs follow the Langmuir adsorption model and mixed type behavior in aggressive media [5,71]. Beyond the traditional monitoring techniques, advanced theoretical based calculation such as density functional theory and molecular simulation (MD/MC simulation) also show good agreement with experimental techniques to prove ILs to be excellent corrosion inhibitors for various metals, such as iron, copper, aluminum, zinc, and their alloys. Ultimately it is concluded that ILs are very effective and green corrosion inhibitors.

Useful links

https://pubs.acs.org/doi/10.1021/jp072262u.
   https://www.mdpi.com/journal/ijms/special_issues/ionic-liquids-2018.
   https://www.journals.elsevier.com/journal-of-molecular-liquids.

Acknowledgments

Author Dr. Dakeshwar Kumar Verma greatly acknowledges the Principal Government Digvijay College Rajnandgaon, Chhattisgarh for the lab and basic facilities.

Conflict of interest

The authors declared no conflict of interest.

References

[1] J. Wang, Y. Lin, A. Singh, W. Liu, Investigation of some porphyrin derivatives as inhibitors for corrosion of N80 steel at high temperature and high pressure in 3.5% NaCl solution containing carbon dioxide, Int. J. Electrochem. Sci. 13 (12) (2018) 11961–11973.

[2] P.P. Kumari, P. Shetty, S.A. Rao, D. Sunil, T. Vishwanath, Synthesis, characterization and anticorrosion behaviour of a novel hydrazide derivative on mild steel in hydrochloric acid medium, Bull. Mater. Sci. 43 (1) (2020) 46. Available from: https://doi.org/10.1007/s12034-019-1995-x.

[3] D.K. Verma, A. Al Fantazi, C. Verma, F. Khan, A. Asatkar, C.M. Hussain, et al., Experimental and computational studies on hydroxamic acids as environmental friendly chelating corrosion inhibitors for mild steel in aqueous acidic medium, J. Mol. Liq. 314 (2020) 113651. Available from: https://doi.org/10.1016/j.molliq.2020.113651.

[4] D.K. Verma, E.E. Ebenso, M.A. Quraishi, C. Verma, Gravimetric, electrochemical surface and density functional theory study of acetohydroxamic and benzohydroxamic acids as corrosion inhibitors for copper in 1 M HCl, Results Phys. 13 (2019) 102194. Available from: https://doi.org/10.1016/j.rinp.2019.102194.

[5] D. Verma, F. Khan, Corrosion inhibition of high carbon steel in phosphoric acid solution by extract of black tea, Adv. Res. 5 (4) (2015) 1–9. Available from: https://doi.org/10.9734/air/2015/18723.

[6] D.K. Verma, Density functional theory (DFT) as a powerful tool for designing corrosion inhibitors in aqueous phase, Advanced Engineering Testing, InTech, 2018. Available from: https://doi.org/10.5772/intechopen.78333.

[7] B. Tan, S. Zhang, Y. Qiang, L. Guo, L. Feng, C. Liao, et al., A combined experimental and theoretical study of the inhibition effect of three disulfide-based flavouring agents for copper corrosion in 0.5 M sulfuric acid, J. Colloid Interface Sci. 526 (2018) 268–280. Available from: https://doi.org/10.1016/j.jcis.2018.04.092.

[8] K. Vinothkumar, M.G. Sethuraman, Corrosion inhibition ability of electropolymerised composite film of 2-amino-5-mercapto-1,3,4-thiadiazole/TiO$_2$ deposited over the copper electrode in neutral medium, Mater. Today Commun. 14 (2018) 27–39. Available from: https://doi.org/10.1016/j.mtcomm.2017.12.007.

[9] D.K. Verma, F. Khan, Corrosion inhibition of mild steel in hydrochloric acid using extract of glycine max leaves, Res. Chem. Intermed. 42 (2016) 3489–3506.

[10] D.K. Verma, F. Khan, Green approach to corrosion inhibition of mild steel in hydrochloric acid medium using extract of spirogyra algae, Green. Chem. Lett. Rev. 9 (1) (2016) 52–60.

[11] D.K. Verma, F. Khan, I. Bahadur, M. Salman, M.A. Quraishi, E.E. Ebenso, et al., Inhibition performance of Glycine max, *Cuscuta reflexa* and Spirogyra extracts for mild steel dissolution in acidic medium: density functional theory and experimental studies, Results Phys. 10 (2018) 665–674.

[12] D. Verma, F. Khan, Corrosion inhibition of mild steel by using sulpha drugs in phosphoric acid medium: a combined experimental and theoretical approach, Am. Chem. Sci. J. 14 (3) (2016) 1–8. Available from: https://doi.org/10.9734/acsj/2016/26282.

[13] M.E. Mashuga, L.O. Olasunkanmi, A.S. Adekunle, S. Yesudass, M.M. Kabanda, E.E. Ebenso, Adsorption, thermodynamic and quantum chemical studies of 1-hexyl-3-methylimidazolium based ionic liquids as corrosion inhibitors for mild steel in HCl, Materials 8 (2015) 3607–3632.

[14] O. Olivares-Xometl, C. López-Aguilar, P. Herrastí-González, N.V. Likhanova, I. Lijanova, R. Martínez-Palou, et al., Adsorption and corrosion inhibition performance by three new ionic liquids on API 5L X52 steel surface in acid media, Ind. Eng. Chem. Res. 53 (2014) 9534−9543.

[15] N.G. Jordana, Biodegradation of ionic liquids—a critical review, Chem. Soc. Rev. 44 (2015) 8200−8237.

[16] J. Estager, J.D. Holbrey, M. Swadźba-Kwaśny, Halometallate ionic liquids—revisited, Chem. Soc. Rev. 43 (2014) 847−886.

[17] M. Sosnowska, M. Barycki, A. Zaborowska, T.P. Rybinska, ILPC: simple chemometric tool supporting the design of ionic liquids, Green Chem. 16 (2014) 4749−4757.

[18] W. Ying, J. Cai, K. Zhou, D. Chen, Y. Ying, Y. Guo, et al., Ionic liquid selectively facilitates CO_2 transport through graphene oxide membrane, ACS Nano 12 (2018) 5385−5393. Available from: https://doi.org/10.1021/acsnano.8b00367.

[19] A. Thomas, M. Prakash, Tuning the CO_2 adsorption by the selection of suitable ionic liquids at ZIF-8 confinement: a DFT study, Appl. Surf. Sci. 491 (2019) 633−639. Available from: https://doi.org/10.1016/j.apsusc.2019.06.130.

[20] J. Qin, Q. Lan, N. Liu, F. Men, X. Wang, Z. Song, et al., A metal-free battery with pure ionic liquid electrolyte, iScience 15 (2019) 16−27.

[21] J. Deng, J. Li, Z. Xiao, S. Song, L. Li, Studies on possible ion confinement in nanopore for enhanced supercapacitor performance in 4V EMIBF4 ionic liquids, Nanomaterials 9 (2019) 1664. Available from: https://doi.org/10.3390/nano9121664.

[22] R. Hajipour, F. Refiee, Recent progress in ionic liquids and their applications in organic synthesis, Org. Prepar. Proc. Inter. 47 (2015) 249−308.

[23] J. Dupont, P.A.Z. Suarez, A.P. Umpierre, R.F.J. De Souza, Pd(II)-Dissolved in ionic liquids: A recyclable catalytic system for the selective biphasic hydrogenation of dienes to monoenes, Braz. Chem. Soc. 11 (2000) 293.

[24] T. Fukushima, A. Kosaka, Y. Ishimura, T. Yamamoto, T. Takigawa, N. Ishii, et al., Molecular ordering of organic molten salts triggered by single-walled carbon nanotubes, Science 300 (2003) 2072−2074.

[25] A.P. Hanzaa, R. Naderib, E. Kowsaric, M. Sayebanid, Corrosion behavior of mild steel in H_2SO_4 solution with 1,4-di[1-methylene-3-methyl imidazolium bromide]-benzene as an ionicliquid, J. Corros. Sci. 107 (2016) 1−11.

[26] J.H. Ha, J.H. Cho, J.H. Kim, B.W. Cho, S.H. Oh, 1-Butyl-1-methylpyrrolidinium chloride as an effective corrosion inhibitor for stainless steel current collectors in magnesium chloride complex electrolytes, J. Power Sources 355 (2017) 90−97.

[27] S.M. Tawfik, Ionic liquids based gemini cationic surfactants as corrosion inhibitors for carbon steel in hydrochloric acid solution, J. Mol. Liq. 216 (2016) 624−635.

[28] Y. Li, S. Zhanga, Q. Dinga, D. Fenga, B. Qina, L. Hua, The corrosion and lubrication properties of 2-mercaptobenzothiazole-functionalized ionic liquids for bronze, Tribol. Int. 114 (2017) 121−131.

[29] A.A. Kityk, D.A. Shaiderov, E.A. Vasil'eva, V.S. Protsenko, F.I. Danilov, Choline chloride based ionic liquids containing nickel chloride: physicochemical properties and kinetics of Ni(II) electroreduction, Electrochim. Acta 245 (2017) 133−145.

[30] Y. Guoa, B. Xub, Y. Liua, W. Yanga, X. Yina, Y. Chena, et al., Corrosion inhibition properties of two imidazolium ionic liquids with hydrophilic tetrafluoroborate and hydrophobic hexafluorophosphate anions in acid medium, J. Ind. Eng. Chem. 56 (2017) 234−247. Available from: https://doi.org/10.1016/j.jiec.2017.07.016.

[31] S.K. Shetty, A.N. Shetty, Eco-friendly benzimidazolium based ionic liquid as a corrosion inhibitor for aluminum alloy composite in acidic media, J. Mol. Liq. 225 (2016) 426−438. Available from: https://doi.org/10.1016/j.molliq.2016.11.037.

[32] Y. Guo, B. Xu, Y. Liu, W. Yang, X. Yin, Y. Chen, et al., Corrosion inhibition properties of two imidazolium ionic liquids with hydrophilic tetrafluoroborate and hydrophobic hexafluorophosphate 3 anions in

acid medium, J. Ind. Eng. Chem. 56 (2017) 234−247. Available from: https://doi.org/10.1016/j.jiec.2017.07.016.

[33] Z. Chen, Y. Guo, Y. Zuo, Y. Chen, W. Yang, B. Xu, Ionic liquids with two typical hydrophobic anions as acidic corrosion inhibitors, Molliq (2018), J. Ind. Eng. Chem. 269 (2018) 1−40.

[34] P.A. Lozada, O.O. Xometl, N.V. Likhanova, I.V. Lijanova, J.R. Vargas-García, R.E. Hernández-Ramírez, Adsorption and performance of ammonium-based ionic liquids as corrosion inhibitors of steel, J. Mol. Liq. 265 (2018) 151−163. Available from: https://doi.org/10.1016/j.molliq.2018.04.153.

[35] Nessim, M.T. Zaky, M.A. Deyab, Three new gemini ionic liquids: synthesis, characterizations and anti-corrosion applications, J. Mol. Liq. 266 (2018) 703−710. Available from: https://doi.org/10.1016/j.molliq.2018.07.001.

[36] F.A. Azeez, O.A. Al-Rashed, A.A. Nazeer, Controlling of mild-steel corrosion in acidic solution using environmentally friendly ionic liquid inhibitors: effect of alkyl chain, J. Mol. Liq. 265 (2018) 654−663.

[37] Y. Ma, F. Han, Z. Li, C. Xia, Acidic-functionalized ionic liquid as corrosion inhibitor for 304 stainless steel in aqueous sulfuric acid, ACS Sustain. Chem. Eng. 4 (2016) 5046−5052. Available from: https://doi.org/10.1021/acssuschemeng.6b01492.

[38] S. Ben Aoun, On the corrosion inhibition of carbon steel in 1 M HCl with a pyridinium-ionic liquid: chemical, thermodynamic, kinetic and electrochemical studies, RSC Adv. 7 (2017) 36688−36696.

[39] A.L. Chong, J.I. Mardel, D.R. MacFarlane, M. Forsyth, A.E. Somers, Synergistic corrosion inhibition of mild steel in aqueous chloride solutions by an imidazolinium carboxylate salt, Chem. Eng. 4 (2016) 1746−1755.

[40] S. Langova, P. Panek, J. Fojtaskova, S. Vicherkova, Alkylimidazolium bromides as corrosion inhibitors for mild steel in acidic medium, Trans. Indian. Inst. Met. 71 (2018) 1−8.

[41] S. Cao, D. Liu, H. Ding, K. Peng, L. Yang, H. Lu, et al., Brönsted acid ionic liquid: electrochemical passivation behavior to mild steel, J. Mol. Liq. 220 (2016) 63−70.

[42] S. Cao, D. Liu, P. Zhang, L. Yang, P. Yang, H. Lu, et al., Green Brönsted acid ionic liquids as novel corrosion inhibitors for carbon steel in acidic medium, Sci. Rep. 7 (2017) 1−14.

[43] A. Bousskri, A. Anejjar, R. Salghi, S. Jodeh, R. Touzani, L. Bazzi, et al., Corrosion control of carbon steel in hydrochloric acid by new eco friendly picolinium based ionic liquids derivative: electrochemical and synergistic studies, J. Mater. Environ. Sci. 7 (2016) 4269−4289.

[44] E.I. Neacsu, V. Constantin, C. Donath, K. Yanushkevich, A. Zhivulka, A. Galyas, et al., Corrosion behavior of some steels in ionic liquids based on choline chloride, Rev. Chim. (Buchar.) 69 (2018) 1−16.

[45] I.M. Gindri, D.A. Siddiqui, C.P. Frizzo, M.A.P. Martins, D.C. Rodrigues, Improvement of tribological and anti-corrosive performance of titanium surfaces coated with dicationic imidazolium-based ionic liquids, RSC Adv. (2016). Available from: https://doi.org/10.1039/C6RA13961B.

[46] A. Khalili Dermani, E. Kowsari, B. Ramezanzadeh, R. Amini, Utilizing imidazole based ionic liquid as an environmentally friendly process for enhancement of the epoxy coating/graphene oxide composite corrosion resistance, J. Ind. Eng. Chem. (2019) 353−363.

[47] A. Yousefi, S. Javadian, N. Dalir, J. Kakemam, J. Akbari, Imidazolium-based ionic liquids as modulators of corrosion inhibition of SDS on mild steel in hydrochloric acíd solutions: experimental and theoretical studies, RSC Adv. (2015) 1−48.

[48] Bhaskaran, P.D. Pancharatna, S. Lata, G. Singh, Imidazolium based ionic liquid as an efficient and green corrosion constraint for mild steel at acidic pH levels, J. Mol. Liq. (2019) 467−476.

[49] C. Verma, L.O. Olasunkanmi, I. Bahadur, H. Lgaz, M.A. Quraishi, J. Haque, et al., Experimental, density functional theory and molecular dynamics supported adsorption behavior of environmental benign imidazolium based ionic liquids on mild steel surface in acidic medium, J. Mol. Liq. 273 (2018) 1−15. Available from: https://doi.org/10.1016/j.molliq.2018.09.139.

[50] D.S. Chauhan, M.A. Quraishi, C. Carriere, A. Seyeux, P. Marcus, A. Singh, Electrochemical, ToF-SIMS and computational studies of 4-amino-5-methyl-4H-1,2,4-triazole-3-thiol as a novel corrosion inhibitor for copper in 3.5% NaCl, J. Mol. Liq. 289 (2019) 111113.

[51] E.K. Ardakani, E. Kowsari, A. Ehsani, Imidazolium-derived polymeric ionic liquid as a green inhibitor for corrosion inhibition of mild steel in 1.0 M HCl: Experimental and computational study, Colloids Surf. 586 (2019) 124195. Available from: https://doi.org/10.1016/j.colsurfa.2019.124195.

[52] F.E.L. Hajjaji, E. Ech-chihbi, N. Rezki, F. Benhiba, M. Taleb, D.S. Chauhan, et al., Electrochemical and theoretical insights on the adsorption and corrosion inhibition of novel pyridinium-derived ionic liquids for mild steel in 1M HCl, J. Mol. Liq. 314 (2020) 113737. Available from: https://doi.org/10.1016/j.molliq.2020.113737.

[53] H. Su, L. Wang, Y. Wu, Y. Zhang, J. Zhang, Insight into inhibition behavior of novel ionic liquids for magnesium alloy in NaCl solution: Experimental and theoretical investigation, Corros. Sci. 165 (2019) 108410. Available from: https://doi.org/10.1016/j.corsci.2019.108410.

[54] L. Feng, S. Zhang, Y. Lu, B. Tan, S. Chen, L. Guo, Synergistic corrosion inhibition effect of thiazolyl based ionic liquids between anions and cations for copper in HCl solution, Appl. Surf. Sci. 483 (2019) 901−911.

[55] M.C. Luna, T.L. Manh, M.R. Romo, M.P. Pardave, E.M.A. Estrada, 1-Ethyl 3-methylimidazolium thiocyanate ionic liquid as corrosion inhibitor of API 5L X52 steel in H_2SO_4 and HCl media, Corros. Sci. 153 (2019) 85−99.

[56] M.C. Luna, T.L. Manh, R.C. Sierra, J.V.M. Flores, L.L. Rojas, E.M.A. Estrada, Study of corrosion behavior of API 5L X52 steel in sulfuric acid in the presence of ionic liquid 1-ethyl 3-methylimidazolium thiocyanate as corrosion inhibitor, J. Mol. Liq. 289 (2019) 111106.

[57] N.V. Likhanova, P.A. Lozada, O.O. Xometl, H.H. Cocoletzi, I.V. Lijanova, J.A. Morales, et al., Effect of organic anions on ionic liquids as corrosion inhibitors of steel in sulfuric acid solution, J. Mol. Liq. 279 (2019) 267−278.

[58] P. Kannan, A. Varghese, K. Palanisamy, A.S. Abousalem, Evaluating prolonged corrosion inhibition performance of benzyltributylammonium tetrachloroaluminate ionic liquid using electrochemical analysis and Monte Carlo simulation, J. Mol. Liq. 297 (2019) 111855. Available from: https://doi.org/10.1016/j.molliq.2019.111855.

[59] S. Cao, D. Liu, P. Zhang, L. Yang, P. Yang, H. Lu, et al., Green Brönsted acid ionic liquids as novel corrosion inhibitors for carbon steel in acidic medium, Sci. Rep. 7 (2017) 8773. Available from: https://doi.org/10.1038/s41598-017-07925-y.

[60] S. Cao, D. Liu, H. Ding, J. Wang, H. Lu, J. Gui, Task-specific ionic liquids as corrosion inhibitors on carbon steel in 0.5 M HCl solution: an experimental and theoretical study, Corros. Sci. (2019) 301−313.

[61] S. Cao, D. Liu, H. Ding, J. Wang, H. Lu, J. Gui, Corrosion inhibition effects of a novel ionic liquid with and without potassium iodide for carbon steel in 0.5 M HCl solution: an experimental study and theoretical calculation, J. Mol. Liq. (2019) 729−740.

[62] T.E. Schmitzhaus, M.R.O. Vega, R. Schroeder, I.L. Muller, S. Mattedi, C. d F. Malfatti, An amino-based protic ionic liquid as a corrosion inhibitor of mild steel in aqueous chloride solutions, Mater. Corros. 71 (2019) 1175−1193. Available from: https://doi.org/10.1002/maco.201911347.

[63] Y. Ma, F. Han, Z. Li, C. Xia, Corrosion behavior of metallic materials in acidic-functionalized ionic liquids, ACS Sustain. Chem. Eng. 4 (2016) 633−639. Available from: https://doi.org/10.1021/acssuschemeng.5b00974.

[64] E. Kowsari, S.Y. Arman, M.H. Shahini, H. Zandi, A. Ehsanie, R. Naderi, et al., In situ synthesis, electrochemical and quantum chemical analysis ofan amino acid-derived ionic liquid inhibitor for corrosion protectionof mild steel in 1M HCl solution, J. Corros. Sci. 112 (2016) 73−85.

[65] F.T. Heakal, A.E. Elkholy, Gemini surfactants as corrosion inhibitors for carbon steel, J. Mol. Liq. 230 (2017) 1−47.

[66] Y. Qiang, S. Zhang, L. Guo, X. Zheng, B. Xiang, S. Chen, Experimental and theoretical studies of four allyl imidazolium-based ionic liquids as green inhibitors for copper corrosion in sulfuric acid, J. Corros. Sci. 119 (2017) 1−29.

[67] C. Verma, I.B. Obot, I. Bahadur, S.M. Sherif, E.E. Ebenso, Choline based ionic liquids as sustainable corrosion inhibitors on mild steel surface in acidic medium: gravimetric, electrochemical, surface morphology, DFT and Monte Carlo simulation studies, J.Apsusc 457 (2018) 1−45.

[68] F.E. Hajjaji, M. Messali, A. Aljuhani, M.R. Aouad, B. Hammouti, M.E. Belghiti, et al., Pyridazinium-based ionic liquids as novel and green corrosion inhibitors of carbon steel in acid medium: electrochemical and molecular dynamics simulation studies, J. Mol. Liq. 249 (2017) 1−26.

[69] G. Vastag, A. Shaban, M. Vranes, A. Tot, S. Belic, S. Gadzuric, Influence of the N-3 alkyl chain length on improving inhibition properties of imidazolium-based ionic liquids on copper corrosion, J. Mol. Liq. 264 (2018) 526−533.

[70] L. Feng, S. Zhang, Y. Qiang, S. Xu, B. Tan, S. Chen, The synergistic corrosion inhibition study of different chain lengths ionic liquids as green inhibitors for X70 steel in acidic medium, J. Mater. Chem. Phys. 215 (2018) 229−241. Available from: https://doi.org/10.1016/j.matchemphys.2018.04.054.

[71] S. Yesudass, L.O. Olasunkanmi, I. Bahadur, M.M. Kabanda, I.B. Obot, E.E. Ebenso, Experimental and theoretical studies on some selected ionic liquids with different cations/anions as corrosion inhibitors for mild steel in acidic medium, J. Taiwan Inst. Chem. Eng. 64 (2016) 1−17. Available from: https://doi.org/10.1016/j.jtice.2016.04.006.

12

Corrosion inhibition by carboxylic acids—an overview

A. Suriya Prabha[1], Panneer Selvam Gayathri[2], R. Keerthana[2],
G. Nandhini[2], N. Renuga Devi[3], R. Dhanalakshmi[2], Susai Rajendran[4],
S. Senthil Kumaran[5]

[1] DEPARTMENT OF CHEMISTRY, MOUNT ZION COLLEGE OF ENGINEERING AND TECHNOLOGY, PUDUKKOTTAI, INDIA [2] PG DEPARTMENT OF CHEMISTRY, M.V. MUTHIAH GOVERNMENT ARTS COLLEGE FOR WOMEN, DINDIGUL, INDIA [3] DEPARTMENT OF ZOOLOGY, GTN ARTS COLLEGE, DINDIGUL, INDIA [4] CORROSION RESEARCH CENTRE, DEPARTMENT OF CHEMISTRY, ST. ANTONY'S COLLEGE OF ARTS AND SCIENCE FOR WOMEN, DINDIGUL, INDIA [5] DEPARTMENT OF MANUFACTURING ENGINEERING, SCHOOL OF MECHANICAL ENGINEERING, VELLORE INSTITUTE OF TECHNOLOGY, VELLORE, INDIA

Chapter outline

Environmentally Sustainable Corrosion Inhibitors. DOI: https://doi.org/10.1016/B978-0-323-85405-4.00010-0

This chapter consists of two sections. Section A deals with the recent trends in the research activities in the field using carboxylic acids as corrosion inhibitors. Section B deals with a case study, where the inhibition of corrosion of mild steel in low chloride medium (an aqueous solution containing 60 ppm of chloride ion as sodium chloride) is discussed.

12.1 Section A

12.1.1 Introduction

Corrosion inhibition of metals and alloys by carboxylic acids has been extensively studied [1−20]. Various metals and alloys have been used in corrosion inhibition research. Different strategies for seeking inhibition efficiency (IE) have been used. Various methods of surface analysis have been used. Significant research results are presented.

12.1.2 Metals

To control corrosion in many metals and alloys, various carboxylic acids have been used. For example, for the corrosion control of mild steel (MS) [5,6], steel [19], aluminum [2,11], carbon steel [1,9,15], copper [7,8], 316 stainless steel [12], and low alloy steel [17], various inhibitors have been used.

12.1.3 Medium

The corrosion of metals in various environments has been studied using chloride medium, seawater, acidic medium, basic medium, and neutral medium. For example, corrosion resistance of metals in acid medium [5,6,9,11,12,17], alkaline medium [19], and sodium chloride medium [7] have been investigated in the presence of carboxylic acids.

12.1.4 Use of various inhibitors (carboxylic acids) as corrosion inhibitors

To monitor corrosion in metals and alloys, several forms in inhibitors have been used. For example, acidic extract of organic carboxylic acids [1]; maleic, malic, succinic, tartaric, citric, tricarboxylic acids, and serine [2]; alkylammonium formate-based protic ionic liquids [3]; surfactant molecules [4]; derivatives of carboxylic acids [5]; carbazole derivatives [6]; cephradine [7]; 1,2,4-triazole-3-carboxylic acid [8]; *Rosmarinus officinalis* and zinc oxide; hydrochloric acid and sulfuric acid [9]; rosemary oil and zinc oxide [10]; phthalic acid [11]; benzonitrile and benzothiazole [12]; dimethylethanolaminen, glutamine, and benzoate [13]; oilfield platform Molikpaq [14]; amide carboxylic acid [15]; benzosulfonazole [16]; chromone-3-acrylic acid (CA) and its derivatives, that is, 6-hydroxy chromone-3-acrylic acid (6-OH-CA) and 7-methoxy chromone-3-acrylic acid (7-Me-CA) [17]; acrylic acid−allylpolyethoxy carboxylate [18]; steel in alkaline media [19]; and water−organic solvent mixture [20] have been used as corrosion inhibitors.

12.1.5 Methods

Various techniques have been used to determine the corrosion activity of metals and alloys in the presence of carboxylic acids, such as the process of weight loss [10,11] and electrochemical experiments such as polarization and AC impedance [2,5,6,7,9,10,12,16,19].

12.1.6 Surface analysis

The formation of a protective film on the metal surface is due to corrosion inhibition. The protective layer usually consists of a metal–inhibitor complex formed on the surface of the metal. They have been analyzed by various spectroscopic techniques. For example, Fourier-transform infrared (FTIR) spectroscopy [1,9,10,11,12,15,16], scanning electron microscopy (SEM) [2,6,7,11,18], X-ray diffraction (XRD) [18], X-ray [1,7,8,18], and transmission electron microscopy [1] have been used to analyze the protective film.

12.1.7 Infrequently used techniques

These are some techniques that are more rarely used. For example, surface analysis by contact angle measurement and energy dispersive X-ray analysis of the metal surface before and after corrosion inhibition have not been employed frequently.

12.1.8 Important findings of the present investigation

From the inhibition tests, several major conclusions were taken. Owing to the creation of a protective layer formed on the metal surface, the key explanation for corrosion resistance is that the defensive layer is made up of a metal–inhibitor complex formed on the surface of the metal. In this work, the as-prepared intelligent coating demonstrates remarkable corrosion protection for MS, which also sheds light on the other metals and alloys' corrosion protection. Molecular dynamics simulations have been performed to illustrate the most conceivable adsorption configuration between the inhibitors and metal surfaces.

If the concentration of different inhibitor increase corrosion inhibitor increases around [1,2,5,10,15,18].

- The first stage in the method of corrosion receptor adsorption on the surface of the metal.
- Electrochemical experiments demonstrate the forming on the metal surface of a protective film.
- As a protective film increases in resistance to shaped liner polarization, change transfer resistance increases, change impedance value decreases, corrosion current value decreases, and double-layer value decreases.

12.1.9 Recent researches on inhibition of corrosion by addition of carboxylic acids in various media

Much research has been carried out on the use of carboxylic acids as corrosion inhibitors. The outcome of the study are summarized in Table 12−1.

Table 12–1 Corrosion inhibition by carboxylic acids.

S. no.	Metal and medium	Inhibitor	Methods	Findings	Reference
1.	Carbon steel	Organic carboxylic acids	Potential measurements used are infrared spectroscopy (FTIR), X-ray diffraction (XRD), transmission electron microscopy (TEM) and zeta	The anodic polarization curves showed that the three modified hydrotalcites act as anodic corrosion inhibitors with an efficiency of 94% for HT-SB, 81% for HT-BZ, and 92% for HT-BTSA at a concentration of 3 g L^{-1}	[1]
2.	Aluminum alloys in alkaline media	Maleic, malic, succinic, tartaric, citric, tricarballylic acids, and serine	Ellipsometry, cyclic polarization, and classic electrochemical impedance spectroscopy (EIS) measurements, whilst the scanning electron microscopy (SEM) and X-ray photoelectron spectroscopy studies worked to assess improvements in the surface topography and chemistry	Inhibition efficacy for tricarballic acid reaches 99% at 6.9 mM, 8.1 mM for citric acid and 12.0 mM for tartaric acid	[2]
3.	Methane hydrate inhibition	Alkylammonium formate-based protic ionic liquids	Industrial applications, including gas purification, lubrication, and catalysis	The THI trend ([EA][Of] ≥ [DMA] [Of] ≥ [DMEA][Of]) reflects the small effect of the structural variation of the cation, with the common formate anion providing the dominant inhibitory action due to presence of a carboxylic group	[3]
4.	Corrosion inhibition to lubrication	Surfactant molecules	Adsorption of surfactants	The results of this study provide insights at the atomic level that help to explain friction outcomes from prior experiments in macroscopic tribology and simulations of classical molecular dynamics	[4]

5.	Mild steel (MS) in 1 M hydrochloric acid	2-(4-Methylbenzylidene)-3-oxo-2,3-dihydro-1H-indene-1-carboxylic acid (MIC), 2-(hydroxymethylene)-3,3-dimethyl-3-oxo-2,3-d hydro-1H-indene-1-one (HIO) and 2-benzylidene-3-oxo-2,3-dihydro-1H-indene-1-carboxylic acid (BIC)	Gravimetric, electrochemical measurements, surface analysis and theoretical studies	With inhibitor concentration, the inhibition performance improved and reached 92%, 87%, and 90% at the maximal MIC, HIO, and BIC concentration, respectively	[5]
6.	MS corrosion in abiotic and biotic environments	Carbazole derivatives	Electrochemical and computational studies and SEM	Carbazole molecules, as demonstrated by the high magnitudes of adsorption energies in the Fe(1 1 0)/ inhibitor/50 H_2O systems, have a tremendous propensity to displace water from the metallic surface and adsorb intensely on the steel surface	[6]
7.	Copper in 0.9% sodium chloride solution	Cephradine	For this inquiry, electrochemical techniques were used, including open circuit potential measurements, potentiodynamic polarization and electrochemical impedance spectroscopy measurements, electron microscopy scanning with energy dispersive X-ray spectroscopy, and quantum chemical calculations	As its concentration increases, the inhibition efficacy of cephradine increases. The SEM with EDAX shows that, due to the adsorption of cephradine on the active sites on the copper surface, a protective coating is formed on the copper surface	[7]
8.	Copper particles on Indium tin oxide	1,2,4-Triazole-3-carboxylic acid	CV using X-ray photoelectron and infrared spectroscopies	For copper safety, double dehydrogenated Trichloroacetic acid-2H (TCA-2H), from the N–H and O–H bond break, plays an important role	[8]

(Continued)

Table 12–1 (Continued)

S. no.	Metal and medium	Inhibitor	Methods	Findings	Reference
9.	Carbon steel in dilute acid solutions hydrochloric acid, sulfuric acid	*Rosmarinus officinalis* and zinc oxide	Potentiodynamic polarization, calculation of open circuit potential, optical microscopy and spectroscopy with attenuated total reflection (ATR)–Fourier-transform infrared (FTIR) spectroscopy	The compound was reported to be more effective in hydrochloric acid solution, with ideal efficiencies of inhibition of 93.26% in hydrochloric acid and 87.7% in sulfuric acid solutions in both acids with mixed form inhibition behavior	[9]
10.	S41003 ferritic steel corrosion in dilute H_2SO_4 and HCl solutions	Rosemary oil and zinc oxide	Potentiodynamic polarization, study of weight loss, calculation of open circuit potential, optical microscopy and spectroscopy with ATR–FTIR	With maximal inhibition efficiencies of 93.68% and 93.26%, 98.93% and 90.5% in hydrochloric acid and sulfuric acid solutions from electrochemical tests with dominant cathodic form inhibition features, the mixed compound is highly efficient	[10]
11.	Aluminum in hydrochloric and tetraoxosulfate (VI) acids	Phthalic acid (PHA)	Gravimetric (weight loss), linear and potentiodynamic polarization methods and FTIR and SEM	PHA prevented aluminum corrosion more in the hydrochloric acid solution than in the sulfuric acid solution	[11]
12.	316 Stainless steel in 6 m hydrochloric acid solution	Benzonitrile and benzothiazole	Polarization of potentiodynamics, coupon calculation, optical microscopy, and study of IR spectroscopy	Maximum inhibition efficiency (IE) value of 95% was obtained	[12]
13.	Delay corrosion in reinforced concrete structures	DMEA, glutamine and benzoate	Potentiodynamic and potentiostatic electrochemical tests	The best mixtures will increase the essential chloride content, so they can reduce the initiation of corrosion on actual reinforced concrete systems as a result	[13]

14.	Chemical structure of water and mineral deposits for formation	Oilfield platform MOLIQPAK	Fluorescence spectrometry with IC, HPLC, GLC, and X-ray	Sulfate-ion concentrations for various wells ranged from $200-1900$ mg L^{-1}, alkalinity (as HCO_3^-) to $450-870$ mg L^{-1}. Among volatile fatty acids, acetic acid was the main component	[14]
15.	Carbon steel in seawater	Amide carboxylic acid	FTIR spectroscopy	The corrosion inhibition rate is 90.43% when the mass concentration of poly(N-maleyl glycine) is 80 mg L^{-1}, which will be effective even if apparent corrosion has already happened for carbon steel	[15]
16.	S40977 stainless steel in 3 M sulfuric acid solution	Benzosulfonazole	Potentiodynamic polarization, calculation of open circuit potential, infrared spectroscopy and optical microscopy	Since the addition of the compound at 0.25% concentration, the corrosion potential value decreased from -0.359 V to -0.306 V, which eventually decreased to -0.278 at 1.25% concentration	[16]
17.	Low alloy steel corrosion in 1 M sulfuric acid	Chromone-3-acrylic acid (CA) and its derivatives viz 6-hydroxy chromone-3-acrylic acid (6-OH-CA) and 7-methoxy chromone-3-acrylic acid (7-Me-CA)	Potentiodynamic polarization and EIS	The order of IE is 7-Me-CA (88.00%) > CA (96.37%) > 6-OH-CA (96.77%)	[17]
18.	Deposits of calcium carbonate in commercial water supplies for recycling	Acrylic acid (AA)–allylpolyethoxy carboxylate (APEL)	SEM, XRD, and TGA static experiments	The maximum scaling IE of 99.1% was achieved by AA–APEL	[18]
19.	Steel in alkaline media	Polyhexamethylene guanidine	Potentiodynamic polarization, open circuit potential measurement	The blocking model is linked to the inhibiting effect of polyhexamethylene guanidine modified by phosphonic groups	[19]

(Continued)

Table 12–1 (Continued)

S. no.	Metal and medium	Inhibitor	Methods	Findings	Reference
20.	Protection of lead substrates	Polymers based on hydrophobic acrylate with built-in group carboxylic acid	In an acetic acid corrosive solution containing tertbutylammonium bromide, odd random phase multisine EIS	The impedance data and electron images of the coatings indicate that the thickness of the coating, pore development, diffusion, and electrolyte absorption profile depend on the AA content of the polymer and the weight of the molecules	[20]

CV, cyclic voltammetry; ([EA]Of), ethylammonium formate; EDAX, energy dispersive X-ray microanalysis; ([DMA]Of), dimethylammonium formate; ([DMEA]Of), dimethylethanolaminen formate; GLC, gas—liquid chromatography; HPLC, high performance liquid chromatography; HT-BTSA, hydrotalcites intercalated by 2-benzothiazolylthio-succinic acid; HT-BZ, hydrotalcites intercalated by benzoate; HT-SB, hydrotalcites intercalated with sebacate; IC, ion chromatography; THI, thermodynamic hydrate inhibition.

12.2 Section B—case study

12.2.1 Introduction: Inhibition of corrosion of MS in chloride medium by malonic acid

In an aqueous solution containing 60 ppm of chloride ion, malonic acid has been used to control corrosion of MS. The effectiveness of inhibition has been calculated using the weight loss method. Electrochemical experiments such as polarization tests and AC impedance spectra have tested mechanistic aspects of corrosion inhibition. The defensive film was studied using UV-visible absorption spectra, fluorescence spectra, and malonic acid inhibition of MS corrosion in the chloride medium.

12.2.2 Experimental methods

12.2.2.1 Weight loss method

Before and after immersion in separate research solutions, the weights of the polished MS specimens were determined. The inhibition efficiencies were determined from the $IE = [(W_1 - W_2)/W_1]100\%$ relationship, where W_1 in the absence of an inhibitor is the corrosion rate and W_2 in the presence of an inhibitor is the corrosion rate.

12.2.2.2 Preparation of malonic acid inhibitor solution

In double distilled water, 1 g of malonic acid was dissolved and make up to 100 mL in a standard measuring flask. As an inhibitor, that solution was used. In 100 mL of distilled water, 1 g of NaCl has been dissolved. Malonic acid was used to control sodium chloride MS corrosion (60 ppm chloride ion).

12.2.2.3 Electrochemical studies

In the existing working corrosion resistance of MS soaked in different research solutions, polarization analysis and AC impedance spectra (21−34) have been calculated. Electrochemical experiments are carried out inside a three-electrode cell assembly. As corrosion resistance increases, linear polarization increases, corrosion current decreases, charge transfer resistance increases, double-layer capacitance decreases, and impedance increases (Schemes 12−1 to 12−3).

12.2.2.4 Polarization study

In a CHI Electrochemical work station/analyzer, model 660A, polarization experiments were performed in the current analysis. An automatic IR payout service has been provided. A cell assembly of three electrodes was used (Scheme 12−1).

MS was the working electrode. The reference electrode was a SCE. The counterelectrode was silver. For the device to reach a steady-state open circuit potential, a time interval of 5−10 min was provided. Corrosion parameters such as corrosion potential (E_{corr}), corrosion

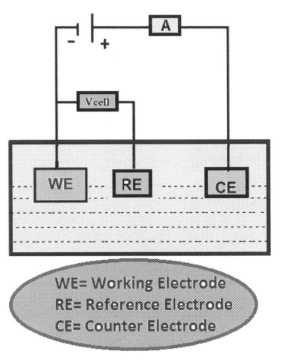

SCHEME 12–1 Three-electrode cell assembly.

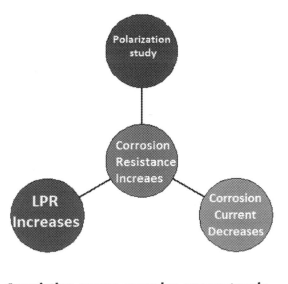

Correlation among corrosion parameters in Polarizaton study

SCHEME 12–2 Correlation among corrosion parameters in polarization study.

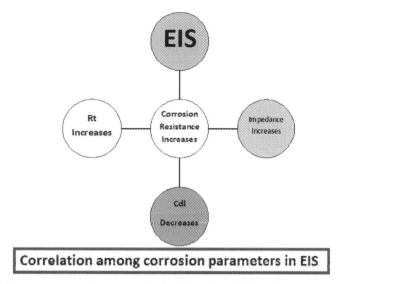

SCHEME 12–3 Correlation among corrosion parameters in electrochemical impedance spectroscopy.

current (I_{corr}), Tafel slopes, anodic = b_a and cathodic = b_c, and LPR (linear polarization resistance) value are taken from polarization studies. The (V/S) scan rate was 0.01.

12.2.2.5 AC impedance spectra
The same instrument and setup used for polarization analysis was also used to document AC impedance spectra in the current research. At different frequencies, the real part (Z') and imaginary part (Z') of cell impedance is calculated in ohms. From the Nyquist map, the values of charge transfer resistance (R_t) and double-layer capacitance (C_{dl}) were computed.

$$R_t = (R_s + R_t) - R_s$$

where R_s, solution resistance.

C_{dl} values were calculated using the relationship

$$C_{dl} = 1/2 \times 3.14 \times R_t \times f_{max}$$

where f_{max} = frequency at maximum imaginary impedance.

12.2.2.6 Surface analysis of protective film
MS samples were immersed for a period of 1 day in the inhibitor system. The sample was extracted after 1 day, dried and subjected to different techniques of surface analysis.

- FTIR spectra were recorded by Perkin–Eleman Spectrum Two.
- SEM images were recorded by Cartizers EVO-18.
- Vickers hardness was recorded by Shimadz HMV-27.

- UV-visible spectra were recorded by Systronics double beam UV-visible spectrophotometer 2202.
- Fluorescence spectra were recorded by Shimadzu spectrofluorophotometer.

12.2.3 Results and discussion

Malonic acid was used in an aqueous solution containing 60 ppm of chloride ion to control the corrosion of MS. The results will be useful in cooling water systems where low chloride ion water is used as a coolant. Cooling water carried by pipelines made of MS may be added to this inhibitor. The weight loss method evaluated the IE of the inhibitor system. The mechanistic aspects were studied by polarization study and AC impedance spectra.

12.2.3.1 Analysis of weight loss method
Inhibition of corrosion of MS in aqueous solution containing chloride ion.

To test the IE of malonic acid in the management of corrosion of MS in the chloride medium, the weight loss procedure was used. The IE of malonic acid in MS corrosion protection is presented in Table 12−2. The rates of corrosion are also given in Table 12−2. Table 12−3 displays the surface coverage values of malonic acid for the regulation of corrosion of MS dissolved in chloride ion.

12.2.3.2 Adsorption isotherm
The corrosion prevention of metals and alloys is due mainly to the adsorption of inhibitor molecules on the metal surface. In the case of dicarboxyl acids, the active concepts of the acids are adsorbed on the metal surface (Table 12−4 and Fig. 12−1).

Table 12−2 Corrosion rates and inhibition efficiency (IE) of inhibitor system (malonic acid) in controlling corrosion of mild steel in chloride medium.

Malonic acid, ppm	Corrosion rate, mdd	IE, %
0	15.76	—
50	7.41	53
100	5.52	65
150	4.09	74
200	2.36	85
250	0.79	95

Table 12−3 Inhibition efficiency (IE) and surface coverage.

Malonic acid, ppm	IE, %	Surface coverage $\theta = IE/100$
50	53	0.53
100	65	0.65
150	74	0.74
200	85	0.85
250	95	0.95

Table 12–4   Parameters for the plot of Langmuir adsorption isotherm, C versus C/θ.

C	C/θ
50	94.33
100	153.84
150	202.70
200	235.29
250	384.61

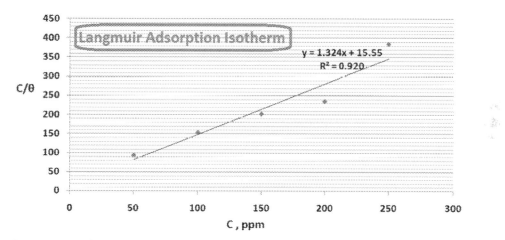

FIGURE 12–1 Langmuir adsorption isotherm.

Table 12–5   Corrosion parameters of mild steel immersed in 60 ppm chloride ion in the absence and presence of malonic acid obtained by polarization study.

System	E_{corr} (V) versus SCE	b_c (V/decade)	b_a (V/decade)	LPR (Ω cm^2)	I_{corr} (A cm^{-2})
Sodium chloride	−0.514	4.989	4.831	2812	1.575×10^{-5}
Sodium chloride + malonic acid	−0.526	4.973	4.845	3219	1.376×10^{-5}

The inhibitor's surface coverage on the metal surface was calculated from the relationship

$$\text{Coverage of surface } \theta = \frac{\text{IE\%}}{100}$$

C/θ was developed as a plot of C. A straight line has been sent. The R_2 value (0.920) was very high. All of these results say that adsorption of inhibitor molecules on the surface of the metal obeys Langmuir adsorption isotherm.

12.2.3.3 Analysis of polarization curves

Corrosion parameters derived from the polarization analysis are given in Table 12−5, namely corrosion potential (E_{corr}), Tafel slopes (b_c, b_a), LPR values, and corrosion current (I_{corr})

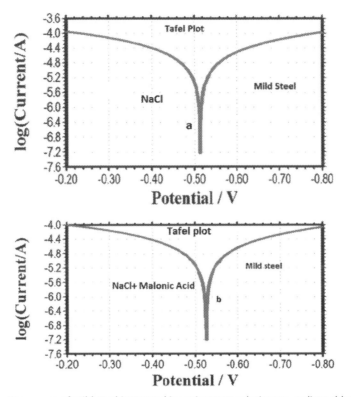

FIGURE 12–2 Polarization curves of mild steel immersed in various test solutions: a, sodium chloride; b, sodium chloride + malonic acid.

values. In the absence and presence of the inhibitor system, Fig. 12–2 indicates the polarization curves of mild steel submerged in sodium chloride.

It is noted from the table that the corrosion potential is −0.514 V versus SCE when MS is submerged in sodium chloride: the LPR value is 2812 Ω cm^2. The current corrosion value is 1.575×10^{-5} A cm^{-2}. According to the table, the corrosion potential in the presence of an agent is moved from −0.514 to −0.526 V versus SCE. This transition is cathodic. It indicates that the cathodic response is mainly regulated. The change is however below 80 mV. This is also reinforced by the fact that, for uninhibited and inhibited processes, the changes in the Tafel slopes are quite tiny. From 2812 to 3219 Ω cm^2, the LPR value rises. The corrosion current value is correspondingly lowered from 1.575×10^{-5} to 1.376×10^{-5} A cm^{-2}. These findings suggest that on the metal surface, a protective film is created. This regulates metal corrosion. A protective film avoids the transfer of electrons from the metal to the medium.

This is due to the fact in presence of inhibitor system, the active principles of the inhibitor system are adsorbed on the metal surface and a protective film is formed. This avoids metal corrosion and thus raises the LPR value (Fig. 12–3) and reduces the existing corrosion value (Scheme 12–2).

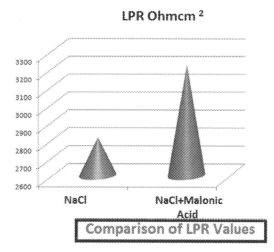

LPR Ohmcm ^{2}

Comparison of LPR Values

FIGURE 12–3 Comparison of linear polarization resistance values.

12.2.3.4 Analysis of AC impedance spectra

The protective film formed on the metal surface is confirmed by the AC impedance spectrum. When a protective film is formed on the metal surface, the charge transfer resistance (R_t) value increases; the double-layer capacitance value (C_{dl}) decreases and the impedance [log(Z/Ω)] value increases (Scheme 12–3).

Fig. 12–4a and b (Nyquist plot) Figs. 12–5 and 12–6 (Bode plots) are seen in the AC impedance spectra of MS submerged in chloride medium in the absence and presence of an agent (malonic acid). Table 12–6 provides the corrosion parameters, namely R_t, C_{dl}, and impedance values.

It is observed from Table 12–6 that, when MS is immersed in NaCl, the R_t value is 1955 Ω cm^2. The C_{dl} value is 2.608×10^{-9} F cm^{-2}. The impedance value is 3.41.

The R_t value rose from 1955 to 2311 Ω cm^2 in the presence of an inhibitor. From 2.608×10^{-9} to 2.206×10^{-9} F cm^{-2} the C_{dl} value decreases. The value of the impedance rose from 3.41 to 3.50.

Implication: Mild steel samples were immersed in chloride medium with malonic acid to test the corrosion prevention efficiency of the dicarboxylic acid. With the rise in inhibitor concentration, the efficacy of inhibition increases. For MS, IE has been achieved by up to 94%. The results suggest that as the concentration of dicarboxylic acid inhibitor rises and the inhibition efficacy of the samples is improved at the same time, the corrosion rate decreases. This result could be important in the cooling water market. To resist corrosion of MS pipes transporting liquids, chloride ion-containing liquids can be used along with malonic acid as a corrosion inhibitor.

12.2.3.5 Analysis of UV-visible absorption spectra

UV-visible absorption spectra have been used in corrosion inhibition study. UV-visible absorption spectrum of malonic acid is shown in Fig. 12–6. A peak appears at 344 nm. The intensity was 0.123. A few crystals of ferrous sulfate were added to the above solution and

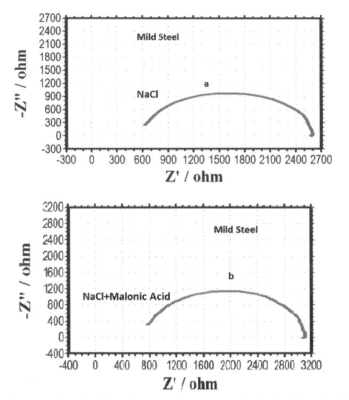

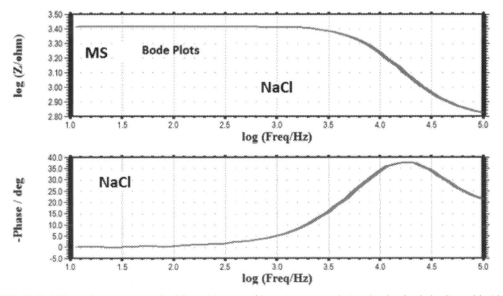

FIGURE 12–4 AC impedance spectra of mild steel immersed in various test solution (Nyquist plot): a, NaCl; b, NaCl + malonic acid.

FIGURE 12–5 AC impedance spectra of mild steel immersed in various test solution (Bode plots): (sodium chloride).

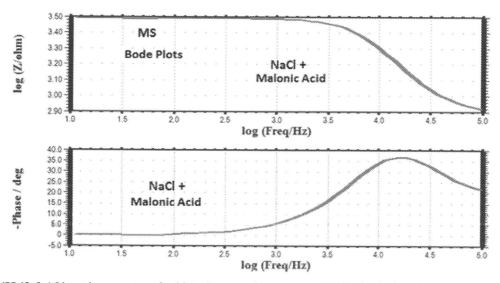

FIGURE 12–6 AC impedance spectra of mild steel immersed in sea water (SW) (Bode plots): sodium chloride + malonic aid.

Table 12–6 Corrosion parameters of mild steel immersed in simulated chloride medium in the absence and presence of malonic acid obtained by AC impedance spectra.

System	R_t (Ω cm^2)	C_{dl} (F cm^{-2})	Impedance log(Z/Ω)
Sodium chloride	1955	2.608×10^{-9}	3.41
Sodium chloride + malonic acid	2311	2.206×10^{-9}	3.50

shaken well. UV-visible absorption spectrum of this solution is shown in Fig. 12–7. Peaks appear at 354.8 and 412.4 nm. The corresponding intensities are 0.267 and 0.436. Red shifts are noticed. There is increase in intensity of the peaks. This spectrum is due to the formation of a complex in solution between Fe^{2+} and malonic acid.

12.2.3.6 Analysis of fluorescence spectra

The Fe^{2+}–malonic acid solution complex was prepared by combining the crystals of ferrous sulfate with malonic acid. Fig. 12–8 shows the fluorescence spectrum of this complex ($\lambda_{max} = 300$ nm). A peak appears at 435.5 nm.

The fluorescence spectrum of the protective film formed on metal surface after immerse in the solution containing malonic acid in NaCl solution is shown in Fig. 12–9. A peak appears at 435.5 nm. This peak is very similar to the peak of Fe^{2+}–malonic acid complex prepared in solution. This confirms that the protective film consist of Fe^{2+}–malonic acid complex formed on the metal surface. It is observed that the intensity of the peak is high.

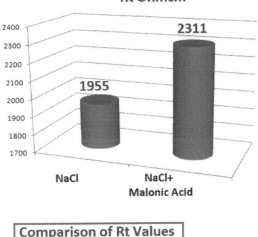

FIGURE 12–7 Comparison of R_t values.

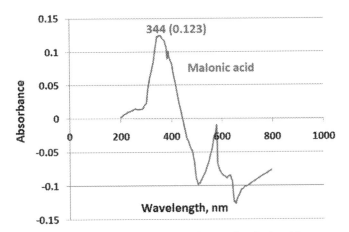

FIGURE 12–8 UV-visible absorption spectrum of an aqueous solution of malonic acid.

Thus it is concluded the protective film consists of Fe^{2+}–malonic acid complex (Figs. 12–10 and 12–11).

12.2.3.7 Analysis of scanning electron microscopy

A scanning electron microscope is a type of electron microscope that, by scanning the surface with a focused electron beam, produces images of a sample. In the sample, electrons communicate with atoms, creating different signals that provide details about the topography of the surface and the composition of the sample. The electron beam is scanned in a raster

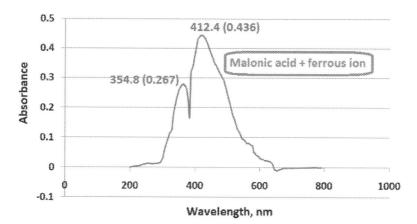

FIGURE 12–9 UV-visible absorption spectrum of an aqueous solution of malonic acid + ferrous ion.

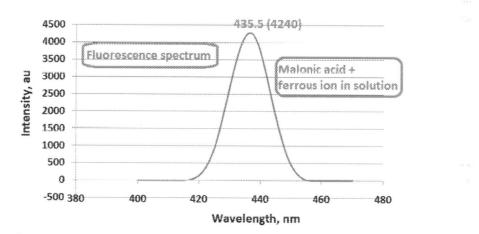

FIGURE 12–10 Fluorescence spectrum of an aqueous solution containing malonic acid + Fe^{2+} complex in solution.

scan pattern, and the beam direction is combined with the intensity of the observed signal to produce an image. In the most common SEM mode, secondary electrons emitted by atoms excited by the electron beam are observed using a secondary electron detector (Everhartz–Thornley detector). Among other factors, the amount of secondary electrons that can be detected, and hence the signal amplitude, depends on the topography of the specimen (Figs. 12–12 and 12–13).

In corrosion inhibition research, SEM analysis has been widely used. SEM images were captured for the polished MS surface (system A), a polished MS surface immersed in corrosive medium (sodium chloride) (system B), and a polished MS surface immersed in corrosive medium (sodium chloride) containing the inhibitor system (malonic acid) (system C). The images are shown in Figs. 12–14 to 12–16.

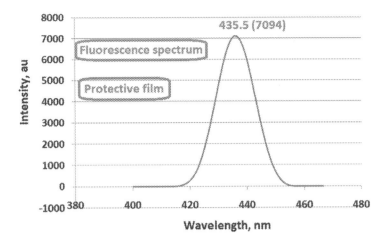

FIGURE 12–11 Fluorescence spectrum of protective film.

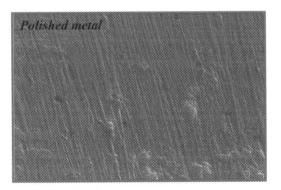

FIGURE 12–12 Scanning electron microscopy image of polished metal.

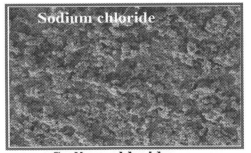

FIGURE 12–13 Scanning electron microscopy image of sodium chloride (corroded metal).

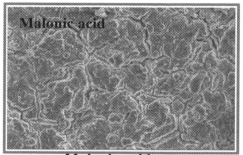

Malonic acid

FIGURE 12–14 Scanning electron microscopy image of protective film (malonic acid).

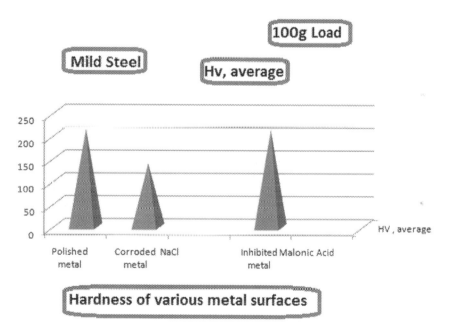

FIGURE 12–15 Comparison of various metal surfaces for 100 g load.

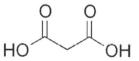

FIGURE 12–16 Structure of malonic acid.

It is observed that the surface is very smooth for system A. The surface is very rugged for system B. Thanks to erosion, pits are observed. The surface is smooth for system C, due to the formation of a protective film. So, SEM is used in the study of corrosion inhibition.

12.2.3.8 Analysis of Vickers hardness

The Vickers test is often easier to use than other hardness tests as the required dimensions are independent of the scale of the indenter, and the indenter can be used on all materials irrespective of hardness. As for all traditional measurements of hardness, the fundamental concept is to observe a material's ability to resist plastic deformation from a standard source. For all metals, the Vickers test may be used and has one of the broadest scales of hardness tests.

Hardness was measured for the surface of polished MS (system A), the surface of polished MS immersed in the corrosive medium (containing chloride ion) (system B), and the surface of polished MS immersed in the corrosive medium containing the inhibitor system (system C). The values are given in Tables 12−7 and 12−8, respectively.

It is observed that the hardness is high for device A. The hardness is poor for system B because the corroded surface contains a porous and amorphous layer of iron oxide. The surface is very rough. There are pits due to corrosion. The surface is smooth for system C, due to the formation of a protective film. In systems A and B, the hardness is in between that of system A and B. That is, lower than that of polished metal but higher than that of corroded surface. Thus the Vickers hardness is used in corrosion inhibition study (Fig. 12−15).

12.2.3.9 Analysis of Fourier-transform infrared spectroscopy

During the corrosion inhibition process, FTIR spectroscopy was used to validate the development of the protective film on the metal surface. In current studies, dicarboxylic acid and malonic acid have been used as corrosion inhibitors (Fig. 12−16). The FTIR spectrum (KBr)

Table 12–7 Vickers hardness (VH) of various surfaces measured along L_1 and L_2.

System (load)	L_1	L_2	VH
Polished metal (100 g)	29.00	30.47	210
Polished metal (100 g)	28.30	29.86	219
Corroded metal (100 g) (NaCl)	36.15	37.39	137
Corroded metal (100 g) (NaCl)	35.58	36.79	142
Inhibited metal (100 g) (malonic acid)	28.74	30.73	210
Inhibited metal (100 g) (malonic acid)	28.07	30.33	218

Table 12–8 Vickers hardness (VH) (average) for 100 gram load.

System	VH, average
Polished metal	214.5
Corroded metal	139.5
Sodium chloride	
Inhibited metal	214
Malonic acid	

of pure malonic acid is shown in Table 12−2 (Fig. 12−17). Fig. 12−18 indicates the FTIR spectrum (KBr) of the protective film formed on the metal surface after immersion for the duration of 1 day in a solution containing sodium chloride and malonic acid solution.

In Fig. 12−17, the FTIR spectroscopy spectrum (potassium bromide) of pure malonic acid is seen. At 1706 cm^{-1}, the C=O stretching frequency occurs. At 3200 cm^{-1}, the O−H stretching frequency occurs. At 2951 cm^{-1}, the aliphatic stretching frequency C−H occurs. The composition of malonic acid is thus confirmed by the spectrum of FTIR spectroscopy.

The FTIR spectrum of the protective film formed on metal surface after immersion in the solution containing 60 ppm of chloride ion and 250 ppm of malonic acid for 1 day is shown

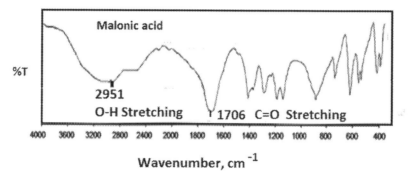

FIGURE 12–17 Fourier-transform infrared spectrum (potassium bromide) of malonic acid.

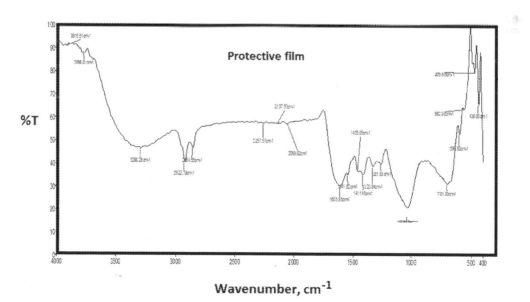

FIGURE 12–18 Fourier-transform infrared spectrum of protective film formed on metal surface.

in Fig. 12−18. It is observed from the figure that the $C{=}O$ stretching has shifted from 1706 to 1605.95 cm^{-1}. The O−H stretching has shifted from 3200 to 3288 cm^{-1}. The aliphatic stretching has shifted from 2951 to 2922 cm^{-1}. The results are summarized in Table 12−9.

In Figs. 12−17 and 12−18, it is found that the stretching frequencies of different functional groups have increased. The outcomes are summarized in Table 12−9. Shifts in the frequencies of different functional groups suggest that on the metal surface, by hydroxyl group and carbonyl group oxygen atoms, the inhibitor has coordinated with Fe^{2+}. The analysis thus contributes to the conclusion that the protective film consists of a complex iron compound produced on the surface of the metal.

12.2.3.10 Summary and conclusion
In the present study and aqueous solution malonic acid has been used as corrosion inhibitor to control corrosion of MS in chloride medium. Methods such as weight loss method, fluorescence study, and AC impedance spectra has been used. The protective film has been analyzed by fluorescence spectra and FTIR spectroscopy. The surface morphology has been analyzed by SEM. The hardness a metal before and after immersion has been determined.

This study leads to the following conclusions:

- Weight loss study reveals that the maximum corrosion IE of 94% is offered by 250 ppm of malonic acid.
- The UV-visible absorption study reveals that the molecules of malonic acid are absorbed on metal surface by Langmuir absorption isotherm.
- Polarization study reveals that the inhibitors function as cathodic inhibitors, because the corrosion potential is shifted to more negative side.
- AC impedance spectra confirms that the formation of a protective film on the metal surface.
- Fluorescence spectra confirms that the protective film consist of Fe^{2+}−malonic acid complex formed on the anodic sites of metal surface.
- This is confirmed by the FTIR spectroscopy spectra.
- Vickers hardness (VH) studies reveal that in the presence of inhibitors, the hardness of the metal increases. In the absence of inhibitor (corrosive medium—chloride ion), the hardness of the metal decreases.
- These findings may be used in cooling water systems where water containing chloride ion can be used as a coolant.

Table 12−9 Stretching frequencies of various functional groups.

	Stretching frequency cm^{-1}	
Various functional groups	Active ingredient of inhibitor (malonic acid)	Protective film formed on the mild steel surface
Hydroxyl group	3200	3288
$C{=}O$	1706	1605
C−H (aliphatic)	2951	2922

12.2.3.11 Scope for further study

In the current research, malonic acid was used to regulate MS corrosion in medium chloride medium as a corrosion inhibitor. Methods such as the polarization analysis of the weight loss method and AC impedance spectra have been used. UV-visible spectra, fluorescence spectra, and FTIR spectra were used to examine the protective film. Scanning electron microscopy has analyzed the surface morphology. The hardness of a metal before and after immersion has been determinded. In future, other metals such as copper and other alloys such as SS316L can been used. Instead of sodium chloride, measurement potassium chloride can be used. Atomic force microscopy can also be used and contact angle measurement can be carried out.

References

[1] D.T. Nguyen, H.T.X. To, J. Gervasi, M. Gonon, M.-G. Olivier, Corrosion inhibition of carbon steel by hydrotalcites modified with different organic carboxylic acids for organic coatings Nguyen, Prog. Org. Coat. 124 (2018) 256−266.

[2] J. Wysocka, M. Cieslik, S. Krakowiak, J. Ryl, Carboxylic acids as efficient corrosion inhibitors of aluminium alloys in alkaline media, Electrochim. Acta 289 (2018) 175−192.

[3] T. Altamash, M. Khraisheh, M.F. Qureshi, S. Aparicio, M. Atilhan, Cost-effective alkylammonium formate-based protic ionic liquids for methane hydrate inhibition, J. Nat. Gas Sci. Eng. 58 (2018) 59−68.

[4] C. Gattinoni, J.P. Ewen, D. Dini, Adsorption of surfactants on α-Fe_2O_3 (0001): a density functional theory study, J. Phys. Chem. C 122 (36) (2018) 20817−20826.

[5] A. Saady, F. El-Hajjaji, M. Taleb, D.S. Chauhan, M.A. Quraishi, Experimental and theoretical tools for corrosion inhibition study of A Density Functional Theory Study solution by new indanones derivatives, Mater. Discov. 12 (2018) 30−42.

[6] H.U. Nwankwo, L.O. Olasunkanmi, E.E. Ebenso, Electrochemical and computational studies of some carbazole derivatives as inhibitors of mild steel corrosion in abiotic and biotic environments, J. Bio Tribo-Corr 4 (2) (2018) 13.

[7] Ž.Z. Tasić, M.B. Petrović Mihajlović, M.B. Radovanović, A.T. Simonović, M.M. Antonijević, Cephradine as corrosion inhibitor for copper in 0.9% NaCl, J. Mol. Struct. 1159 (2018) 46−54.

[8] F.-H. Chang, T.-Y. Chen, S.-H. Lee, Y.-J. Chen, J.-L. Lin, Corrosion inhibition of copper particles on ITO with 1,2,4-triazole-3-carboxylic acid, Surf. Interfaces 10 (2018) 162−169.

[9] R.T. Loto, Surface coverage and corrosion inhibition effect of *Rosmarinus officinalis* and zinc oxide on the electrochemical performance of low carbon steel in dilute acid solutions, Res. Phys. 8 (2018) 172−179.

[10] R.T. Loto, Inhibition studies of the synergistic effect of Rosemary oil and zinc oxide on S41003 ferritic steel corrosion in dilute sulphuric and hydrochloric acid solutions, Orient. J. Chem. 34 (1) (2018) 240−254.

[11] P.O. Ameh, N.O. Eddy, Experimental and computational chemistry studies on the inhibition efficiency of phthalic acid (PHA) for the corrosion of aluminum in hydrochloric and tetraoxosulphate (VI) acids, Prot. Met. Phys. Chem. Surf. (2018).

[12] R.T. Loto, C.A. Loto, A. McPepple, G. Olanrewaju, A. Olaitan, Synergistic effect of benzonitrile and benzothiazole on the corrosion inhibition of 316 stainless steel in 6M HCL solution, Minerals, Metals & Materials Series Part F12, 2018, pp. 901−908.

[13] A. Brenna, M. Pedeferri, F. Bolzoni, M. Ormellese, Chloride-induced corrosion inhibition by binary mixtures in reinforced concrete structures, 2018 NACE—International Corrosion Conference Series, April 2018.

[14] N.V. Polyakova, P.A. Zadorozhny, I.S. Trukhin, S.V. Sukhoverkhov, A.N. Markin, V.A. Avramenko, A.V. Brikov, Determination of the chemical composition of formation and sea waters, inorganic deposits sampled at oilfield platform MOLIQPAK, Neftyanoe Khozyaystvo—OIJ (4)(2018) 43−47.

[15] R.T. Loto, Corrosion polarization behaviour and inhibition of S40977 stainless steel in benzosulfonazole/3M H_2SO_4 solution, S. Afr. J. Chem. Eng. 24 (2017) 148−155.

[16] R. Kumar, S. Chahal, S. Kumar, R. Salghi, S. Jodeh, Corrosion inhibition performance of chromone-3-acrylic acid derivatives for low alloy steel with theoretical modeling and experimental aspects, J. Mol. Liq. 243 (2017) 439−450.

[17] G. Liu, M. Xue, Q. Liu, Y. Zhou, Double-hydrophilic block copolymer as an effective and green scale inhibitor in industrial recycling water systems, Water Science and Technology: Water Supply 17 (4) (2017) 1193−1200.

[18] E.D. Rubl'ova, V.B. Obraztsov, F.I. Danylov, Influence of the PH value of the medium on the inhibition of corrosion in steel modified by polyhexamethylene guanidines, Mater. Sci. 52 (2017) 620−626.

[19] M. De Keersmaecker, T. Hauffman, O. van den Berg, F. Du Prez, A. Adriaens, Acrylate-based coatings to protect lead substrates, Electrochim. Acta 229 (2017) 8−21.

[20] J.A. Thangakani, S. Rajendran, J. Sathiabama, R.M. Joany, R.J. Rathis, S.S. Prabha, Int. J. Nano Corros. Sci. Eng. 1 (2014) 50.

[21] A. Nithya, P. Shanthy, N. Vijaya, R.J. Rathish, S.S. Prabha, R.M. Joany, S. Rajendran, Int. J. Nano Corros. Sci. Eng. 2 (2015) 1.

[22] A.C.C. Mary, S. Rajendran, H. Al-Hashem, R.J. Rathish, T. Umasankareswari, J. Jeyasundari, Int. J. Nano Corros. Sci. Eng. 1 (2015) 42.

[23] A. Anandan, S. Rajendran, J. Sathiyabama, D. Sathiyaraj, Influence of some tablets on corrosion resistance of orthodontic wire made of SS 316L alloy in artificial saliva, Int. J. Corros. Scale Inhib 6 (2) (2017) 132−141. Available from: https://doi.org/10.17675/2305-6894-2017-6-2-3.

[24] C.O. Akalezi, C.E. Ogukwe, E.A. Ejele, E.E. Oguzie, Int. J. Corros. Scale Inhib 5 (2) (2016) 132−146. Available from: https://doi.org/10.17675/2305-6894-2016-5-2-3.

[25] T.A. Onat, D. Yiğit, H. Nazır, M. Güllü, G. Dönmez, Int. J. Corros. Scale Inhib 5 (no. 3) (2016) 273−281. Available from: https://doi.org/10.17675/2305-6894-2016-5-3-7.

[26] A.S. Fouda, M.A. El-Morsy, A.A. El-Barbary, L.E. Lamloum, Study on corrosion inhibition efficiency of some quinazoline derivatives on stainless steel 304 in hydrochloric acid solutions, Int. J. Corros. Scale Inhib 5 (2) (2016) 112−131. Available from: https://doi.org/10.17675/2305-6894-2016-5-2-2.

[27] V.I. Vigdorovich, L.E. Tsygankova, E.D. Tanygina, A.Y. Tanygin, N.V. Shel, Int. J. Corros. Scale Inhib 5 (1) (2016) 59−65. Available from: https://doi.org/10.17675/2305-6894-2016-5-1-5.

[28] P.N. Devi, J. Sathiyabama, S. Rajendran, Int. J. Corros. Scale Inhib 6 (1) (2017) 18−31. Available from: https://doi.org/10.17675/2305-6894-2017-6-1-2.

[29] S. Rajendran, M.K. Devi, A.P.P. Regis, A.J. Amalraj, J. Jeyasundari, M. Manivannan, Electroplating using environmental friendly garlic extract—a case study, Zaštita Materijala 50 (broj 3) (2009). 131−14.

[30] S. Rajendran, M. Agasta, R.B. Devi, B.S. Devi, K. Rajam, J. Jeyasundari, Corrosion inhibition by an aqueous extractof Henna leaves (*Lawsonia inermis* L), Zaštita Materijala 50 (broj 2) (2009) 77−84.

[31] S. Rajendran, P. Chitradevi, S. Johnmary, A. Krishnaveni, S. Kanchana, L. Christy, R. Nagalakshmi, B. Narayanasamy, Corrosion behaviour of SS 316 L in artificial saliva in presence of electoral, Zaštita Materijala 51 (broj 3) (2010) 149−158.

[32] S. Rajendran, T.S. Muthumegala, M. Pandiarajan, P. Nithya Devi, A. Krishnaveni, J. Jeyasundari, N. Samy, H.N. Beevi, Corrosion resistance of SS316L in simulated concretepore solution in presence of tri-sodiumcitrate, Zaštita Materijala 52 (broj 2) (2011) 85–89.

[33] S. Gowri, B. Jaslin Lara, B. Kanagamani, A. Kavitha, A. Maria Belciya, C. Muthunayaki, C. Pandeeswari, J. Maria Praveena, T. Umasankareswari, L. Jerald Majellah, S. Rajendran, Influence of metformin hydrochloride-250 mg (MFH) tablet on corrosion resistance of orthodontic wire made of NiCr alloy in artificial saliva, J. Chem. Sci. Chem. Eng. 1 (1) (2020) 1–10.

13 ▦

Chemical modification of epoxy prepolymers as anticorrosive materials: a review

Omar Dagdag[1], Eno E. Ebenso[2], Chandrabhan Verma[3], Mustapha El Gouri[1]

[1]LABORATORY OF INDUSTRIAL TECHNOLOGIES AND SERVICES, DEPARTMENT OF PROCESS ENGINEERING, HEIGHT SCHOOL OF TECHNOLOGY, SIDI MOHAMMED BEN ABDALLAH UNIVERSITY, FEZ, MOROCCO [2]INSTITUTE FOR NANOTECHNOLOGY AND WATER SUSTAINABILITY, COLLEGE OF SCIENCE, ENGINEERING AND TECHNOLOGY, UNIVERSITY OF SOUTH AFRICA, JOHANNESBURG, SOUTH AFRICA [3]INTERDISCIPLINARY RESEARCH CENTER FOR ADVANCED MATERIALS, KING FAHD UNIVERSITY OF PETROLEUM AND MINERALS, DHAHRAN, SAUDI ARABIA

Chapter outline

Environmentally Sustainable Corrosion Inhibitors. DOI: https://doi.org/10.1016/B978-0-323-85405-4.00013-6

Abbreviations

ER1   4,4′-isopropylidenediphenol oxirane

ER2   4,4′-isopropylidene tetrabromodiphenol oxirane

ER3   2,2′-(((sulfonyl bis(4,1-phenylene)) bis(oxy))bis(methylene)) bis(oxirane)

ER4   2,2′-bis(oxiran-2-ylmethoxy)-1,1′-biphenyl

ER5   diglycidyl amino benzene

ER6   2-(oxiran-2-yl-methoxy)-N,N-bis(oxiran-2-yl-methyl)aniline

ER7   N, N-bis(oxiran-2-ylmethyl)-2-((oxiran-2-ylmethyl) thio)aniline

ER8 N^1,N^1,N^2,N^2-tetrakis (oxiran-2-ylmethyl)bbenzene-1,2-diamine

ER9   4-methyl- N^1,N^1,N^2,N^2-tetrakis (oxiran-2-ylmethyl) benzene-1,2-diamine

ER10   tetraglycidyl-1,2-aminobenzamide

ER11   4,4′-(ethane-1, 2-diyl) bis (N, N-bis (oxiran-2-ylmethyl) aniline)

ER12   4,4′-oxybis(N,N-bis(oxiran-2-ylmethyl)aniline)

ER13   tetraglycidyl ethylenedianiline

ER14   pentaglycidyl ether pentabisphenol A of phosphorus

ER15   hexaglycidyl tris (p-ethylene dianiline) phosphoric triamide

ER16   hexaglycidyl cyclotriphosphazene

ER17   decaglycidyl phosphorus penta methylene dianiline

ER18   triglycidyl ether tribisphenol A of ethylene

ER19   hexaglycidyl trimethylene dianiline of ethylene

ER20   triglycidyl ether of triethoxy triazine

ER21   5,6-anhydro-3-O-methyl-1,2-O-isopropylidene-α-D-glucofuranose)

ER22   5,6-anhydro-3-O-hexadecyl-1,2-O-isopropylidene-α-D-glucofuranose

ER23   5,6-anhydro-3-O-dodecyl-1,2-O-isopropylidene-α-D-glucofuranose

ER24   5,6-anhydro-3-O-hexyl-1,2-O-isopropylidene-α-D-glucofuranose

17.1 Introduction

Epoxy resins (ER) are a unique class of material with a large number of industrial applications. Among the ER, those made from the condensation of bisphenol A and epichlorohydrin have received the most attention due to their superior mechanical, rheological, and anticorrosive properties. Therefore they have been widely used in industry in applications such as adhesives, coatings, laminates, encapsulating materials, electronics, and in making composites [1−4]. The synthetic methods and characterization procedures of bisphenol A and epichlorohydrin type of resins have been covered thoroughly in the literature [5−10]. Apart from the above, ERs are highly compressive materials that possess excellent corrosion resistance, high tensile strength, resistance to physical abuses and superior fatigue properties [11].

In this chapter, we present some synthesized ER produced in recent years by different researchers in our laboratory to try to highlight their use as corrosion inhibitors in aqueous as well as in coating phase.

17.2 Epoxy resins

The term "epoxy resins" designates a wide variety of prepolymers containing one or more epoxy groups (or oxirane) (Fig. 13−1) [12,13].

These resins are most often prepared from epichlorohydrin in two stages:

1. *First step:* condensation of epichlorohydrin and formation of α-chlorohydrin (Fig. 13−2).
2. *Second step:* dehydrohalogenation of α-chlorohydrins and regeneration of epoxy cycles by the action of an alkali metal (Fig. 13−3).

Epoxy resins were discovered almost simultaneously by the Swiss P. Castan (1939) and by the American S.O. Greenlee (1939).

There are several epoxy prepolymers on the market (bisphenol formaldehyde, phenol novolaks, cresol-novolak, etc.), but DGEBA is the most widespread epoxy prepolymer, with production representing 95% of the world's tonnage of epoxy prepolymers [14].

17.3 Reaction mechanism of epoxy/amine systems

The amine functions of the cross-linking agent can react with two epoxy groups. Their reactivity depends essentially on their basicity. This is because an aliphatic amine is much more reactive than an aromatic amine. First, the primary amine reacts with an epoxy group, creating a secondary amine which in turn can react with another epoxy group (Fig. 13−4).

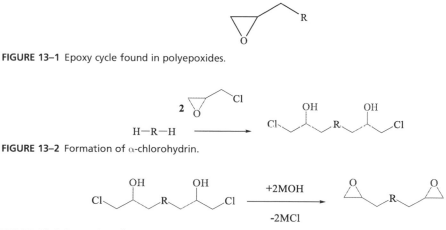

FIGURE 13–1 Epoxy cycle found in polyepoxides.

FIGURE 13–2 Formation of α-chlorohydrin.

FIGURE 13–3 Formation of epoxy rings.

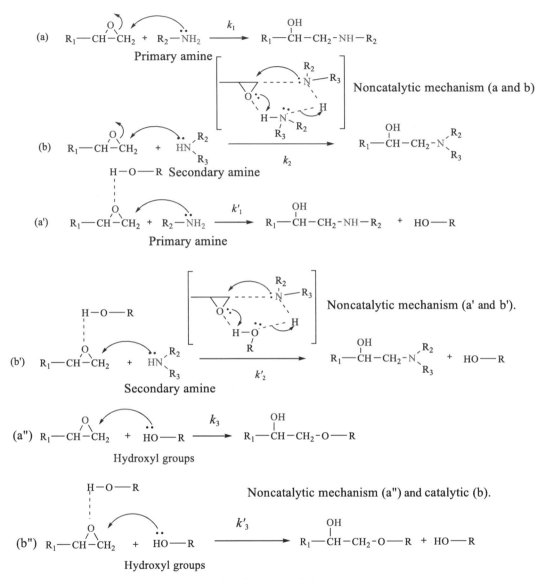

FIGURE 13–4 Main chemical reactions taking place during cross-linking.

The reactivity of primary and secondary amines to epoxies cannot be differentiated due to the observation of a unique activation energy (E_a) and reaction enthalpy (ΔH_a) [15]. The reactivity ratio between these two reactions (k_2/k_1) varies between 0.1 and 1 [16]. The reaction mechanism accepted so far in the literature (Fig. 13–4) involves reactive complexes [17] in which a hydrogen bridge is formed between the nitrogen atom of the amine and the oxygen atom of the epoxy group.

However, the actual form of these complexes is still debated. [15]. These complexes perform two functions; on the one hand, the hydrogen bond weakens the C−O bond of the epoxy group, the carbon therefore becomes more electrophilic, thus facilitating the nucleophilic attack by the nitrogen atom [17]. The formation of such a complex allows the nucleophile to be close enough to the electrophile for enough time for the reaction to take place [17]. A reaction following such a mechanism is considered uncatalyzed [18]. During the reaction between the amine (primary or secondary) and the epoxy group, a hydroxyl function (secondary alcohol) is also formed. These hydroxyl functions have a catalytic effect on the epoxide-amine reaction (Fig. 13−4). In fact, the presence of hydrogen bonds with the oxygen atoms of epoxides facilitates nucleophilic attack [17,19−22]. In this case, a self-catalyzed reaction is then mentioned because it is the groups resulting from the reaction which provide the catalysis.

However, the hydroxyl functions are also likely to initiate an etherification reaction with the epoxy functions [(a") and (b")]. This is then the homopolymerization of the epoxy resin. This last reaction is in competition with the two preceding ones [16]. It is catalyzed by tertiary amines formed during cross-linking. This reaction has been highlighted very little because of the special conditions it requires [23]. It can take place in the case of aromatic amines [23] due to the low reactivity of secondary amines compared to primary amines [16]. However, it does not occur below 200°C [15]. The mechanism proposed to explain the action of tertiary amine breaks down into three stages: initiation, propagation, and termination [22].

17.4 Applications of epoxy resins

Epoxy polymers are important thermosetting materials and can be used in the form of varnishes, but also as paints, adding colorants, additives, or pigments, for their exceptional physical, mechanical, or chemical properties. They can be obtained by adding fillers, in particular zinc phosphates, iron oxides, or metal powder. Finally, the hardeners used in the polymerization of epoxy prepolymers are of great importance. They are the ones who determine how and in which application the finished product is best used.

17.5 Epoxy resins as effective polymeric corrosion inhibitors

17.5.1 Epoxy resins based on hydroxyl

In our previous work [5,6], we have synthesized four ER based on hydroxyl (ER1, ER2, ER3, and ER4). Their structures are shown in Fig. 13−5. The corrosion inhibition activities for four compounds were evaluated using electrochemical methods. This study has shown that these compounds provide good protection for CS against corrosion inhibition in the 1 M HCl medium. The result of the electrochemical study shows that these four compounds have a remarkable corrosion inhibiting effect and that the inhibitory efficiency increases with the increase in concentrations for the inhibitors. The highest inhibition efficiency values at 10^{-3} M (optimum concentration) are 98.1%, 96.5%, 95.8%, and 95.6% for ER3, ER2, ER4, and ER1,

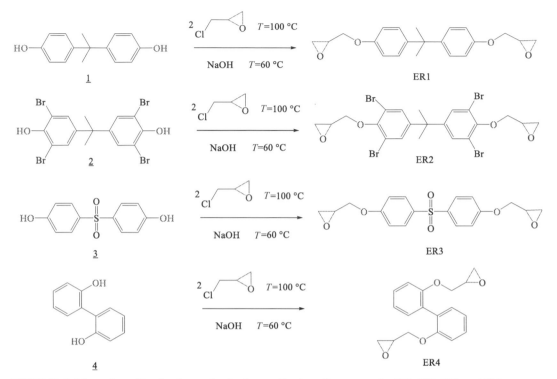

FIGURE 13–5 Schematic outline for the synthesis of aromatic phenolic epoxy resins ER1, ER2, ER3, and ER4.

respectively (Table 13–1). Their high inhibition efficiency is attributed to their association with larger molecular size and the presence of extensive conjugation in the form of two aromatic rings and heteroatoms. Electrochemical results showed that ER1 and ER2 behave as predominantly anodic type and cathodic type of corrosion inhibitors, respectively. Electrochemical results showed that ER3 and ER4 act as anodic type and interfacial type corrosion inhibitors. Thermodynamic studies showed that ERs adsorb spontaneously using their chemisorption mechanism.

17.5.2 Epoxy resins based on amines

In our previous work [9,24–26], we have also synthesized six ER based on amines (ER5, ER6, ER7, ER8, ER9, and ER10). Their structures are shown in Fig. 13–6.

The corrosion inhibition activities for six compounds were evaluated using electrochemical methods. This study has shown that these compounds provide good protection for CS against corrosion inhibition in the 1 M HCl medium. The result of the electrochemical study shows that these six compounds have a remarkable corrosion inhibiting effect and that the inhibitory efficiency increases with the increase in concentrations for the inhibitors. Electrochemical results demonstrate that ERs act as reasonably good inhibitors for carbon

Table 13–1 Inhibition efficiency comparisons for some epoxy resins as effective polymeric corrosion inhibitors.

Epoxy resins	Type of solution	C (M)	Type of substrate	η_{Max} (%)
ER1	1M HCl	10^{-3}	CS	95.6
ER2	1M HCl	10^{-3}	CS	96.5
ER3	1M HCl	10^{-3}	CS	98.1
ER4	1M HCl	10^{-3}	CS	95.8
ER5	1M HCl	10^{-3}	CS	97.3
ER6	1M HCl	10^{-3}	CS	94.3
ER7	1M HCl	10^{-3}	CS	95.4
ER8	1M HCl	10^{-3}	CS	91.7
ER9	1M HCl	10^{-3}	CS	92.9
ER10	1M HCl	10^{-3}	CS	95.0
ER11	1M HCl	10^{-3}	CS	96.5
ER12	1M HCl	10^{-3}	CS	91.3
ER13	3% NaCl	ER13-MDA	CS	93.0
ER14	1M HCl	10^{-3}	CS	94.2
ER15	3% NaCl	ER15-MDA	CS	98.8
ER16	1M HCl	10^{-3}	CS	95.0
ER17	1M HCl	10^{-3}	CS	91.0
ER18	1M HCl	10^{-3}	CS	93.0
ER19	1M HCl	10^{-3}	CS	95.0
ER20	1M HCl	10^{-3}	CS	88.0
ER21	1M HCl	10^{-3}	CS	94.0
ER22	1M HCl	10^{-3}	CS	95.0
ER23	1M HCl	10^{-3}	CS	94.4
ER24	1M HCl	10^{-3}	CS	94.3

steel in 1 M HCl medium and their effectiveness followed the sequence: ER5 (97.3%) > ER7 (95.4%) > ER10 (95.0%) > ER6 (94.3%) > ER9 (92.9%) > ER8 (91.7%) (Table 13–1). The Tafel polarization study showed that ER5 acts as a mixed-type inhibitor with predominant cathodic effectiveness; ER6, ER7, and ER10 act as mixed-type inhibitors with slight anodic predominance; and ER8 and ER9 act as mixed type inhibitors.

In our previous work [3,7,8], we have synthesized other ER derived from tetrafunctional aromatic amines ER11, ER12, and ER13. Their structures are shown in Fig. 13–7. The anticorrosive properties of ER11 and ER12 compounds for CS corrosion in 1M HCl solution were evaluated using electrochemical methods. Electrochemical studies demonstrate that ER11 and ER12 act as sensibly good inhibitors for CS in 1 M HCl medium and showed highest efficiency as high as 96.5% and 91.3% at 10^{-3} M for the ER11 and ER12 compounds, respectively (Table 13–1). potentiodynamic polarization (PDP) study suggested that ER11 behaves with a slight anodic predominance and ER12 behaves as a predominantly cathodic type of inhibitor.

Herein, an epoxy resin ER13 cured with a methylene dianiline was used as an effective anticorrosive coating for CS corrosion/3% NaCl system. The anticorrosive performance of

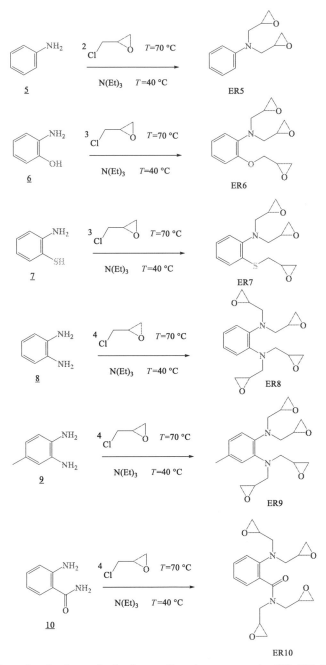

FIGURE 13–6 Schematic outline for the synthesis of aromatic amines epoxy resins ER5, ER6, ER7, ER8, ER9, and ER10.

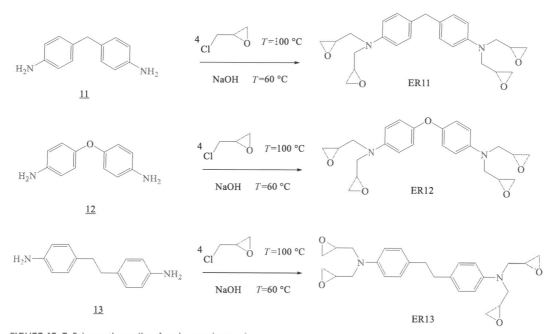

FIGURE 13–7 Schematic outline for the synthesis of epoxy resins derived from tetrafunctional aromatic amines ER11, ER12, and ER13.

coating was investigated using PDP and electrochemical impedance spectroscopy (EIS) methods. EIS study showed an excellent performance of the ER13-MDA as an effective anti-corrosive coating at metal−electrolyte interfaces. The PDP showed that protection efficiency for the ER13-MDA of about 93%.

17.5.3 Epoxy resins based on phosphorus

In our previous work [27−31], we have prepared other ER based on phosphorus ER14, ER15, ER16, and ER17. Their structures are shown in Fig. 13−8. The corrosion inhibition activities for four compounds were evaluated using electrochemical methods. This study has shown that these compounds provide good protection for CS against corrosion inhibition in the 1 M HCl medium for the compounds ER14, ER16, and ER17 and in the 3% NaCl medium for the compound ER15. Electrochemical results demonstrate that ERs (ER14, ER16, and ER17) act as reasonably good inhibitors for CS in 1 M HCl medium and their effectiveness followed the sequence: ER16 (95%)>ER14 (94,18%)>ER17 (91%) (Table 13−1).

The epoxy resin ER15 cured with a MDA was used as an effective anticorrosive coating for CS corrosion/3% NaCl system. The anticorrosive performance of the coating was investigated using the EIS method. EIS study showed an excellent performance of the ER15-MDA

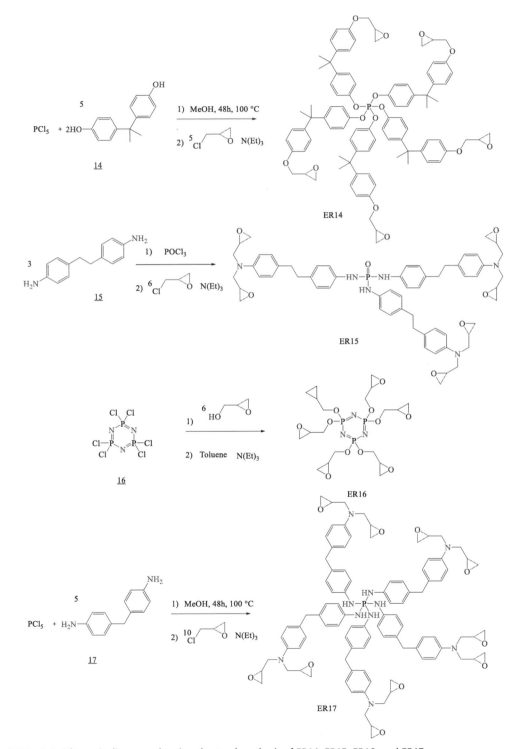

FIGURE 13–8 Schematic diagrams showing the total synthesis of ER14, ER15, ER16, and ER17.

as an effective anticorrosive coating at metal–electrolyte interfaces. The EIS showed protection efficiency for the ER15-MDA of about 98,81%.

17.5.4 Epoxy resins based on trichloroethylene

Hsissou et al. [32,33] for their part studied the electrochemical behavior of two ER based on tricholoroethylene, one trifunctional (ER18) and the other hexafunctional (ER19), as corrosion inhibitors for CS in 1M HCl medium. The chemical structures of ER18 and ER19 are presented in Fig. 13–9.

Electrochemical studies demonstrated that ER18 and ER19 act as sensibly good inhibitors for CS in 1 M HCl medium and showed highest efficiency as high as 93% and 95% at 10^{-3} M for the ER18 and ER19 compounds, respectively.

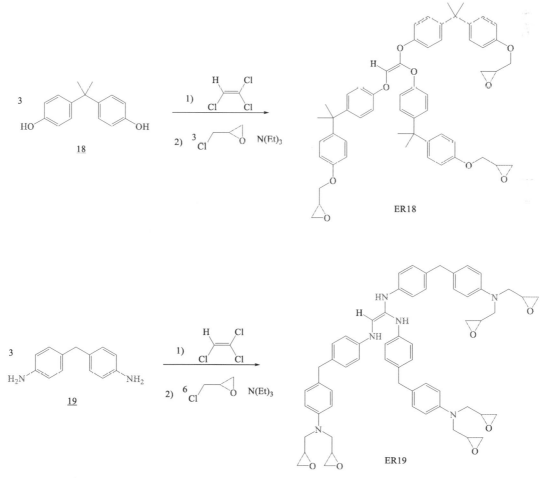

FIGURE 13–9 Chemical structures of epoxy resins ER18 and ER19.

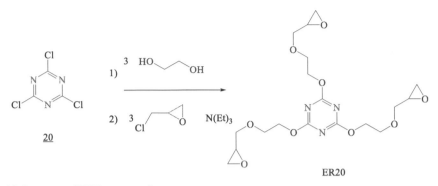

FIGURE 13–10 Structure of ER20 epoxy resin.

17.5.5 Epoxy resins based on 2,4,6-trichloro-1,3,5-triazine

Hsissou et al. [34] reported the preparation of the epoxy resin (ER20) synthesized based on 2,4,6-trichloro-1,3,5-triazine. It was evaluated and investigated as a corrosion inhibitor of E24 in 1 M HCl solution. The synthetic route of ER20 is shown in Fig. 13–10.

The results showed that ER20 has better inhibition performance. The thermodynamic kinetic parameters showed that the adsorption of the ER20 compound formed a protective layer on the E24 surface obeyed the Langmuir isotherm model and the adsorption at the metal–electrolyte interface involved physisorption.

17.5.6 Epoxy resins based on glucose

Rbaa et al. [35,36] reported the preparation of the ERs (ER21, ER22, ER23, and ER24) based on glucose derivatives as ecological corrosion inhibitors for CS in 1M HCl medium. The synthetic route of ER20 is shown in Fig. 13–11. The corrosion inhibition activities for four compounds were evaluated using electrochemical methods. This study has shown that these compounds provide good protection for CS against corrosion inhibition in the 1 M HCl medium, and that the inhibitory efficacy reaches an optimal value of 95%, 94.4%, 94.3%, and 94 % for the ER22, ER23, ER24, and ER21 compounds, respectively. The results of the PDP show that two ER22 and ER21 inhibitors act as cathode-type inhibitors and both ER23 and ER24 inhibitors act on the metal surface as inhibitors of mixed type. Very high negative magnitude of the standard Gibbs free energy of adsorption (-44.41 to -42.68 kJmol^{-1}) values showed that these compounds interact strongly with the metallic surface.

17.6 Epoxy resins as aqueous phase corrosion inhibitors

Most of the ER are macromolecules that possess limited solubility in the polar electrolytic media, including HCl and NaCl solutions. However, few of them that are relatively

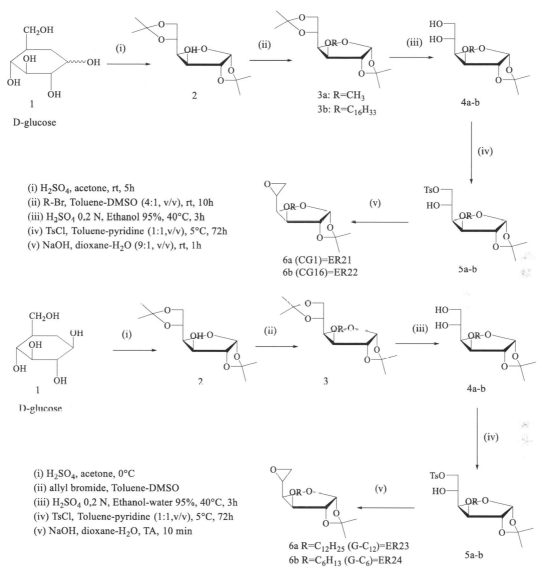

(i) H$_2$SO$_4$, acetone, rt, 5h
(ii) R-Br, Toluene-DMSO (4:1, v/v), rt, 10h
(iii) H$_2$SO$_4$ 0,2 N, Ethanol 95%, 40°C, 3h
(iv) TsCl, Toluene-pyridine (1:1,v/v), 5°C, 72h
(v) NaOH, dioxane-H$_2$O (9:1, v/v), rt, 1h

6a (CG1)=ER21
6b (CG16)=ER22

(i) H$_2$SO$_4$, acetone, 0°C
(ii) allyl bromide, Toluene-DMSO
(iii) H$_2$SO$_4$ 0,2 N, Ethanol-water 95%, 40°C, 3h
(iv) TsCl, Toluene-pyridine (1:1,v/v), 5°C, 72h
(v) NaOH, dioxane-H$_2$O, TA, 10 min

6a R=C$_{12}$H$_{25}$ (G-C$_{12}$)=ER23
6b R=C$_6$H$_{13}$ (G-C$_6$)=ER24

FIGURE 13–11 Synthesis path for ecological inhibitors ER21, ER22, ER23, and ER24.

small in molecular size and/or contain several peripheral polar functional groups show remarkable solubility in such electrolytic media [37]. In an acidic medium the epoxide ring can undergo a ring-opening reaction to form an open chain structure, as shown in Fig. 13–12 [38].

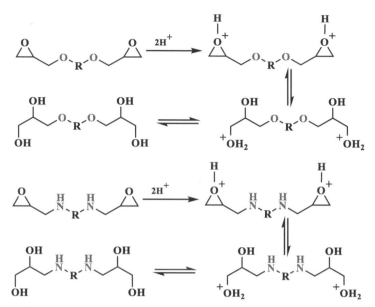

FIGURE 13–12 Mechanism of ring-opening reaction of epoxy resins in acid solution.

17.7 Conclusion

ER represent a special group of highly reactive prepolymers or polymers that contain epoxide groups. Their molecular structures contain several peripheral polar functional groups including hydroxyl (-OH), amino (-NH$_2$) groups and amide (-CONH$_2$) through which they can get easily adsorbed over the metallic surface and act as effective anticorrosive materials in coating as well as solution phase. Several ERs in pure and cured forms have been used as anticorrosive coating materials especially for CS in 1M HCl medium and 3% NaCl medium. Electrochemical (EIS and PDP methods) demonstrations showed that most of the ERs acted as interface type and mixed type anticorrosive materials.

Declaration of interest statement

The authors declare that they have no known competing financial interests or personal relationships that could have appeared to influence the work reported in this paper.

Conflict of interest statement

The authors declare that they have no known conflict of interest.

Author's contributions

Authors collectively contribute in the designing, framing and writing of the chapter.

References

[1] R. Hsissou, A. Bekhta, O. Dagdag, A. El Bachiri, M. Rafik, A. Elharfi, Rheological properties of composite polymers and hybrid nanocomposites, Heliyon 6 (2020) e04187.

[2] O. Dagdag, M. El Gouri, A. El Mansouri, A. Outzourhit, A. El Harfi, O. Cherkaoui, et al., Rheological and electrical study of a composite material based on an epoxy polymer containing cyclotriphosphazene, Polymers 12 (2020) 921.

[3] O. Dagdag, R. Hsissou, A. El Harfi, A. Berisha, Z. Safi, C. Verma, et al., Fabrication of polymer based epoxy resin as effective anti-corrosive coating for steel: computational modeling reinforced experimental studies, Surf. Interfaces 18 (2020) 100454.

[4] O. Dagdag, A. Berisha, Z. Safi, S. Dagdag, M. Berrani, S. Jodeh, et al., Highly durable macromolecular epoxy resin as anticorrosive coating material for carbon steel in 3% NaCl: computational supported experimental studies, J. Appl. Polym. Sci. 137 (2020) 49003.

[5] O. Dagdag, Z. Safi, Y. Qiang, H. Erramli, L. Guo, C. Verma, et al., Synthesis of macromolecular aromatic epoxy resins as anticorrosive materials: computational modeling reinforced experimental studies, ACS Omega 5 (2020) 3151−3164.

[6] O. Dagdag, Z. Safi, H. Erramli, N. Wazzan, I. Obot, E. Akpan, et al., Anticorrosive property of heterocyclic based epoxy resins on carbon steel corrosion in acidic medium: electrochemical, surface morphology, DFT and Monte Carlo simulation studies, J. Mol. Liq. 287 (2019) 110977.

[7] O. Dagdag, Z. Safi, H. Erramli, N. Wazzan, L. Guo, C. Verma, et al., Epoxy prepolymer as a novel anti-corrosive material for carbon steel in acidic solution: electrochemical, surface and computational studies, Mater. Today Commun. 22 (2020) 100800.

[8] O. Dagdag, Z. Safi, N. Wazzan, H. Erramli, L. Guo, A.M. Mkadmh, et al., Highly functionalized epoxy macromolecule as an anti-corrosive material for carbon steel: computational (DFT, MDS), surface (SEM-EDS) and electrochemical (OCP, PDP, EIS) studies, J. Mol. Liq. 302 (2020) 112535.

[9] R. Hsissou, F. Benhiba, O. Dagdag, M. El Bouchti, K. Nouneh, M. Assouag, et al., Development and potential performance of prepolymer in corrosion inhibition for carbon steel in 1.0 M HCl: outlooks from experimental and computational investigations, J. Colloid Interface Sci. 575 (2020) 43−60.

[10] O. Dagdag, A. El Harfi, Z. Safi, L. Guo, S. Kaya, C. Verma, et al., Cyclotriphosphazene based dendrimeric epoxy resin as an anti-corrosive material for copper in 3% NaCl: experimental and computational demonstrations, J. Mol. Liq. (2020) 113020.

[11] M. Singla, V. Chawla, Mechanical properties of epoxy resin−fly ash composite, J. Miner. Mater. Charact. Eng. 9 (2010) 199−210.

[12] F.-L. Jin, X. Li, S.-J. Park, Synthesis and application of epoxy resins: a review, J. Ind. Eng. Chem. 29 (2015) 1−11.

[13] S.J. Park, F.L. Jin, Epoxy resins: fluorine systems, in: Wiley Encyclopedia of Composites. Wiley, 2011, pp. 1−6

[14] P. Bardonnet, Monographies. Résines époxydes composants et propriétés, J. Tech. de. l'ingénieur. Plastiques et. Compos. 3 (1992), pp. A3465. 1-A3465. 16.

[15] B.A. Rozenberg, Kinetics, thermodynamics and mechanism of reactions of epoxy oligomers with aminesed Epoxy Resins and Composites II, Springer, 1986pp. 113−165.

[16] C. Barrere, F. Dal Maso, Résines époxy réticulées par des polyamines: structure et propriétés, Rev. de. l'Institut Français du. Pétrole 52 (1997) 317−335.

[17] R. Vinnik, V. Roznyatovsky, Kinetic method by using calorimetry to mechanism of epoxy-amine cure reaction, J. Therm. Anal. Calorim. 74 (2003) 29.

[18] L. Shechter, J. Wynstra, R.P. Kurkjy, Glycidyl ether reactions with amines, Ind. Eng. Chem. 48 (1956) 94−97.

[19] H. Flammersheim, Kinetics and mechanism of the epoxide−amine polyaddition, Thermochim. Acta 310 (1998) 153−159.

[20] I.T. Smith, The mechanism of the crosslinking of epoxide resins by amines, Polymer 2 (1961) 95−108.

[21] R. Mezzenga, L. Boogh, J.-A.E. Månson, B. Pettersson, Effects of the branching architecture on the reactivity of epoxy − amine groups, Macromolecules 33 (2000) 4373−4379.

[22] N. Enikolopiyan, New aspects of the nucleophilic opening of epoxide ringsed Polymerization of Heterocycles (Ring Opening), Elsevier, 1977pp. 317−328.

[23] J.A. Ramos, N. Pagani, C.C. Riccardi, J. Borrajo, S.N. Goyanes, I. Mondragon, Cure kinetics and shrinkage model for epoxy-amine systems, Polymer 46 (2005) 3323−3328.

[24] O. Dagdag, Z. Safi, H. Erramli, O. Cherkaoui, N. Wazzan, L. Guo, et al., Adsorption and anticorrosive behavior of aromatic epoxy monomers on carbon steel corrosion in acidic solution: computational studies and sustained experimental studies, RSC Adv. 9 (2019) 14782−14796.

[25] O. Dagdag, Z. Safi, R. Hsissou, H. Erramli, M. El Bouchti, N. Wazzan, et al., Epoxy pre-polymers as new and effective materials for corrosion inhibition of carbon steel in acidic medium: computational and experimental studies, Sci. Rep. 9 (2019) 1−14.

[26] O. Dagdag, A. El Harfi, O. Cherkaoui, Z. Safi, N. Wazzan, L. Guo, et al., Rheological, electrochemical, surface, DFT and molecular dynamics simulation studies on the anticorrosive properties of new epoxy monomer compound for steel in 1 M HCl solution, RSC Adv. 9 (2019) 4454−4462.

[27] R. Hsissou, S. Abbout, R. Seghiri, M. Rehioui, A. Berisha, H. Erramli, et al., Evaluation of corrosion inhibition performance of phosphorus polymer for carbon steel in [1 M] HCl: computational studies (DFT, MC and MD simulations), J. Mater. Res. Technol. (2020).

[28] O. Dagdag, A. El Harfi, A. Essamri, M. El Gouri, S. Chraibi, M. Assouag, et al., Phosphorous-based epoxy resin composition as an effective anticorrosive coating for steel, Int. J. Ind. Chem. 9 (2018) 231−240.

[29] O. Dagdag, A. El Harfi, L. El Gana, Z.S. Safi, L. Guo, A. Berisha, et al., Designing of phosphorous based highly functional dendrimeric macromolecular resin as an effective coating material for carbon steel in NaCl: computational and experimental studies, J. Appl. Polym. Sci. (2020) 49673.

[30] M. Galai, M. El Gouri, O. Dagdag, Y. El Kacimi, A. Elharfi, M.E. Touhami, New hexa propylene glycol cyclotiphosphazene as efficient organic inhibitor of carbon steel corrosion in hydrochloric acid medium, J. Mater. Environ. Sci. 7 (2016) 1562−1575.

[31] R. Hsissou, O. Dagdag, M. Berradi, M. El Bouchti, M. Assouag, A. Elharfi, Development rheological and anti-corrosion property of epoxy polymer and its composite, Heliyon 5 (2019) e02789.

[32] R. Hsissou, F. Benhiba, S. Abbout, O. Dagdag, S. Benkhaya, A. Berisha, et al., Trifunctional epoxy polymer as corrosion inhibition material for carbon steel in 1.0 M HCl: MD simulations, DFT and complexation computations, Inorg. Chem. Commun. 115 (2020) 107858.

[33] R. Hsissou, B. Benzidia, M. Rehioui, M. Berradi, A. Berisha, M. Assouag, et al., Anticorrosive property of hexafunctional epoxy polymer HGTMDAE for E_{24} carbon steel corrosion in 1.0 M HCl: gravimetric, electrochemical, surface morphology and molecular dynamic simulations, Polym. Bull. 77 (2020) 3577−3601.

[34] R. Hsissou, O. Dagdag, S. Abbout, F. Benhiba, M. Berradi, M. El Bouchti, et al., Novel derivative epoxy resin TGETET as a corrosion inhibition of E_{24} carbon steel in 1.0 M HCl solution. Experimental and computational (DFT and MD simulations) methods, J. Mol. Liq. 284 (2019) 182−192.

[35] M. Rbaa, F. Benhiba, P. Dohare, L. Lakhrissi, R. Touir, B. Lakhrissi, et al., Synthesis of new epoxy glucose derivatives as a non-toxic corrosion inhibitors for carbon steel in molar HCl: experimental, DFT and MD simulation, Chem. Data Collect. 27 (2020) 100394.

[36] M. Rbaa, P. Dohare, A. Berisha, O. Dagdag, L. Lakhrissi, M. Galai, et al., New Epoxy sugar based glucose derivatives as eco friendly corrosion inhibitors for the carbon steel in 1.0 M HCl: experimental and theoretical investigations, J. Alloy. Compd. 833 (2020) 154949.

[37] I. Yarovsky, E. Evans, Computer simulation of structure and properties of crosslinked polymers: application to epoxy resins, Polymer 43 (2002) 963−969.

[38] E.N. Jacobsen, F. Kakiuchi, R.G. Konsler, J.F. Larrow, M. Tokunaga, Enantioselective catalytic ring opening of epoxides with carboxylic acids, Tetrahedron Lett. 38 (1997) 773−776.

14

Green corrosion inhibitors derived through one-step multicomponent reactions: recent developments

Chandrabhan Verma, M.A. Quraishi

INTERDISCIPLINARY RESEARCH CENTER FOR ADVANCED MATERIALS, KING FAHD UNIVERSITY OF PETROLEUM AND MINERALS, DHAHRAN, SAUDI ARABIA

Chapter outline

14.1 Introduction

14.1.1 Corrosion and corrosion inhibitors

Metallic materials undergo chemical and/or electrochemical reactions with the species of the surrounding environment. The process of degradation of metallic components by their reaction with the environment is known as corrosion [1,2]. Corrosion is a highly damaging phenomenon that causes huge safety and safety losses. According to NACE (National Association of Corrosion Engineers), the annual global cost of corrosion is about US$ 2.5 trillion, which constitutes 3%—5% of GDP [3,4]. Along with direct and indirect economic losses, several accidents have been reported because of corrosion. In view of the very high economic and fatality losses associated with corrosion, numerous methods have been developed and implemented for corrosion mitigation [5,6]. Some of the common methods are coatings, painting, alloying and dealloying, synthetic inorganic and organic compounds. Use of synthetic compounds is a most effective and economic method of corrosion prevention. The inhibitors may be of organic, inorganic, or natural types. Fig. 14—1 represents some common examples of each kind of corrosion inhibitors. However, most of the inorganic compounds, mainly chromates, molybdates, and nitrates, are highly toxic, and therefore their current use is highly restricted because of increasing ecological consciousness. Nevertheless, the use of synthetic organic compounds, especially heterocyclic compounds are effective, economical, and easy methods of corrosion mitigation. Generally, organic compounds adsorb through their electron-rich centers, called active sites, on the metallic surface and build a film which isolates the metallic structures from an aggressive environment, thereby protecting from corrosive degradation. The most common actives sites are polar functional groups, such as $-OH$ (hydroxyl), $-NO_2$ (nitro), $-CONH_2$ (amide), $-NH_2$ (primary amine), $-COCl$ (acid chloride), $>NH$ (secondary amine), $>N-$ (tertiary amine) $-COOC_2H_5$ (ester), $-O-$ (ether) $>C=O$ (carbonyl), $>C=S$ (thio-carbonyl), $>C=N-$ (imine) etc. Apart from polar functional groups, homo-, such as $>C=C<$, $-N=N-$, $-C\equiv C-$, and hetero-atomic, such as $-N=O$, $>S=O$, $-C\equiv N$, multiple bonds act as adsorption or active sites [7,8].

Organic corrosion inhibitors adsorb using either physical or chemical or mixed modes (physiochemical) of adsorption mechanisms. The physisorption is mainly attributed to the electrostatic force of attraction between metal surface and inhibitor molecules.

Chemisorption mainly takes place through sharing or transfer of the charges (electrons), whereas mixed adsorption takes place by the combination of both. The electron-rich centers transfer their electrons into the d-orbital of surface metallic atoms and form coordination bonding. Polar functional groups and multiple bonds increase the conjugation and increase the effectiveness of compounds. However, polar functional groups do not increase protection efficiency in all cases. In general, the presence of polar functional groups that are electron releasing in nature increases the protection effectiveness. On the other hands,

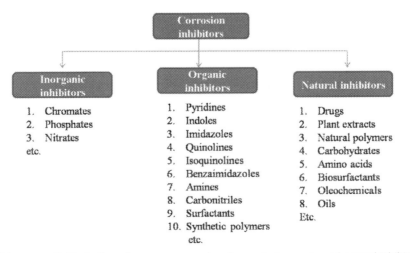

FIGURE 14–1 Diagrammatic illustration of common examples of organic, inorganic, and natural inhibitors.

polar functional substituents that are electron withdrawing in nature decrease the protection effectiveness of the compounds. The effect of substituents can be explained on the basis of Hammett substituent constants. The different forms of Hammett equations are presented below [6,7]:

$$\log \frac{K_R}{K_H} = \sigma\rho \tag{14.1}$$

$$\log \frac{1 - \eta\%_R}{1 - \eta\%_H} = \sigma\rho \tag{14.2}$$

$$\log \frac{\eta\%_R}{\eta\%_H} = \frac{C_{rH}}{C_{rR}} = \sigma\rho - \log\frac{\theta_R}{\theta_H} \tag{14.3}$$

In the above equations, K_H, $\eta\%_H$, C_{rH}, and θ_H are the equilibrium constant, inhibition efficiency, corrosion rate, and surface coverage, respectively [7]. In the above equations, "R" represents a substituent. ρ is the reaction parameter, and σ is the Hammett constant that reflects the total electron density around an active site. A detailed description of these symbols can be found elsewhere. By keeping other things constant, it is expected that the presence of electron-donating groups such as $-OH$, $-NH_2$, $-OMe$, $-NMe_2$, and $-NHMe$, etc. increases the protection effectiveness, whereas electron-withdrawing substituents, such as $-CN$ and $-NO_2$, are expected to decrease the protection efficiency [7,8].

14.1.2 Environmental aspects of corrosion inhibition

According to OSPAR (Oslo Paris Commission) and TEACH (Registration, Evaluation, Authorization, and Restriction of the chemical), a corrosion inhibitor which is associated with high biodegradability, nonbioaccumulation behavior, and negligible environmental toxicity would be considered as environmentally friendly [9,10]. There are two indices, namely, ethical concentration (EC_{50}) and lethal concentration (LC_{50}), that can be used for the measurement of toxicity. LC_{50} and EC_{50} represent the concentrations of compounds which cause death of 50% living population and which adversely affect the growth of living population, respectively [11,12]. For a chemical to be treated as environmentally friendly its EC_{50}/LC_{50} magnitude should be more than 10 mgL^{-1} [13]. According to OSPAR, the biodegradation rate of an environment-friendly chemical would be 60% or more in 28 days [14]. The partition coefficient (log K_{OW} or D^{OW}) is used as a gage for the measurement of bioaccumulation. A K_{OW} or D^{OW} with magnitude of less than 3 establishes the nonbioaccumulative behavior of the chemical [15,16]. This chapter describes an array of different environment-friendly alternatives being used as corrosion inhibitors. To the best of our knowledge this is the first article that collectively describes the all kinds of environment-friendly corrosion inhibitors. The green corrosion inhibitors may be of synthetic type, which includes compounds derived through multicomponent reactions (MCRs), solid phase reactions (SPRs), mechanochemical mixing (MCM), microwave (MW), and

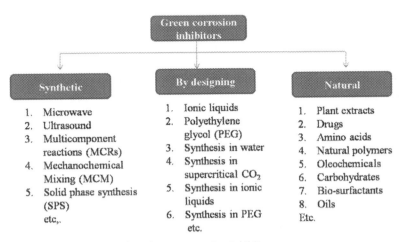

FIGURE 14–2 Some representative examples of green corrosion inhibitors.

ultrasound (US) irradiation [17,18]. Natural environment-friendly alternatives include plant extracts, drugs, carbohydrates, amino acids, biopolymers, and biosurfactants [19,20]. Ionic liquids, polyethylene glycol (PEG), and compounds derived using ionic liquids, PEG, water, and supercritical CO_2 can also be regarded as environment-friendly [21]. A summary of each type of green corrosion inhibitor is presented in Fig. 14–2. Different environment-friendly alternatives being used for traditional toxic corrosion inhibitors with their salient features are described herein.

14.1.3 Green corrosion inhibitors and multicomponent reactions

It is important to mention that most of the traditional organic corrosion inhibitors that are toxic in nature are derived using multistep reactions. Because of increasing ecological consciousness, the use of these compounds is highly restricted. Generally, multistep reactions are associated with numerous disadvantages including a huge release of toxic chemicals and solvents, being highly expensive, tedious, and giving a lower yield. On the other hands, one-step MCRs have emerged as a powerful tool for the synthesis of environmentally benign corrosion inhibitors. The MCRs are connected with several advantages including high synthetic yield and atom economy, ease of performance, shorter reaction time, and lower number of purification steps. It is also reported that MCRs are highly selective compared to the multistep reactions. Because of the reduction in the number of purification steps, synthetic yields are significantly higher for MCRs compared to the multistep reactions. More so, because of lower number of purification steps, a smaller amount of solvents is used during the workup process, and therefore MCRs are associated with a minimum discharge of toxic solvents that later adversely affect living organisms and the surrounding environment. Several compounds derived from MCRs are widely used as corrosion inhibitors [17]. This chapter describes the findings of some major publications on corrosion inhibitors derived from MCRs.

14.2 Corrosion inhibitors derived from multicomponent reactions: literature study

A literature study showed that numerous organic compounds, especially heterocyclic derivatives, are widely used as corrosion inhibitors. Different series of heterocyclic compounds such as pyridine, imidazole, quinoline, isoquinoline, triazole, tetraazole, etc. are widely used. It is observed that in such compounds the presence of polar substituents increases the protection effectiveness compared to the compounds without such substituents. However, this increase in the inhibition effectiveness is more pronounced in the presence of electron-donating substituents. In fact, there are numerous reports available in which compounds containing electron-withdrawing substituents, such as −CN, −NO$_2$ and −COOH, show lower protection effectiveness as compared to the compounds without such substituents. However, an inverse order of protection effectiveness has also been reported in a few studies. A literature study showed that pyridine-based heterocyclic compounds derived from MCRs are widely used as corrosion inhibitors. In pyridine derivatives synthesized through MCRs, the effect of substituents is widely studied. Singh et al. [22] investigated the four benzene-fused pyridine (quinoline) derivatives differing in the nature of substituents for mild steel corrosion in acidic electrolyte. Different chemical, electrochemical, and surface methods were employed to demonstrate the anticorrosive effect of the tested compounds. Results showed that inhibition effectiveness of the tested compounds followed the order: Q-4 (-NMe$_2$) > Q-3 (-OMe) > Q-2 (-Me) > Q-1 (-H). Results showed that inhibition effect of the four tested compounds increased on increasing their concentrations and decreased with temperature. Adsorption of the tested compounds followed the Langmuir adsorption isotherm model. Polarization analyses showed that Q-3 (-OMe), Q-2 (-Me), and Q-1 (-H) acted as mixed-type inhibitors, whereas Q-4 (-NMe$_2$) behaved as a cathodic type inhibitor. Interfacial behavior of Q1−4 was observed using electrochemical impedance spectroscopy (EIS) studies. Adsorption mechanism of corrosion inhibition was supported using surface morphological characterization through scanning electron microscopy (SEM), atomic force microscopy (AFM), and X-ray photoelectron spectroscopy (XPS) methods. SEM and AFM studies showed that the presence of Q1−4 in the corrosive electrolyte causes significant improvement in the surface morphology which is attributed to the adsorption of these compounds over the metallic surface.

A similar observation was also observed by these authors while studying the anticorrosion effect of three naphthyridine derivatives on mild steel corrosion in 1 M HCl using experimental and computational methods [23]. Results showed that inhibition efficiencies of tested naphthyridine derivatives followed the order: 94.28% (N-1; -H) < 96.66% (N-2;-Me) < 98.09% (N-3;-OMe). The protection effectiveness of the N1−3 increases on increasing their concentration and maximum protection efficiency was observed at 6.54×10^{-5} M concentration. Authors observed that an increase in temperature causes a subsequent decrease in the inhibition effect of tested compounds. Polarization studies revealed that tested compounds behaved with mixed-type behavior with slight cathodic dominance. EIS studies showed that N1−3 behaved as interface-type inhibitors. Adsorption of the tested compounds followed the Langmuir adsorption isotherm model. Adsorption behavior of N1−3 was supported by SEM and AFM studies, where significant smoothness in the surface morphologies of metallic specimens occurs in the presence of inhibitors. Density functional theory (DFT) analyses showed that studied compounds interact with the metal surface through donor−acceptor interactions. Strong and spontaneous behavior of inhibitor adsorption on a metallic surface was observed using Monte Carlo (MC) simulations. It was observed that N1−3 adsorbs on a metallic surface using a flat orientation. Similarly, other compounds derived from MCRs are widely used as corrosion inhibitors. Synthetic schemes, the nature of metals, and the electrolytes of some major reports are presented in Table 14.1. Generally, pyridine derivatives derived thorough MCRs act as mixed-type corrosion inhibitors with some cathodic dominance and their adsorption follows the Langmuir adsorption isotherm model. EIS studies showed that such compounds adsorb at the interface of the environment and metal surface and acted as interface-type inhibitors.

Highly concentrated acidic solutions (mostly 15% HCl), are extensively employed in different industrial cleaning processes, such as acid descaling, acid cleaning, oil well acidification, and acid pickling. The acid cleaning and pickling processes are conducted to remove the surface impurities during metallurgical processing of metallic ores in order to get pure metals. However, because of their highly aggressive nature, these processes are associated with exhaustive metallic dissolution in the form of metallic corrosion. Therefore these processes need the addition of some external additives called corrosion inhibitors. Organic compounds containing heteroatoms and aromatic rings are established as one of the most effective class of corrosion inhibitors. These compounds adsorb on metallic surface and form inhibitive film. Adsorption of these compounds is influenced by numerous factors, including their molecular structures and concentration, temperature, and immersion time. Compounds derived through MCRs are also extensively reported as metallic corrosion inhibitors in 15% HCl (pickling) solution. Ansari et al. [59] investigated the anticorrosion property of two chromenopyridine derivatives (PPCs) on N80 steel in 15% HCl using experimental methods. Weight loss analysis showed that PPCs acted as good corrosion inhibitors and showed the highest protection power of 92.4% (PPC-1) and 84.1% (PPC-2) at 200 mgL^{-1} concentration. PPC-1 and PPC-2 acted as mixed-type corrosion inhibitors as derived from polarization investigation. Adsorption of the PPCs on metallic surface obeyed the Langmuir adsorption isotherm model. Adsorption mechanism of corrosion inhibition on PPC-1 and PPC-2 was further supported by SEM investigation. Synthetic schemes, the nature of metals and electrolytes of some major reports are presented in Table 14.2.

Table 14.1 Synthetic schemes for some major corrosion inhibitors derived through multicomponent reactions for low-concentration acidic media.

MCR synthetic scheme	System	Ref.	MCR synthetic scheme	System	Ref.
	Fe/ 1M HCl	[22]		Fe/ 1M HCl	[23]
	Fe/ 1M HCl	[24]		Fe/ 1M HCl	[25]
	Fe/ 1M HCl	[26]		Fe/ 1M HCl	[27]
	Fe/ 1M HCl	[28]		Fe/ 1M HCl	[29]
	Cu/0.5 M HCl	[30]		Fe/ 1M HCl	[31]
	Fe/ 1M HCl	[32]		Fe/ 1M HCl	[33]
	Fe/ 1M HCl	[34]		Fe/ 1M HCl,	[35]

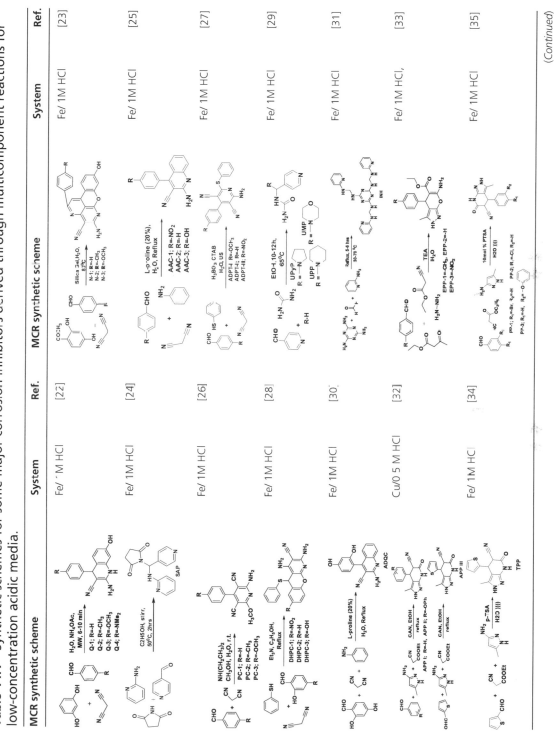

(Continued)

Table 14.1 (Continued)

MCR synthetic scheme	System	Ref.	MCR synthetic scheme	System	Ref.

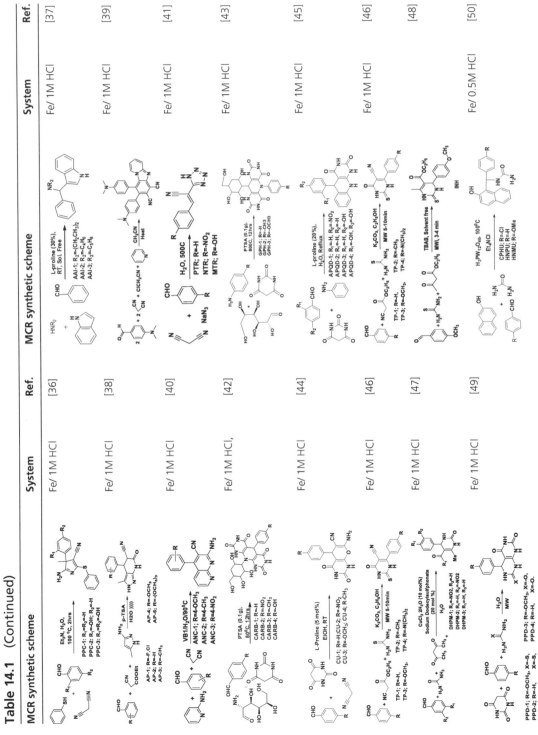

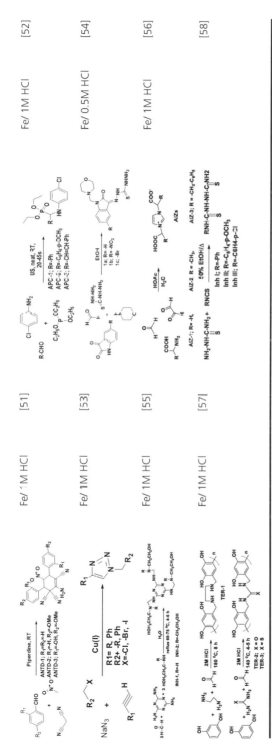

[51] [52]

[53] [54]

[55] [56]

[57] [58]

Fe/ 1M HCl

Fe/ 1M HCl

Fe/ 1M HCl

Fe/ 1M HCl

Fe/ 1M HCl

Fe/ 0.5M HCl

Fe/ 1M HCl

Fe/ 1M HCl

MCRs, multicomponent reactions.

Table 14.2   Synthetic schemes for some major corrosion inhibitors derived through multicomponent reactions for N80 steel in 15% HCl.

MCR synthetic scheme	System	Ref.	MCR synthetic scheme	System	Ref.
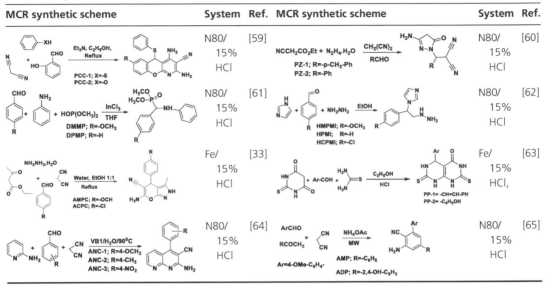	N80/ 15% HCl	[59]		N80/ 15% HCl	[60]
	N80/ 15% HCl	[61]		N80/ 15% HCl	[62]
	Fe/ 15% HCl	[33]		Fe/ 15% HCl,	[63]
	N80/ 15% HCl	[64]		N80/ 15% HCl	[65]

MCRs, multicomponent reactions.

14.3 Conclusion

On the basis of ongoing discussions it is established that one-step MCRs act as a tool for the synthesis of environment-friendly corrosion inhibitors. The MCRs are associated with several advantages such as lower number of purification and workup process that make them cost-effective and easy to perform. Several classes of organic compounds, especially heterocyclic compounds, derived through MCRs are used as valuable corrosion inhibitors for academic as well as industrial purposes. Most of the compounds derived through MCRs act as mixed- and interface-type inhibitors with cathodic predominance in a few studies. Their adsorption on metallic surface obeyed the Langmuir adsorption isotherm model. The adsorption of compounds derived through MCRs on a metallic surface was supported using several surface analysis methods including SEM, AFM, and XPS. Theoretical analyses showed that such compounds interact with a metallic surface using a donor—acceptor mechanism. They spontaneously adsorb and acquire the flat or nearly flat orientation, thereby covering a larger surface area and behave as excellent corrosion inhibitors.

Useful links

https://www.sciencedirect.com/science/article/abs/pii/S0167732218346737.
https://chemistry-europe.onlinelibrary.wiley.com/doi/abs/10.1002/ejoc.201300166.
https://www.mdpi.com/1420-3049/14/12/4936.

References

[1] U. Chatterjee, S.K. Bose, S.K. Roy, Environmental Degradation of Metals: Corrosion Technology Series/14, CRC Press, 2001.

[2] A. Balamurugan, S. Rajeswari, G. Balossier, A. Rebelo, J. Ferreira, Corrosion aspects of metallic implants —an overview, Mater. Corros. 59 (2008) 855−869.

[3] C. Verma, H. Lgaz, D. Verma, E.E. Ebenso, I. Bahadur, M. Quraishi, Molecular dynamics and Monte Carlo simulations as powerful tools for study of interfacial adsorption behavior of corrosion inhibitors in aqueous phase: a review, J. Mol. Liq. 260 (2018) 99−120.

[4] A. Mishra, J. Aslam, C. Verma, M. Quraishi, E.E. Ebenso, Imidazoles as highly effective heterocyclic corrosion inhibitors for metals and alloys in aqueous electrolytes: a review, J. Taiwan Inst. Chem. Eng. 114 (2020) 341−358.

[5] M.G. Fontana, Corrosion Engineering, Tata McGraw-Hill Education, 2005.

[6] V.S. Sastri, Green Corrosion Inhibitors: Theory and Practice, John Wiley & Sons, 2012.

[7] C. Verma, L. Olasunkanmi, E.E. Ebenso, M. Quraishi, Substituents effect on corrosion inhibition performance of organic compounds in aggressive ionic solutions: a review, J. Mol. Liq. 251 (2018) 100−118.

[8] C. Verma, E.E. Ebenso, M. Quraishi, Molecular structural aspects of organic corrosion inhibitors: influence of−CN and−NO_2 substituents on designing of potential corrosion: inhibitors for aqueous media, J. Mol. Liq. 316 (2020) 113874.

[9] M.S. Wong, The convention for the protection of the marine environment of the North-East Atlantic (the 'OSPAR Convention')(and Annexes I, II, III, IV), Elgar Encyclopedia of Environmental Law, Edward Elgar Publishing Limited, 2017, pp. 189−198.

[10] V. Bakir, Policy agenda setting and risk communication: greenpeace, shell, and issues of trust, Harv. Int. J. Press/Polit. 11 (2006) 67−88.

[11] N. van der Hoeven, A.A. Gerritsen, Effects of chlorpyrifos on individuals and populations of *Daphnia pulex* in the laboratory and field, Environ. Toxicol. Chem. Int. J. 16 (1997) 2438−2447.

[12] S. Hasenbein, S.P. Lawler, J. Geist, R.E. Connon, The use of growth and behavioral endpoints to assess the effects of pesticide mixtures upon aquatic organisms, Ecotoxicology 24 (2015) 746−759.

[13] T. Tišler, J. Zagorc-Končan, Acute and chronic toxicity of arsenic to some aquatic organisms, Bull. Environ. Contam. Toxicol. 69 (2002) 421−429.

[14] D. Hasson, H. Shemer, A. Sher, State of the art of friendly "green" scale control inhibitors: a review article, Ind. Eng. Chem. Res. 50 (2011) 7601−7607.

[15] M. Nendza, R. Kühne, A. Lombardo, S. Strempel, G. Schüürmann, PBT assessment under REACH: screening for low aquatic bioaccumulation with QSAR classifications based on physicochemical properties to replace BCF in vivo testing on fish, Sci. Total Environ. 616 (2018) 97−106.

[16] K. Bittermann, L. Linden, K.-U. Goss, Screening tools for the bioconcentration potential of monovalent organic ions in fish, Environ. Sci. Process. Impacts 20 (2018) 845−853.

[17] C. Verma, J. Haque, M. Quraishi, E.E. Ebenso, Aqueous phase environmental friendly organic corrosion inhibitors derived from one step multicomponent reactions: a review, J. Mol. Liq. 275 (2019) 18−40.

[18] C. Verma, M. Quraishi, E.E. Ebenso, Microwave and ultrasound irradiations for the synthesis of environmentally sustainable corrosion inhibitors: an overview, Sustain. Chem. Pharm. 10 (2018) 134−147.

[19] C. Verma, E.E. Ebenso, I. Bahadur, M. Quraishi, An overview on plant extracts as environmental sustainable and green corrosion inhibitors for metals and alloys in aggressive corrosive media, J. Mol. Liq. 266 (2018) 577−590.

[20] S.A. Umoren, U.M. Eduok, Application of carbohydrate polymers as corrosion inhibitors for metal substrates in different media: a review, Carbohydr. Polym. 140 (2016) 314−341.

[21] C. Verma, E.E. Ebenso, M. Quraishi, Ionic liquids as green and sustainable corrosion inhibitors for metals and alloys: an overview, J. Mol. Liq. 233 (2017) 403−414.

[22] P. Singh, V. Srivastava, M. Quraishi, Novel quinoline derivatives as green corrosion inhibitors for mild steel in acidic medium: electrochemical, SEM, AFM, and XPS studies, J. Mol. Liq. 216 (2016) 164−173.

[23] P. Singh, E.E. Ebenso, L.O. Olasunkanmi, I. Obot, M. Quraishi, Electrochemical, theoretical, and surface morphological studies of corrosion inhibition effect of green naphthyridine derivatives on mild steel in hydrochloric acid, J. Phys. Chem. C. 120 (2016) 3408−3419.

[24] M. Jeeva, M. susai Boobalan, G.V. Prabhu, Adsorption and anticorrosion behavior of 1-((pyridin-2-yla-mino)(pyridin-4-yl) methyl) pyrrolidine-2, 5-dione on mild steel surface in hydrochloric acid solution, Res. Chem. Intermed. 44 (2018) 425−454.

[25] C. Verma, M. Quraishi, L. Olasunkanmi, E.E. Ebenso, L-Proline-promoted synthesis of 2-amino-4-aryl-quinoline-3-carbonitriles as sustainable corrosion inhibitors for mild steel in 1 M HCl: experimental and computational studies, RSC Adv. 5 (2015) 85417−85430.

[26] K. Ansari, M. Quraishi, A. Singh, Corrosion inhibition of mild steel in hydrochloric acid by some pyridine derivatives: an experimental and quantum chemical study, J. Ind. Eng. Chem. 25 (2015) 89−98.

[27] M.A. Quraishi, 2-Amino-3, 5-dicarbonitrile-6-thio-pyridines: new and effective corrosion inhibitors for mild steel in 1 M HCl, Ind. Eng. Chem. Res. 53 (2014) 2851−2859.

[28] C. Verma, L.O. Olasunkanmi, I. Obot, E.E. Ebenso, M. Quraishi, 2, 4-Diamino-5-(phenylthio)-5 H-chromeno [2, 3-b] pyridine-3-carbonitriles as green and effective corrosion inhibitors: gravimetric, electrochemical, surface morphology and theoretical studies, RSC Adv. 6 (2016) 53933−53948.

[29] M. Jeeva, G.V. Prabhu, M.S. Boobalan, C.M. Rajesh, Interactions and inhibition effect of urea-derived Mannich bases on a mild steel surface in HCl, J. Phys. Chem. C. 119 (2015) 22025−22043.

[30] C. Verma, M. Quraishi, 2-Amino-4-(2, 4-dihydroxyphenyl) quinoline-3-carbonitrile as sustainable corrosion inhibitor for SAE 1006 steel in 1 M HCl: electrochemical and surface investigation, J. Assoc. Arab. Univ. Basic Appl. Sci. 23 (2017) 29−36.

[31] C. Verma, M. Quraishi, E. Ebenso, Green ultrasound assisted synthesis of N 2, N 4, N 6-tris ((pyridin-2-ylamino) methyl)-1, 3, 5-triazine-2, 4, 6-triamine as effective corrosion inhibitor for mild steel in 1 M hydrochloric acid medium, Int. J. Electrochem. Sci. 8 (2013) 10864−10877.

[32] M. Quraishi, The corrosion inhibition effect of aryl pyrazolo pyridines on copper in hydrochloric acid system: computational and electrochemical studies, RSC Adv. 5 (2015) 41923−41933.

[33] M. Yadav, L. Gope, N. Kumari, P. Yadav, Corrosion inhibition performance of pyranopyrazole derivatives for mild steel in HCl solution: gravimetric, electrochemical and DFT studies, J. Mol. Liq. 216 (2016) 78−86.

[34] P. Singh, M. Quraishi, S. Gupta, A. Dandia, Investigation of the corrosion inhibition effect of 3-methyl-6-oxo-4-(thiophen-2-yl)-4, 5, 6, 7-tetrahydro-2H-pyrazolo [3, 4-b] pyridine-5-carbonitrile (TPP) on mild steel in hydrochloric acid, J. Taibah Univ. Sci. 10 (2016) 139−147.

[35] S. Gupta, A. Dandia, P. Singh, M. Qureishi, Green synthesis of pyrazolo [3, 4-b] pyridine derivatives by ultrasonic technique and their application as corrosion inhibitor for mild steel in acid medium, J. Mater. Environ. Sci. 6 (2015) 168−177.

[36] C. Verma, E. Ebenso, I. Bahadur, I. Obot, M. Quraishi, 5-(Phenylthio)-3H-pyrrole-4-carbonitriles as effective corrosion inhibitors for mild steel in 1 M HCl: experimental and theoretical investigation, J. Mol. Liq. 212 (2015) 209−218.

[37] C. Verma, M. Quraishi, E. Ebenso, I. Obot, A. El Assyry, 3-Amino alkylated indoles as corrosion inhibitors for mild steel in 1M HCl: experimental and theoretical studies, J. Mol. Liq. 219 (2016) 647−660.

[38] A. Dandia, S. Gupta, P. Singh, M. Quraishi, Ultrasound-assisted synthesis of pyrazolo [3, 4-b] pyridines as potential corrosion inhibitors for mild steel in 1.0 M HCl, ACS Sustain. Chem. Eng. 1 (2013) 1303−1310.

[39] C. Verma, M. Quraishi, Adsorption behavior of 8, 9-bis (4 (dimethyl amino) phenyl) benzo [4, 5] imidazo [1, 2-a] pyridine-6, 7-dicarbonitrile on mild steel surface in 1 M HCl, J. Assoc. Arab. Univ. Basic Appl. Sci. 22 (2017) 55–61.

[40] P. Singh, M. Makowska-Janusik, P. Slovensky, M. Quraishi, Nicotinonitriles as green corrosion inhibitors for mild steel in hydrochloric acid: electrochemical, computational and surface morphological studies, J. Mol. Liq. 220 (2016) 71–81.

[41] C. Verma, M. Quraishi, A. Singh, 5-Substituted 1H-tetrazoles as effective corrosion inhibitors for mild steel in 1 M hydrochloric acid, J. Taibah Univ. Sci. 10 (2016) 718–733.

[42] C. Verma, L.O. Olasunkanmi, E.E. Ebenso, M.A. Quraishi, I.B. Obot, Adsorption behavior of glucosamine-based, pyrimidine-fused heterocycles as green corrosion inhibitors for mild steel: experimental and theoretical studies, J. Phys. Chem. C 120 (2016) 11598–11611.

[43] C. Verma, M.A. Quraishi, K. Kluza, M. Makowska-Janusik, L.O. Olasunkanmi, E.E. Ebenso, Corrosion inhibition of mild steel in 1M HCl by D-glucose derivatives of dihydropyrido [2, 3-d: 6, 5-d′] dipyrimidine-2, 4, 6, 8 (1H, 3H, 5H, 7H)-tetraone, Sci. Rep. 7 (2017) 44432.

[44] D.K. Yadav, M.A. Quraishi, Application of some condensed uracils as corrosion inhibitors for mild steel: gravimetric, electrochemical, surface morphological, UV–visible, and theoretical investigations, Ind. Eng. Chem. Res. 51 (2012) 14966–14979.

[45] C. Verma, L. Olasunkanmi, I. Obot, E.E. Ebenso, M. Quraishi, 5-Arylpyrimido-[4, 5-b] quinoline-diones as new and sustainable corrosion inhibitors for mild steel in 1 M HCl: a combined experimental and theoretical approach, RSC Adv. 6 (2016) 15639–15654.

[46] P. Singh, A. Singh, M. Quraishi, Thiopyrimidine derivatives as new and effective corrosion inhibitors for mild steel in hydrochloric acid: electrochemical and quantum chemical studies, J. Taiwan Inst. Chem. Eng. 60 (2016) 588–601.

[47] D.K. Yadav, B. Maiti, M. Quraishi, Electrochemical and quantum chemical studies of 3, 4-dihydropyrimidin-2 (1H)-ones as corrosion inhibitors for mild steel in hydrochloric acid solution, Corros. Sci. 52 (2010) 3586–3598.

[48] R. Korde, C.B. Verma, E. Ebenso, M. Quraishi, Electrochemical and thermo dynamical investigation of 5-ethyl 4-(4-methoxyphenyl)-6-methyl-2-thioxo-1, 2, 3, 4 tetrahydropyrimidine-5-carboxylate on corrosion inhibition behavior of aluminium in 1M hydrochloric acid medium, Int. J. Electrochem. Sci. 10 (2015) 1081–1093.

[49] K. Ansari, Sudheer, A. Singh, M. Quraishi, Some pyrimidine derivatives as corrosion inhibitor for mild steel in hydrochloric acid, J. Dispers. Sci. Technol. 36 (2015) 908–917.

[50] M. Bahrami, S. Hosseini, P. Pilvar, Experimental and theoretical investigation of organic compounds as inhibitors for mild steel corrosion in sulfuric acid medium, Corros. Sci. 52 (2010) 2793–2803.

[51] C. Verma, M. Quraishi, A. Singh, 2-Amino-5-nitro-4, 6-diarylcyclohex-1-ene-1, 3, 3-tricarbonitriles as new and effective corrosion inhibitors for mild steel in 1 M HCl: experimental and theoretical studies, J. Mol. Liq. 212 (2015) 804–812.

[52] N.K. Gupta, C. Verma, R. Salghi, H. Lgaz, A. Mukherjee, M. Quraishi, New phosphonate based corrosion inhibitors for mild steel in hydrochloric acid useful for industrial pickling processes: experimental and theoretical approach, N. J. Chem. 41 (2017) 13114–13129.

[53] R. González-Olvera, V. Román-Rodríguez, G.E. Negrón-Silva, A. Espinoza-Vázquez, F.J. Rodríguez-Gómez, R. Santillan, Multicomponent synthesis and evaluation of new 1, 2, 3-triazole derivatives of dihydropyrimidinones as acidic corrosion inhibitors for steel, Molecules 21 (2016) 250.

[54] A.K. Singh, M. Quraishi, Inhibiting effects of 5-substituted isatin-based Mannich bases on the corrosion of mild steel in hydrochloric acid solution, J. Appl. Electrochem. 40 (2010) 1293–1306.

[55] C. Verma, M. Quraishi, E. Ebenso, Mannich bases derived from melamine, formaldehyde alkanolamines as novel corrosion inhibitors for mild steel in hydrochloric acid medium, Int. J. Electrochem. Sci. 8 (2013) 10851–10863.

[56] V. Srivastava, J. Haque, C. Verma, P. Singh, H. Lgaz, R. Salghi, et al., Amino acid based imidazolium zwitterions as novel and green corrosion inhibitors for mild steel: experimental, DFT and MD studies, J. Mol. Liq. 244 (2017) 340−352.

[57] C.B. Verma, M. Quraishi, E. Ebenso, Electrochemical and thermodynamic investigation of some soluble terpolymers as effective corrosion inhibitors for mild steel in 1M hydrochloric acid solution, Int. J. Electrochem. Sci. 8 (2013) 12894−12906.

[58] S. Kumar, D. Sharma, P. Yadav, M. Yadav, Experimental and quantum chemical studies on corrosion inhibition effect of synthesized organic compounds on N80 steel in hydrochloric acid, Ind. Eng. Chem. Res. 52 (2013) 14019−14029.

[59] K. Ansari, M. Quraishi, A. Singh, Chromenopyridin derivatives as environmentally benign corrosion inhibitors for N80 steel in 15% HCl, J. Assoc. Arab. Univ. Basic Appl. Sci. 22 (2017) 45−54.

[60] K. Ansari, M. Quraishi, A. Singh, S. Ramkumar, I.B. Obote, Corrosion inhibition of N80 steel in 15% HCl by pyrazolone derivatives: electrochemical, surface and quantum chemical studies, RSC Adv. 6 (2016) 24130−24141.

[61] M. Yadav, D. Sharma, S. Kumar, I. Bahadur, E. Ebenso, Electrochemical and theoretical studies on amino phosphonates as efficient corrosion inhibitor for N80 steel in hydrochloric acid solution, Int. J. Electrochem. Sci. 9 (2014) 6580−6593.

[62] M. Yadav, P. Yadav, U. Sharma, Substituted imidazoles as corrosion inhibitors for N80 steel in hydrochloric acid, Indian J. Chem. Technol. 20 (2013) 363−370.

[63] J. Haque, K. Ansari, V. Srivastava, M. Quraishi, I. Obot, Pyrimidine derivatives as novel acidizing corrosion inhibitors for N80 steel useful for petroleum industry: a combined experimental and theoretical approach, J. Ind. Eng. Chem. 49 (2017) 176−188.

[64] K. Ansari, M. Quraishi, Experimental and computational studies of naphthyridine derivatives as corrosion inhibitor for N80 steel in 15% hydrochloric acid, Phys. E Low-Dimen. Syst. Nanostruct. 69 (2015) 322−331.

[65] K. Ansari, M. Quraishi, A. Singh, Pyridine derivatives as corrosion inhibitors for N80 steel in 15% HCl: electrochemical, surface and quantum chemical studies, Measurement 76 (2015) 136−147.

15

Ultrasound and microwave heating for the synthesis of green corrosion inhibitors: a literature study

Chandrabhan Verma[1], Eno E. Ebenso[2], M.A. Quraishi[1]

[1]INTERDISCIPLINARY RESEARCH CENTER FOR ADVANCED MATERIALS, KING FAHD UNIVERSITY OF PETROLEUM AND MINERALS, DHAHRAN, SAUDI ARABIA [2]INSTITUTE FOR NANOTECHNOLOGY AND WATER SUSTAINABILITY, COLLEGE OF SCIENCE, ENGINEERING AND TECHNOLOGY, UNIVERSITY OF SOUTH AFRICA, JOHANNESBURG, SOUTH AFRICA

Chapter outline

Abbreviations

PEG	polyethylene glycol
PTC	phase transfer catalyst
VOCs	volatile organic compounds
WL	weight loss
HE	hydrogen evolution
AAs	amino acids

Environmentally Sustainable Corrosion Inhibitors. DOI: https://doi.org/10.1016/B978-0-323-85405-4.00015-X

GDP	gross domestic product
%IE	percentage inhibition efficiency
C_R	corrosion rate
EIS	electrochemical impedance spectroscopy
PDP	potentiodynamic polarization
LPR	linear polarization resistance
OCP	open circuit potential
θ	surface coverage
SEM	scanning electron microscopy
EDX	energy dispersive X-ray
XRD	X-ray powder diffraction
XPS	X-ray photoelectron spectroscopy
AFM	atomic force microscopy
DFT	density functional theory
MDS	molecular dynamics simulation
MCS	Monte Carlo simulation
REACH	registration, evaluation, authorization and restriction of the chemical
OSPAR	Oslo Paris Commission
LC_{50}	lethal concentration
EC_{50}	effective concentration
$\log K_{OW}$ or D^{OW}	partition coefficient
MCRs	multicomponent reactions
MW	microwave
US	ultrasound
CH: TS-Cht	chitosan

15.1 Introduction

15.1.1 Corrosion and corrosion inhibitors: basic information

Corrosion is defined as degradation of metal by the electrochemical reactions with the components of the environment. Corrosion causes immense economic, safety, and fatality losses in both developed and developing countries. According to National Association of Corrosion Engineers, the recent corrosion cost is about 3.5% of the world's GDP [1,2]. In view of very high economic loss, several methods of corrosion mitigation have been developed depending upon the nature of environment and metallic materials [3–5]. Among the available methods of corrosion protection, the use of organic, inorganic, and mixed coatings, painting, alloying, dealloying, galvanization, and synthetic organic compounds are the most common (Fig. 15–1).

Recently, the implementation of organic (heterocyclic) compounds as corrosion inhibitors is one of the most cost-effective, easiest, and most effective methods of corrosion mitigation [6–8]. Generally, most of the organic compounds contain several electron-rich centers including polar functional groups, such as $-OH$, $-CONH_2$, $-OMe$, $>C=O$, $-NMe_2$, $-N=N-$, $-NHMe$, $-NH_2$, $>C=S$, $-N=O$, and multiple bonds, such as $>C=C<$, $-C\equiv C-$, $-C\equiv N$, through which they can adsorb effectively on a metallic surface. This type of adsorption results in the formation of protective films which separate metal from an

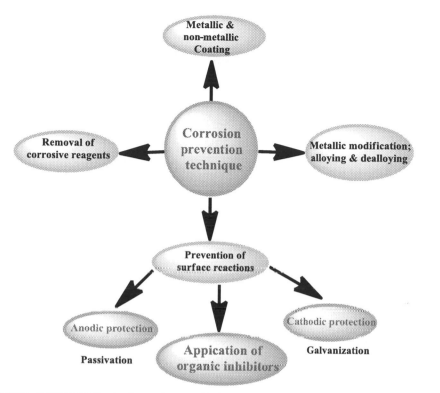

FIGURE 15–1 Some common methods of corrosion monitoring.

aggressive environment [9,10]. Organic inhibitors can be adsorbed physically or chemically. However, most of the organic compounds adsorb on a metallic surface using a physiochemisorption mechanism [11,12]. Obviously, physisorption takes place through electrostatic interactions between a charged metallic surface and charged inhibitor molecules, while chemisorption occurs through sharing of charges between metallic d-orbitals and inhibitor molecules.

15.1.2 Green corrosion inhibitors

Organic compounds are established as one of the most effective and economic methods of corrosion mitigation. However, their current implementation is restricted because of their synthesis using toxic chemicals, solvents, and catalysts in multistep reactions. The increasing ecological and environmental awareness do not permit the use of such chemicals. In view of this, several environment-friendly alternatives (Fig. 15–2) have been developed and used as corrosion inhibitors for different metals and alloys in versatile electrolytes [5,13,14]. In this direction the use of plant extracts as environment-friendly corrosion inhibitors is gaining particular attention. Because of their biological (plant) origin, plant extracts are regarded as one of the most significant environment-friendly alternatives to the traditional toxic

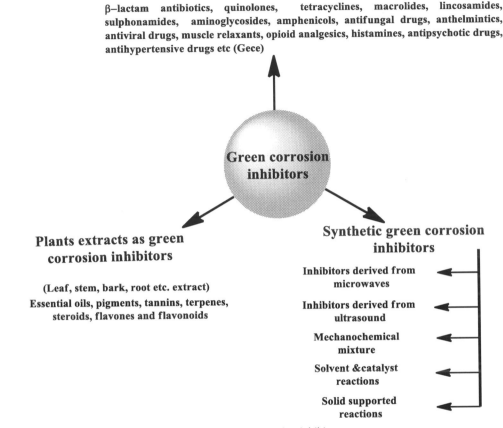

Drugs as green corrosion inhibitors

β–lactam antibiotics, quinolones,   tetracyclines, macrolides, lincosamides, sulphonamides, aminoglycosides, amphenicols, antifungal drugs, anthelmintics, antiviral drugs, muscle relaxants, opioid analgesics, histamines, antipsychotic drugs, antihypertensive drugs etc (Gece)

Green corrosion inhibitors

Plants extracts as green corrosion inhibitors

(Leaf, stem, bark, root etc. extract)
Essential oils, pigments, tannins, terpenes, steroids, flavones and flavonoids

Synthetic green corrosion inhibitors

Inhibitors derived from microwaves

Inhibitors derived from ultrasound

Mechanochemical mixture

Solvent &catalyst reactions

Solid supported reactions

FIGURE 15–2 Examples of synthetic and natural green corrosion inhibitors.

corrosion inhibitors [15]. In view of this several research and review articles have been recently published dealing with the corrosion inhibition effect of plant extracts. Another environment-friendly alternative is chemical medicines (drugs) that are widely used as corrosion inhibitors [16,17]. Due to their complex structures, the presence of polar functional groups, and multiple bonds, they offer strong anticorrosion activity. Several review and research articles have been published dealing with the corrosion inhibition effect of drugs [16,18].

Nowadays, microwave (MW) and ultrasound (US) irradiation have emerged as powerful energy heating sources [19–21]. Because of their several advantages, such as shorter reaction time, high synthetic efficiency, high selectivity, uniform heating, cost-effectivity, and high synthetic yields, MW and US catalyzed reactions can be regarded as environment-friendly approaches for the synthesis of corrosion inhibitors [22–24]. MW and US radiation coupled with multicomponent reactions (MCRs) offer some of the greenest synthetic approaches. The organic compounds derived from US and MW irradiation with and without MCRs are widely used as corrosion inhibitors [25,26]. Apart from that, ionic liquids and natural polymers can also be used as environment-friendly alternatives to traditional toxic corrosion inhibitors [27,28]. Carbohydrates, amino acid, and their derivatives are also treated as environmentally benign alternatives [29]. Compounds synthesized in green solvents such as water, ionic liquids, and supercritical CO_2 are also regarded as green alternatives. This chapter deals with the collection of works published on the corrosion inhibition effectiveness of organic compounds derived using nonconventional MW and US heating.

15.1.3 Microwave irradiation in organic synthesis

Use of MW irradiation has emerged as a powerful alternative heating source for the synthesis of various heterocycles, polymeric, nanomaterials and inorganic molecules [30]. Generally, in conventional heating methods (heating on hot-plate), chemical transformations occur via slow activation of reactant molecules in which energy (heat) first passes through the wall of the reaction vessel in order to reach the reactant and solvent molecules [31]. Therefore conventional heating is a slow and inefficient method of transferring thermal and kinetic energy to the solvent and reactant molecules. On the other hand MW radiation directly coupled with reactant molecules can cause a sudden and random increase of the reaction temperature. The rate and selectivities of the MW-catalyzed reactions can be ensured by appropriately selecting the MW indices. MW heating has several advantages over conventional heating, such as rapid deep inside heating, high temperature homogeneity, and preferential selective heating [32]. In the last 30 years, several MW-assisted chemical transformations have been carried out (Fig. 15−3). MW-mediated heating has the following major advantages over a conventional heating method [32]:

1. Uniform heating throughout the reaction mixture.
2. High reaction rate (reaction rate increases up to 1000-fold).
3. High efficiency of heating.
4. Reduces the production of unwanted side products (waste prevention).
5. High purity of the synthesized products.
6. Improved reproducibility.
7. No environmental heat loss during MW heating.
8. Avoids unwanted heating of the reaction vessel.
9. Low operating cost.

15.1.3.1 How does microwave irradiation works?

Since MW irradiation causes instantaneous heating of the reaction mixture, the chemical reactions that generally require several hours or even days for their completion by conventional heating can be performed in a few minutes or even in seconds with the aid of MW irradiation (Fig. 15−4). The rate of MW-catalyzed reaction can be described using the Arrhenius equation [33,34]:

$$K = A\,e^{-Ea/RT} \tag{1}$$

where T, R, A, and k denote the absolute temperature, the gas constant, the Arrhenius preexponential factor and the rate constant, respectively. E_a is the activation energy. The kinetics of any chemical transformation is mainly determined by temperature effect. Under

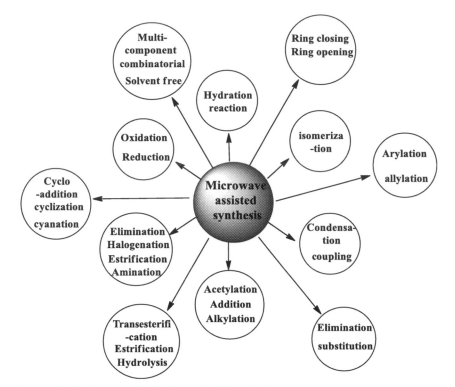

FIGURE 15–3 Examples of microwave irradiation-assisted chemical transformation.

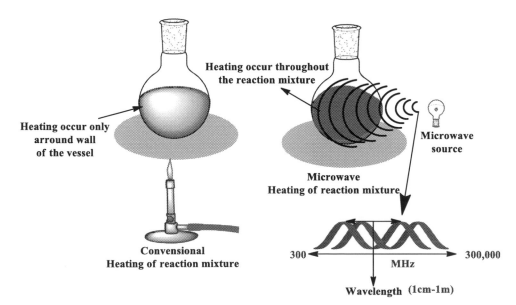

FIGURE 15–4 Diagrammatic representation of conventional and microwave heating.

conventional heating methods, absolute temperature can be defined as the bulk temperature (T_B) at which all the reacting molecules undergo heating with respect to their polar or non-polar nature. However, MW heating is totally different from conventional heating, which instantaneously activates all the reacting molecules that have either permanent dipole and/or have ionic conductance. In MW influenced reactions, heating of the reactant mixtures occurs in nanoseconds $(10^{-9}\,\text{s})$ and therefore excited reactant molecules are incapable to be relaxed to ground state (relaxation requires $10^{-5}\,\text{s}$).

In other words, in MW-irradiated reactions, excited molecules are unable to attain the state of equilibrium; this results in a state of nonequilibrium which generally leads to an instantaneous high temperature (T_I). Generally, T_I is a function of the MW power input [35]. Obviously, the greater the intensity of the MW radiation supplied to the reaction mixture, the greater will be value of T_I. Although T_I cannot be measured directly, its magnitude is much higher for most of the chemical transformations than that of T_B which is a measurable quantity. The following two theories have been proposed to explain the mechanism of MW-induced heating:

15.1.3.1.1 Dielectric or dipolar heating mechanism

The dipole polarization is a process through which polar molecules interact with the MW radiation and produce heat. Whenever, polar molecules interact with electromagnetic radiation of an appropriate frequency, they try to align themselves in a phase depending upon the direction of the applied electromagnetic radiation [36]. However, at the same time a few molecules are unable to follow the field because of the intermolecular force of attraction and repulsion acting between them [37]. The two different but adverse phenomena result in the development of phase difference, as shown in Fig. 15−5. These two opposite effects offer random motion of molecules that generates heat due to molecular friction and dielectric losses. In other words, during dielectric or dipolar heating, the MW energy is converted into

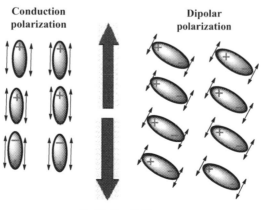

FIGURE 15−5 Mechanism of dipolar and conduction polarization.

thermal and kinetic energy. The electromagnetic radiation of 300–30,000 MHz (3–30 GHz) range constitutes MW radiation. During, MW heating it should be noted that at very high applied frequency of MW radiation (>30 GHz), reacting molecules are unable to follow the field due to their intermolecular (van der Waals) force of attraction that restricts the quick motion of the molecules, which ultimately results in inadequate interparticle interactions [37]. Moreover, at very low frequency of applied MW radiation (<0.3 GHz), the polar molecules align themselves with the field due to availability of sufficient time, resulting in no random motion of the reactant molecules, which is essential for the conversion of electrical energy of MW radiation into kinetic and thermal energy. The greatest advantage of the MW irradiation is that energy of the MW range lies at around 0.037 kJ mol^{-1}, which is very low compared to the bond dissociation energy of the common chemical bonds; those bond dissociation energies lie at 80–120 kJ mol^{-1}. Therefore MW irradiation does not affect the structure of molecules, and therefore interaction between MW irradiation and reacting molecules is purely kinetic.

15.1.3.1.2 Ionic conductance heating mechanism

In the ionic conductance heating method, it is considered that whenever charged particles such as ions are subject to interaction with an oscillating electromagnetic field, they start to move in a back-and-forth fashion [38]. During this movement these charged particles collide with nearby molecules (charged or neutral), as a result of which heat is produced. Although the ability to produce heat is much greater via ionic conductance compared to dielectric polarization, the ionic conductance heating is limited only to the molecules that have not very high electrical conductivity because highly conductive molecules reflect most of the electromagnetic radiation falling upon them [38]. The diagrammatic mechanism of ionic conductance heating is presented in Fig. 15–5.

15.1.4 Ultrasound irradiation in organic syntheses

US irradiation has several advantages in the field of synthetic organic chemistry compared to conventional heating methods [39]. For example, chemical transformations using US irradiation are generally more efficient, more economical, and less time-consuming. A literature survey reveals that US irradiation can increase the chemical reactivity beyond a million-fold [39]. Recently, various chemical transformations (Fig. 15–6), such as esterification, saponification, hydrolysis/solvolysis, substitution, addition, alkylation, oxidation, reduction, and condensation, have been carried out using US irradiation [39]. Generally the process of US activation occurs through acoustic cavitation, which involves the formation, rapid growth, and collapse of the bubbles in the liquids.

15.1.4.1 What is cavitation and how does it occurs?

The complete US range consists of three distinct parts: power US (20–100 kHz), which is the most important region of US from the chemical transformation point of view; high frequency US (100 kHz–1 MHz), which is being utilized for the detection of cracks on the solid surfaces and animal navigation; and diagnostic US (1–500 MHz), which is used for

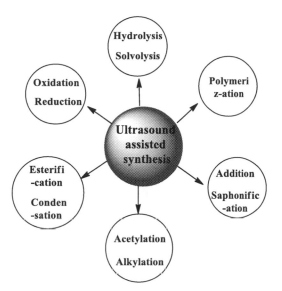

FIGURE 15–6 Some common examples of ultrasound-assisted syntheses.

diagnostic purposes. Whenever US irradiation is allowed to pass through liquids, cyclic expansion and compression starts due to mechanical vibrations. During the cyclic vibration, the compression cycle applies a positive pressure and ultimately impels the liquid molecules together. However, the expansion cycle vibrations apply a negative pressure and therefore push the liquid molecules apart from each other [40]. When the magnitude of the pressure during the expansion cycle exceeds the textile strength of liquid, the intermolecular van der Waals force of attraction between the liquid molecules is not sufficiently strong to sustain cohesion, which leads to the formation of gas-filled microbubbles or small cavities [40]. It is also important to note that pure liquids associated with high tensile strength and therefore US irradiation are incapable of producing enough negative pressure to cause cavitation [40]. However, very fortunately most of the liquids are associated with some impurities such as small suspended particles, dissolved solid materials, and other contaminants that reduce the tensile strength of the liquid molecules. For example, pure or distilled water requires nearly 1000 atm pressure, while tap water requires just a few atm of pressure for the same [41]. During the growth of microbubbles, the spherical microbubbles undergo radial and tangential deformation before their collapse. In general, collapse of the microbubbles occurs in an adiabatic manner in a relatively small volume which results in the transfer of the collapsed energy to only a few reacting molecules. This phenomenon of energy transfer is responsible for sonoluminescence, that is, the process through which liquid molecules interact with the US wave and produce UV–visible electromagnetic radiation (200–700 nm) [41]. The energy produced due to a collapse of microbubbles is sufficient to excite and/or dissociate the reacting molecules in order to emit light during their relaxation in ground state [42].

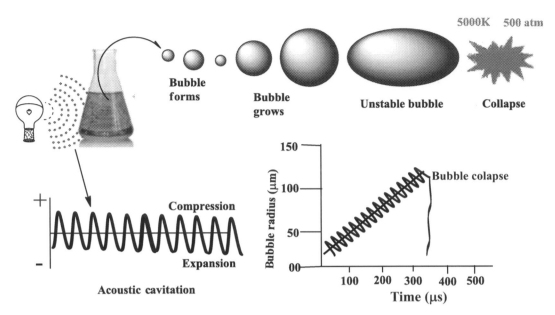

FIGURE 15-7 Mechanism of acoustic cavitation.

15.1.4.2 How does ultrasound irradiation works?

The cavitation process consists of the repetition of the three steps, that is, the formation, rapid growth, and violent collapse of the microbubbles in the liquid, as shown in Fig. 15−7. During their collapse each microbubble acts as a "hotspot" and produces energy that enormously increases the temperature and pressure up to 5000 K and 500 atm, respectively [43]. The US irradiation for chemical transformation has several other advantages over conventional and even MW heating, for example, the cooling rate is very fast (109 K s^{-1}) [44]. The collapse of the liquid microbubbles produces extremely high energy which can drive most of the chemical transformations reported in the literature. However, there are several factors that affect the rate of US catalyzed (rate of bubbles formation and their collapse) reactions and these are listed below.

15.2 Green corrosion inhibitors from microwave and ultrasound irradiation

Several classes of organic compounds derived through MW and US irradiation are effectively used as corrosion inhibitors for metals and alloys. Palomar-Pardavé et al. [45] demonstrated the inhibition effect of 2-amino-5-alkyl-1,3,4-thiadiazole derivatives differing in the nature of alkyl chain for carbon steel in 1 M H_2SO_4 system. The developed compounds were synthesized using spectral characterization methods and their protection efficiencies were determined using electrochemical and scanning electron microscopy (SEM) analyses. All compounds differing in the nature of alkyl chain length became effective by the adsorption mechanism that followed the Langmuir isotherm model. Synthesis of the compounds is presented in Scheme 15−1.

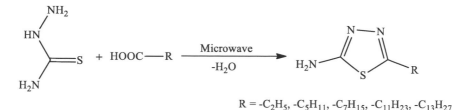

$$R = -C_2H_5, -C_5H_{11}, -C_7H_{15}, -C_{11}H_{23}, -C_{13}H_{27}$$

SCHEME 15–1 Synthetic scheme of 2-amino-5-alkyl-1,3,4-thiadiazole derivatives differing in the nature of alkyl chain length and tested as corrosion inhibitors for steel/ 1 M H_2SO_4 system.

Synthesis of the compounds was carried out using a conventional hot-plate as well as MW heating methods and the results showed that compounds derived through MW heating acted as better corrosion inhibitors than the compounds derived through hot-plate heating. Results showed that the alkyl chain length played a significant role in determining the protection efficiency of the developed compounds.

It was observed that up to undecyl carbon chain spacer, protection efficiency increases rapidly, but after that a sudden drop in the protection power of the compounds was observed on further increase in carbon chain length. Ansari et al. [46] reported the protection efficiency of two isatin derivatives (TZ-1 and TZ-2) for mild steel in 20% H_2SO_4 using numerous methods. The compounds were derived using the MW irradiation method as per Scheme 15–2.

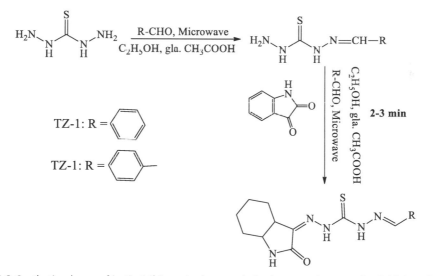

SCHEME 15–2 Synthetic scheme of isatin-β-thiosemicarbazone derivatives tested as corrosion inhibitors for mild steel 20% H_2SO_4 system.

Polarization study showed TZs acted as predominantly cathodic-type corrosion inhibitors. Adsorption of the TZs on MS surface obeyed the Langmuir adsorption isotherm model. Values of Gibb's free energies showed that both TZ molecules adsorb on metallic surface using the physisorption mechanism. EIS study showed that TZs act as interface-type corrosion inhibitors as their adsorption increase the values of charge transfer resistance.

Dutta et al. [47] synthesized four bis-benzimidazole (BBI) using MW heating and evaluated them as corrosion inhibitors for a MS/1 M HCl system using experimental and computational methods. Tested BBIs acted as effective corrosion inhibitors even for a long time exposure. Polarization results showed that BBIs behaved as mixed-type corrosion inhibitors and their adsorption obeyed the Langmuir isotherm model. Inhibition effectiveness of the BBIs followed the order: BBP > BBMS > BBMO > BBE. Synthesis of the BBIs is presented in Scheme 15−3. Computational studies showed that BBIs interact strongly with the metallic surface using the donor−acceptor mechanism and acquire flat or horizontal orientations.

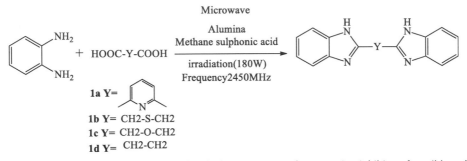

SCHEME 15–3 Synthetic scheme of bis-benzimidazole derivatives tested as corrosion inhibitors for mild steel in a 1 M HCl system.

Our research [48] team synthesized three 2-aminobenzene-1,3-dicarbonitrile derivatives (ABDNs) differing in the nature of substituents using MW irradiation and tested them as corrosion inhibitors for MS in 1 M HCl system using experimental and computational methods. Results showed that the compounds containing a higher number of electron releasing

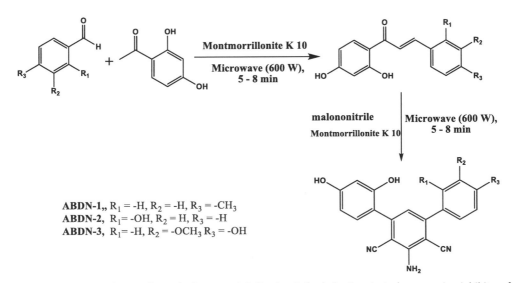

ABDN-1,, R$_1$ = -H, R$_2$ = -H, R$_3$ = -CH$_3$
ABDN-2, R$_1$= -OH, R$_2$ = H, R$_3$ = -H
ABDN-3, R$_1$= -H, R$_2$ = -OCH$_3$ R$_3$ = -OH

SCHEME 15–4 Synthetic scheme of 2-aminobenzene-1,3-dicarbonitrile derivatives tested as corrosion inhibitors for mild steel in 1 M HCl system.

substituents behaved as better corrosion inhibitors. All studied compounds behaved as mixed-type corrosion inhibitors and their adsorption obeyed the Langmuir isotherm model. EIS analyses showed that the presence of the ABDNs in corrosive electrolyte increased the values of charge transfer resistance by forming the barrier for charge transfer through their adsorption. The donor–acceptor mechanism of corrosion inhibition was observed using density functional theory (DFT) analyses. The scheme for the synthesis of ABDNs is shown in Scheme 15–4.

The corrosion inhibition effect of compounds derived through MW irradiation has also been reported in other studies.

Similar to MWs, US irradiation is also used extensively to synthesize the compounds that are used as corrosion inhibitors. We reported the synthesis of a chalcone by the reaction of aromatic aldehyde and acetophenone and three heterocyclic compounds by the chemical reactions of chalcone with urea, thiourea, and hydroxyl amine to form three heterocyclic compounds [49]. All synthesized compounds were tested as corrosion inhibitors for mild steel in 1 M HCl using different methods. The compounds acted as effective corrosion inhibitors for mild steel in an acidic medium and they behaved as mixed-type corrosion inhibitors. The tested compounds inhibited corrosion by an adsorption mechanism that followed the adsorption isotherm model. Adsorption of the compounds on metallic surface was supported by SEM and Energy dispersive X-ray analyses. All tested compounds showed the highest protection power at 25 ppm concentration. EIS analysis showed that tested compounds behaved as interface-type corrosion inhibitors. Synthesis of the compounds is presented in Scheme 15–5.

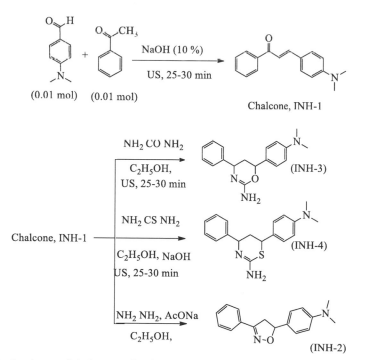

SCHEME 15.5 Synthetic scheme of chalcone and its heterocyclic derivatives tested as corrosion inhibitors for mild steel in 1 M HCl system.

Our research team synthesized another triazine-based heterocyclic compound (INH) using the US irradiation method and tested it as a corrosion inhibitor for mild steel in 1 M HCl. Results showed that INH acted as an effective anticorrosive material and its protection efficiency increased with its increasing concentration [50]. INH acted as a mixed- and interface-type corrosion inhibitor. The scheme for the synthesis of INH is shown in Scheme 15−6. Compounds derived through US irradiation are also used in other studies.

SCHEME 15–6 Synthetic scheme of triazine-based heterocyclic compound tested as corrosion inhibitors for mild steel in 1 M HCl system.

15.3 Conclusions

Organic compounds are established as the most effective class of corrosion inhibitors. Because of their ease of synthesis, cost-effectiveness, and high protection efficiency they are used widely. However, they are mostly developed using multistep reactions using conventional heating sources. These reactions are expensive, time-consuming, tedious and not environmentally friendly as they utilize traditional heating methods that require minutes, hours, days, or even weeks to activate the reactant molecules. In view of this, several nonconventional heating methods including MW and US heating have emerged as new and efficient heating sources. Numerous organic compounds with different industrial and biological applications including corrosion inhibitors have been developed using MW and US irradiation. However, the literature on MW and US irradiation as sources for the synthesis of corrosion inhibitors is relatively scarce and should be further explored.

Conflict of interests

There is no conflict of interest in any way.

Author's contributions

All authors collectively contributed in the designing and write-up of the book chapter.

Important links

1. https://www.sciencedirect.com/science/article/abs/pii/S2352554118301001
2. https://pubs.acs.org/doi/pdf/10.1021/sc400155u

References

[1] C. Verma, L. Olasunkanmi, E.E. Ebenso, M. Quraishi, Substituents effect on corrosion inhibition performance of organic compounds in aggressive ionic solutions: a review, J. Mol. Liq. 251 (2018) 100–118.

[2] B. Hou, X. Li, X. Ma, C. Du, D. Zhang, M. Zheng, et al., The cost of corrosion in China, NPJ Mater. Degrad. 1 (2017) 1–10.

[3] P. Marcus, Corrosion Mechanisms in Theory and Practice, CRC press, 2011.

[4] M.G. Fontana, Corrosion Engineering, Tata McGraw-Hill Education, 2005.

[5] V.S. Sastri, Green Corrosion Inhibitors: Theory and Practice, John Wiley & Sons, 2012.

[6] G. Bahlakeh, B. Ramezanzadeh, A. Dehghani, M. Ramezanzadeh, Novel cost-effective and high-performance green inhibitor based on aqueous *Peganum harmala* seed extract for mild steel corrosion in HCl solution: detailed experimental and electronic/atomic level computational explorations, J. Mol. Liq. 283 (2019) 174–195.

[7] M. Finšgar, J. Jackson, Application of corrosion inhibitors for steels in acidic media for the oil and gas industry: a review, Corros. Sci. 86 (2014) 17–41.

[8] S. Kumar, H. Vashisht, L.O. Olasunkanmi, I. Bahadur, H. Verma, G. Singh, et al., Experimental and theoretical studies on inhibition of mild steel corrosion by some synthesized polyurethane tri-block co-polymers, Sci. Rep. 6 (2016) 30937.

[9] O. Olivares-Xometl, N. Likhanova, M. Domínguez-Aguilar, J. Hallen, L. Zamudio, F. Arce, Surface analysis of inhibitor films formed by imidazolines and amides on mild steel in an acidic environment, Appl. Surf. Sci. 252 (2006) 2139–2152.

[10] R. Ding, W. Li, X. Wang, T. Gui, B. Li, P. Han, et al., A brief review of corrosion protective films and coatings based on graphene and graphene oxide, J. Alloy Compd. 764 (2018) 1039–1055.

[11] O. Dagdag, Z. Safi, H. Erramli, N. Wazzan, I. Obot, E. Akpan, et al., Anticorrosive property of heterocyclic based epoxy resins on carbon steel corrosion in acidic medium: electrochemical, surface morphology, DFT and Monte Carlo simulation studies, J. Mol. Liq. 287 (2019) 110977.

[12] C. Verma, I. Obot, I. Bahadur, E.-S.M. Sherif, E.E. Ebenso, Choline based ionic liquids as sustainable corrosion inhibitors on mild steel surface in acidic medium: gravimetric, electrochemical, surface morphology, DFT and Monte Carlo simulation studies, Appl. Surf. Sci. 457 (2018) 134–149.

[13] L.K.M. Goni, M.A. Mazumder, Green Corrosion Inhibitors, Corrosion Inhibitors, IntechOpen, 2019.

[14] A. McMahon, D. Harrop, Green Corrosion Inhibitors: An Oil Company Perspective, NACE International, Houston, TX, 1995.

[15] C. Verma, E.E. Ebenso, I. Bahadur, M. Quraishi, An overview on plant extracts as environmental sustainable and green corrosion inhibitors for metals and alloys in aggressive corrosive media, J. Mol. Liq. 266 (2018) 577–590.

[16] G. Gece, Drugs: a review of promising novel corrosion inhibitors, Corros. Sci. 53 (2011) 3873–3898.

[17] M. Abdallah, Rhodanine azosulpha drugs as corrosion inhibitors for corrosion of 304 stainless steel in hydrochloric acid solution, Corros. Sci. 44 (2002) 717–728.

[18] R. Pathak, P. Mishra, Drugs as corrosion inhibitors: a review, Int. J. Sci. Res. 5 (2016) 671–677.

[19] G. Cravotto, M. Beggiato, A. Penoni, G. Palmisano, S. Tollari, J.-M. Lévêque, et al., High-intensity ultrasound and microwave, alone or combined, promote Pd/C-catalyzed aryl–aryl couplings, Tetrahedron Lett. 46 (2005) 2267–2271.

[20] J.-M. Leveque, G. Cravotto, Microwaves, power ultrasound, and ionic liquids. A new synergy in green organic synthesis, Chim. Int. J. Chem. 60 (2006) 313–320.

[21] C. Leonelli, T.J. Mason, Microwave and ultrasonic processing: now a realistic option for industry, Chem. Eng. Process. Process Intensif. 49 (2010) 885–900.

[22] E. Martinez-Guerra, V.G. Gude, Transesterification of used vegetable oil catalyzed by barium oxide under simultaneous microwave and ultrasound irradiation, Energy Convers. Manage. 88 (2014) 633–640.

[23] A. Barge, S. Tagliapietra, L. Tei, P. Cintas, G. Cravotto, Pd-catalyzed reactions promoted by ultrasound and/or microwave irradiation, Curr. Org. Chem. 12 (2008) 1588–1612.

[24] P. Cintas, S. Tagliapietra, E.C. Gaudino, G. Palmisano, G. Cravotto, Glycerol: a solvent and a building block of choice for microwave and ultrasound irradiation procedures, Green Chem. 16 (2014) 1056–1065.

[25] C. Verma, M. Quraishi, E.E. Ebenso, Microwave and ultrasound irradiation for the synthesis of environmentally sustainable corrosion inhibitors: an overview, Sustain. Chem. Pharm. 10 (2018) 134–147.

[26] C. Verma, J. Haque, M. Quraishi, E.E. Ebenso, Aqueous phase environmental friendly organic corrosion inhibitors derived from one step multicomponent reactions: a review, J. Mol. Liq. 275 (2019) 18–40.

[27] C. Verma, E.E. Ebenso, M. Quraishi, Ionic liquids as green and sustainable corrosion inhibitors for metals and alloys: an overview, J. Mol. Liq. 233 (2017) 403–414.

[28] S. Umoren, Polymers as corrosion inhibitors for metals in different media - a review, Open Corros. J. 2 (2009).

[29] S.A. Umoren, U.M. Eduok, Application of carbohydrate polymers as corrosion inhibitors for metal substrates in different media: a review, Carbohydr. Polym. 140 (2016) 314–341.

[30] M.B. Gawande, S.N. Shelke, R. Zboril, R.S. Varma, Microwave-assisted chemistry: synthetic applications for rapid assembly of nanomaterials and organics, Acc. Chem. Res. 47 (2014) 1338–1348.

[31] B.L. Hayes, Recent advances in microwave-assisted synthesis, Aldrichim. Acta 37 (2004) 66–76.

[32] M. Nüchter, B. Ondruschka, W. Bonrath, A. Gum, Microwave assisted synthesis—a critical technology overview, Green Chem. 6 (2004) 128–141.

[33] A. de la Hoz, A. Diaz-Ortiz, A. Moreno, Microwaves in organic synthesis, Therm. Non-Thermal Microw. Effects Chem. Soc. Rev. 34 (2005) 164–178.

[34] K.D. Raner, C.R. Strauss, Influence of microwaves on the rate of esterification of 2, 4, 6-trimethylbenzoic acid with 2-propanol, J. Org. Chem. 57 (1992) 6231–6234.

[35] A. Stadler, B.H. Yousefi, D. Dallinger, P. Walla, E. Van der Eycken, N. Kaval, et al., Scalability of microwave-assisted organic synthesis. From single-mode to multimode parallel batch reactors, Org. Process. Res. Dev. 7 (2003) 707–716.

[36] C.O. Kappe, Microwave dielectric heating in synthetic organic chemistry, Chem. Soc. Rev. 37 (2008) 1127–1139.

[37] C. Gabriel, S. Gabriel, E.H. Grant, B.S.J. Halstead, D.M.P. Mingo, Dielectric parameters relevant to microwave dielectric heating, Chem. Soc. Rev. 27 (1998) 213–223.

[38] J.D. Moseley, C.O. Kappe, A critical assessment of the greenness and energy efficiency of microwave-assisted organic synthesis, Green Chem. 13 (2011) 794–806.

[39] T.J. Mason, Ultrasound in synthetic organic chemistry, Green Chem. 26 (1997) 443–451.

[40] A. Brotchie, R. Mettin, F. Griesera, M. Ashokkumar, Cavitation activation by dual-frequency ultrasound and shock waves, Phys. Chem. Chem. Phys. 11 (2009) 10029–10034.

[41] H. Frenzel, H. Schultes, LuminescenzimultraschallbeschicktenWasser, Z. Phys. Chem. Abt. B 27 (1934) 421.

[42] G. Cravotto, P. Cintas, Power ultrasound in organic synthesis: moving cavitational chemistry from academia to innovative and large-scale applications, Chem. Soc. Rev. 35 (2006) 180–196.

[43] K.S. Suslick, Sonochemistry, Science 247 (1990) 1439–1445.

[44] . J. L P. Cintas, in: T.J. Mason (Ed.), Advances in Sonochemistry, 5, JAI Press, London, 1999, pp. 147–174.

[45] M. Palomar-Pardavé, M. Romero-Romo, H. Herrera-Hernández, M.A. Abreu-Quijano, N.V. Likhanova, J. Uruchurtu, et al., Influence of the alkyl chain length of 2 amino 5 alkyl 1,3,4 thiadiazole compounds on the corrosion inhibition of steel immersed in sulfuric acid solutions, Corros. Sci. 54 (2010) 231−243.

[46] K.R. Ansari, M.A. Quraishi, A. Singh, Isatin derivatives as a non-toxic corrosion inhibitor for mild steel in 20% H_2SO_4, Corros. Sci. 95 (2015) 62−70.

[47] A. Dutta, S.K. Saha, P. Banerjee, D. Sukul, Correlating electronic structure with corrosion inhibition potentiality of some bis-benzimidazole derivatives for mild steel in hydrochloric acid: combined experimental and theoretical studies, Corros. Sci. 98 (2015) 541−550.

[48] C. Verma, M.A. Quraishi, A. Singh, 2-Aminobenzene-1,3-dicarbonitriles as green corrosion inhibitor for mild steel in 1 M HCl: electrochemical, thermodynamic, surface and quantum chemical investigation, J. Taiwan Inst. Chem. Eng. 49 (2015) 229−239.

[49] C.B. Verma, M.J. Reddy M.A. Quraishi, Ultrasound assisted green synthesis of 3-(4-(dimethylamino) phenyl)-1-phenylprop-2-en-1-one and its heterocyclics derived from hydrazine, urea and thiourea as corrosion inhibitor for mild steel in 1M HCl. Anal. Bioanal. Electrochem. 6 (204) 515−534.

[50] C. Verma, M.A. Quraishi, E.E. Ebenso, Green ultrasound assisted synthesis of N 2, N 4, N 6-tris ((Pyridin-2-ylamino) methyl)-1, 3, 5-triazine-2, 4, 6-triamine as effective corrosion inhibitor for mild steel in 1 M hydrochloric acid medium, Int. J. Electrochem. Sci. 8 (2013) 10864−10877.

16

Polyethylene glycol and its derivatives as environmental sustainable corrosion inhibitors: A literature survey

Chandrabhan Verma[1], Chaudhery Mustansar Hussain[2]

[1]INTERDISCIPLINARY RESEARCH CENTER FOR ADVANCED MATERIALS, KING FAHD UNIVERSITY OF PETROLEUM AND MINERALS, DHAHRAN, SAUDI ARABIA [2]DEPARTMENT OF CHEMISTRY AND ENVIRONMENTAL SCIENCE, NEW JERSEY INSTITUTE OF TECHNOLOGY, NEWARK, NJ, UNITED STATES

Chapter outline

Environmentally Sustainable Corrosion Inhibitors. DOI: https://doi.org/10.1016/B978-0-323-85405-4.00005-7

Abbreviations

PEG	polyethylene glycol
PTC	phase transfer catalyst
VOCs	volatile organic compounds
WL	weight loss
HE	hydrogen evolution
AAs	amino acids
GDP	gross domestic product
%IE	percentage inhibition efficiency
C_R	corrosion rate
EIS	electrochemical impedance spectroscopy
PDP	potentiodynamic polarization
LPR	linear polarization resistance
OCP	open circuit potential
θ	surface coverage
SEM	scanning electron microscopy
EDX	energy dispersive X-ray
XRD	X-ray powder diffraction
XPS	X-ray photoelectron spectroscopy
AFM	atomic force microscopy
DFT	density functional theory
MDS	molecular dynamics simulation
MCS	Monte Carlo simulation
REACH	registration, evaluation, authorization and restriction of the chemical
OSPAR	Oslo Paris Commission
LC_{50}	lethal concentration
EC_{50}	effective concentration
$\log K_{OW}$ or D^{OW}	partition coefficient
MCRs	multicomponent reactions
MW	microwave
US	ultrasound
CH: TS-Cht	Chitosan

16.1 Introduction

16.1.1 Polyethylene glycol: Basics and applications

Increasing ecological awareness is increasing pressure on the synthesis, development, and use of nonhazardous alternatives to traditional toxic corrosion inhibitors [1,2]. There are numerous potential benefits to replacing the traditional toxic corrosion inhibitors with polyethylene glycol (PEG) and several other kinds of green and environment-friendly alternatives [3–5]. The most significant benefits are minimized flammability, cost-effectiveness, minimized toxicity, and minimum environmental risk [6]. PEG and its derivatives may represent an increasingly noteworthy selection for the replacement of conventional toxic corrosion inhibitors [7−9]. Toxicity information is available for wide range of molecular weights of PEG and several of them are already approved for internal consumption by the US FDA [10,11]. Other leading toxic corrosion inhibitor alternatives include ionic liquids [12], plant extracts [2], chemical medicines [1], and chemicals derived using multicomponent reactions [13], especially using microwave (MW) and ultrasound heating [14]. A literature survey shows that PEG and its solutions are used as solvents and phase transfer catalysts (PTC) for various chemical transformations.

PEG is available in wide range of molecular weights ranging from 200 to tens of thousands [6,15]. At ambient temperature, PEG up to molecular weight 600 represents a hydroscopic, highly viscous, and water-soluble polymer. PEG with molecular weight of 600−800 exists as a waxy material [6,15]. PEG acquires the white solid property when its molecular weight exceeds 800. It is important to mention that increased PEG molecular weight results in the subsequent decrease in its solubility in water, for example, PEG-2000 has water solubility of 60% at 20°C [6]. PEG is recognized as a safe chemical and can be used for internal consumption. PEG-derived aqueous solutions are biocompatible and are extensively used as media for tissue culture and in the preservation of organs [6,15].

16.1.2 Advantages of using polyethylene glycol over volatile organic compounds corrosion inhibitors

Recently, several review articles have been published dealing with PEG chemistry and its uses in medicine and biotechnology, such as PEG-derived aqueous biphasic systems (ABS), and PEG and PEG-reinforced catalysts as alternative separation media, ABS in bioconversion, PEG and its solutions as alternative solvents for chemical transformations, and PEG and its derivatives as PTC in synthesis [6,16−18]. However, not one of these review articles has paid attention to PEG and its derivatives acting as anticorrosion materials. Here we develop a comprehensive review of PEG and its derivatives as green corrosion inhibitors. PEG and its derivatives acquire the unique ability to coordinate with metallic cations [19]. Because of their ability to coordinate with metal cations, PEG and its derivatives can be used to separate and protect them from corrosion in aggressive electrolytes.

Unlike traditional volatile organic compounds (VOCs), PEGs are nonvolatile alternatives [20−22]. As per MSDS data, low-molecular-weight PEG acquires the vapor density of greater than unity (relative to the density of air), which is reliable for the industry standard for the assortment of substitute chemicals to VOCs. PEG and its derivatives also acquire the ability of biodegradability and low flammability that offer the PEG and its derivatives as environmental sustainable alternatives to be used as corrosion inhibitors [23,24]. PEG and its derivatives are stable in acids and bases, therefore they can be used as long-term corrosion inhibitors. They can also be used as high-temperature corrosion inhibitors as they are stable up to high temperature. The PEG is stable for H_2O_2 oxidation and hydride reduction ($NaBH_4$) [25−27]. One of the greatest advantages of using PEG and its derivatives as corrosion inhibitors is that they can be recovered from aqueous media through an extraction process using a suitable solvent or by direct distillation of solvent or water [28,29].

16.2 Literature study: Polyethylene glycol and its derivatives as corrosion inhibitors

16.2.1 Polyethylene glycol and its derivatives as corrosion inhibitors in H_2SO_4

Sulfuric acid, having the molecular formula of H_2SO_4, and also known as oil of vitriol, is the most commonly used acidic solution for pickling processes in United States and Europe. The efficiency of acid pickling by H_2SO_4 greatly depends upon the nature of the metallic materials and the pickling temperature. Generally, 5%−28% solutions of H_2SO_4 are used for descaling and pickling processes. Lower concentrations (1−3 M) of H_2SO_4 are widely used for academic purposes. PEG and PEG derivatives are expansively used as corrosion inhibitors in such electrolytes. Ashassi-Sorkhabi and Ghalebsaz-Jeddi [30] studied the inhibition properties of PEG of different molecular weights ranging from 200 to10,000 g mol^{-1} on carbon steel corrosion in 3 N H_2SO_4 solution using weight (mass) loss, potentiodynamic polarization (PDP), and electrochemical impedance spectroscopy (EIS) methods. Results demonstrated that PEGs showed more than 90% efficiency and they adsorbed on a metallic surface using the physisorption mechanism. EIS studies showed that an increase in molecular weight of PEG results in the corresponding increase in the protection efficiency with a subsequent increase in the charge transfer resistance (R_{ct}). An increase in PEGs concentrations causes an increase in the corrosion inhibition efficiency. All tested PEGs showed highest protection efficiency at 10^{-1} M concentration. The same authors also reported the inhibition effect of PEG of different molecular weights (from 400 up to10,000 g mol^{-1}) for carbon steel corrosion in 3 N H_2SO_4 solution [31]. The study was designed to demonstrate the effect of cavitation on the inhibition effect of PEG. Similar to previous studies, an increase in inhibition efficiency was observed with an increase in molecular weight and concentration of PEG. SEM analyses showed that the presence of PEG as an inhibitor prevented the propagation of pits and cracks developing on the metallic surface. SEM analyses showed that an increase in the molecular weight of PEG results in the corresponding increase in the surface morphology (smoothness). SEM images of a metallic surface corroded in 3 N H_2SO_4 after 2.5 h ultrasonication in the presence of different molecular weights of PEG are shown in Fig. 16−1. It can be seen that surface smoothness increases when increasing the molecular weight of the PEG. Umoren et al. [32] studied the corrosion inhibition property of polyvinyl alcohol (PVA) and PEG for aluminum in an acidic medium using weight loss (WL) (gravimetric), gasometric, and thermometric methods. Corrosion inhibition studies were conducted at different temperatures ranging from 30°C−60°C. Protection efficiencies of PVA and PEG increase with their concentrations and decrease with increase in temperature. Adsorption of PVA and PEG on aluminum surface obeyed the Temkin adsorption isotherm. Inhibition efficiency values followed the order: PEG > PVA. The same group of authors also reported the inhibition effect of a natural polymer (gum Arabic; GA) and PEG as synthetic polymer for mild steel

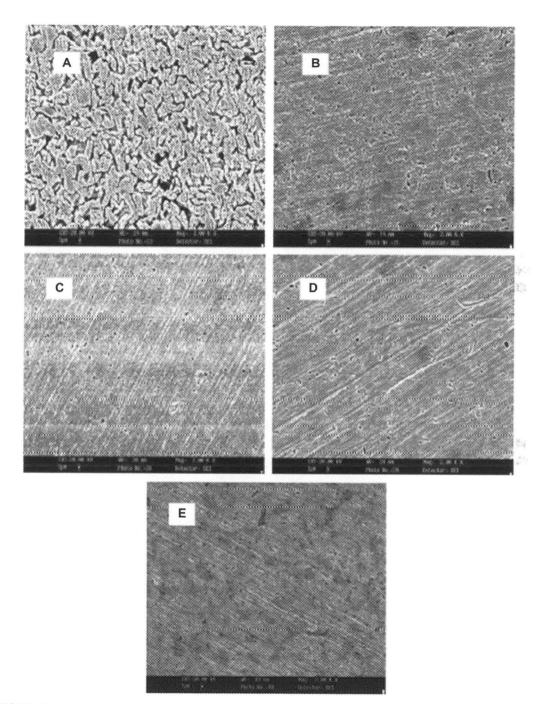

FIGURE 16–1 Scanning electron microscopy images of carbon steel surfaces corroded in 3 N H_2SO_4 after 2.5 h ultrasonication in the presence of 10^{-4} M concentration of (A) PEG-400, (B) PEG-1000, (C) PEG-4000, (D) PEG-10,000, and (E) PEG-10,000 without ultrasonication [31].

corrosion in H_2SO_4 medium using gravimetric, gasometric (hydrogen evolution), and thermometric methods [33]. Corrosion inhibition studies were conducted at different temperatures ranging from 30°C−60°C. The study was also designed to demonstrate the effect of halides (KI, KBr, KCl) on the corrosion inhibition properties of GA and PEG. Results showed that PEG showed better protection than that of the GA and their protection efficiencies increases by increasing the size of halide ions, that is, I < Br < Cl. Adsorption of GA and PEG on aluminum surface obeyed the Temkin adsorption isotherm. The increase in radii and decrease in electronegativity values of halide ions were consistent with the order of inhibition efficiency.

The synergistic effect of halides (KI, KBr, and KCl) on mild steel corrosion in H_2SO_4 medium was also described by these authors in another study [34]. Similar conclusions were drawn. Umoren demonstrated the corrosion inhibition efficiency of PEG and polyvinyl pyrrolidone (PVP) and their blends for mild steel corrosion in sulfuric acid medium using WL and hydrogen evolution methods [35]. Results showed that the protection efficiency of PEG and PVP increases with their concentration. Their blends showed better protection efficiencies than that of the homopolymers and maximum efficiency was derived for (PEG:PVP) blending ratio of 1:3. Both PEG and PVP adsorb on metallic surface using physisorption mechanism. Adsorption of PEG and PVP obeyed the Temkin adsorption isotherm. A relative study of PEG and PVP on mild steel corrosion in 0.5 M H_2SO_4 and 0.5 M HCl was described using PDP, linear polarization resistance (LPR), EIS, and SEM methods [36].

Results demonstrated that both PVP and PEG acted as good anticorrosive materials for mild steel in both electrolytes. PDP analyses showed that PVP and PEG acted as mixed-type corrosion inhibitors. EIS studies showed that the presence of PVP and PEG increased the charge transfer resistance for metallic corrosion. Adsorption of PVP and PEG on a metallic surface obeyed the Langmuir adsorption isotherm. SEM analyses showed that the presence of PEG and PVP showed significant improvement in the surface morphologies. Fig. 16−2 shows the SEM images of mild steel corroded in 0.5 M H_2SO_4 and 0.5 M HCl solution with and without PE and PVP.

The inhibition effect of PEG in the presence of two surfactants, namely sodium dodecylbenzene sulfonate and cetyltrimethyl ammonium bromide, on mild steel corrosion in 0.1 M H_2SO_4 was reported by Mobin and Khan using different methods [37]. Inhibition efficiency of PEG increases with an increase in its concentration, and maximum efficiency was observed at 25 ppm concentration. SEM, EFX, and atomic force microscopy analyses were conducted to demonstrate the adsorption of PEG on metallic surface. The corrosion inhibition effects of PEG and PEG derivatives in H_2SO_4 media are also reported in other studies [37−39].

16.2.2 Polyethylene glycol and its derivatives as corrosion inhibitors in hydrochloric acid

Hydrochloric acid (molecular formula HCl) is an inorganic chemical which is also known as hydronium chloride and is associated with a specific smell. Hydrochloric acid is widely used

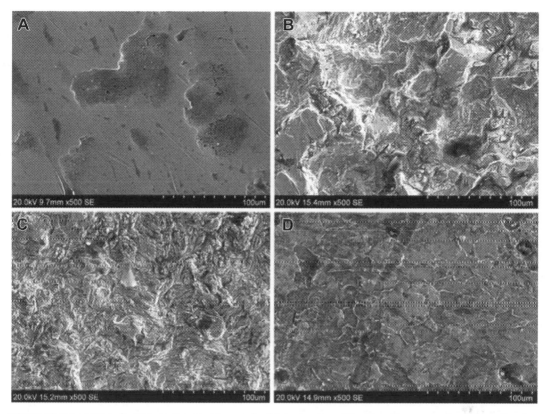

FIGURE 16–2 Scanning electron microscopy images of mild steel surface (A) without immersion in acid solution, (B) immersion in 0.5 M H_2SO_4 and 0.5 M HCl in the absence of inhibitors, (C) immersion in 0.5 M H_2SO_4 and 0.5 M HCl in the presence of polyethylene glycol, and (D) immersion in 0.5 M H_2SO_4 and 0.5 M HCl in the presence of polyvinyl pyrrolidone [36].

for descaling, oil well acidification, and acid pickling processes. Lower concentrations of HCl (1−3 M) are widely used for educational purposes, whereas higher concentrations are useful for industrial applications. A literature study showed that PEG and PEG derivatives are widely used as corrosion inhibitors for metals and alloys in different electrolytes. Ashassi-Sorkhabi et al. [9] studied the corrosion inhibition property of a few PEGs on carbon steel in 0.5 M HCl solution using WL, PDP, and EIS methods. The inhibition effect of PEGs having different molecular weights, 400, 1000, 4000, and 10,000 g mol^{-1} was evaluated. Results showed that protection efficiency increases with an increase in molecular weight and concentration. PDP study showed that PEGs acted as mixed-type corrosion inhibitors. Their adsorption on a metallic surface obeyed the Langmuir adsorption isotherm model.

Relative corrosion inhibition effect of PEG and ciprofloxacin on aluminum corrosion in HCl medium were investigated in another study [40]. Analyses showed that significant enhancement in the protection power of PEG was observed in the presence of ciprofloxacin.

PEG with and without ciprofloxacin spontaneously adsorbed on the metallic surface. Adsorption of PEG and ciprofloxacin on an aluminum surface obeyed the Langmuir adsorption isotherm model. SEM analyses showed that PEG and ciprofloxacin form well-ordered inhibitive layers over the metallic surface that led to higher protection efficiency. The ciprofloxacin enhances corrosion inhibition efficiency of PEG through synergism. An inhibition effect of PEG (PEG-300−600) on aluminum corrosion in HCl medium was also reported by Awad et al. [41]. The study was conducted using WL and PDP methods. PEG exhibited good protection efficiency for aluminum corrosion in hydrochloric acid solution and its efficiency increases with its concentration. PDP analyses showed that PEG behaved as a mixed-type corrosion inhibitor, and its presence remarkably affected both anodic and cathodic polarization curves. The experimental results were corroborated using density functional theory (DFT) and MDS studies. DFT studies showed that PEGs interact with a metallic surface using the donor−acceptor method. MD simulation studies showed that different PEGs adsorb on a metallic surface using flat or horizontal orientations.

Recently, chitosan-functionalized PEG, designated as chitosan−PEG, was evaluated as a corrosion inhibitor for mild steel in 1 M HCl medium using WL, electrochemical, and DFT methods [8]. The chitosan−PEG showed highest protection efficiency of 93.9% at 200 mgL^{-1} concentration. Adsorption of chitosan−PEG on a mild steel surface obeyed the Langmuir adsorption isotherm model. PDP studies showed that chitosan−PEG acted as a mixed−type corrosion inhibitor. DFT analyses showed that chitosan−PEG interacts with a metallic surface using donor−acceptor interactions. A literature study revealed that there are numerous other studies published in which PEG and its derivatives have been evaluated as corrosion inhibitors.

16.2.3 Polyethylene glycol and its derivatives as corrosion inhibitors in NaCl

Natural (seawater) and synthetic or simulated NaCl solutions are widely used by industry and academics as electrolytic media. A literature survey reveals that several classes of organic compounds are widely used as corrosion inhibitors for NaCl solution. Nevertheless, there are few reports are available in which PEG is used as a corrosion inhibitor in NaCl media. The NaCl media of different concentrations have been used as electrolytes. Boudellioua et al. [42] demonstrated the corrosion resistance property of ASTM A915 mild steel in the absence and presence of a combination of cerium(III) and PEG in 0.1 M NaCl and in 0.1 M Na$_2$SO$_4$ media. The corrosion inhibition effect of the combinations of cerium(III) and PEG was tested using cyclic voltammetry, LPR, EIS, and PDP methods. Results showed that the presence of PEG and cerium nitrate inhibited the formation of corrosion products over the metallic surface. Electrochemical analyses showed that the formulations behaved as mixed-type and interface-type corrosion inhibitors. In another study [43], these authors described the effect of cerium(III) ions on the corrosion inhibition effect of PEG for ASTM A915 in chloride solution. Similar conclusions were derived. Božović and coworkers developed and employed an environment-friendly synergistic mixture of different components for steel corrosion in a chloride medium [44]. The mixture consisted of sodium benzoate, tannin, propolis, PEG400, and starch. Studies conducted by LPR probe and EIS methods indicated that the mixture formed a protective and inhibitive layer that inhibited the diffusion of oxygen and protected from corrosion. SEM examination was carried out to demonstrate the surface morphologies of the metallic surfaces. Several other reports dealing with the anticorrosive effect of PEG in NaCl media have been published [45].

16.2.4 Polyethylene glycol and its derivatives as corrosion inhibitors in metal hydroxides

Metal hydroxides (MOH) solutions are extensively used as electrolytes for processing or different metals and alloys. Therefore corrosion inhibitors in such electrolytes are highly essential. A literature study showed that PEG and PEG derivatives are also used as corrosion inhibitors in KOH media. Wanees and coworkers [46] described the effect of PEG on polarization behavior and pitting corrosion of cadmium using cyclic voltammetry and PDP methods. The effect of halide ions (Br and I) at different concentrations on the anticorrosion effect of PEG was also demonstrated. Results showed that the presence of PEG-200, PEG-400, and PEG-1540 shifted the corrosion potential toward the noble direction indicating that PEGs increase resistance for pitting attacks. In another study, the authors demonstrated the electrochemical and corrosion behavior of zinc in 8.5 M KOH solution in the absence and presence of organic compounds [47]. The organic compounds include PEG (molecular weight 600) and polyoxyethylen alkyl phosphate ester acid (GAFAC RA600). Results showed that PEG acted as a more efficient anticorrosive material than GAFAC RA600. High-resolution SEM analyses showed that in the presence of PEG and GAFAC RA600, the surface morphology of the metallic surface changed remarkably, which provided a clue for metallic corrosion inhibition.

Umoren et al. [47] studied the corrosion inhibition effect on aluminum corrosion in 1 M NaOH solution using WL and thermometric methods. The corrosion inhibition studies were carried out at different temperatures (30°C-60°C). The effect of different halides, namely chloride, bromide, and iodide ions, were evaluated on the anticorrosion property of PEG. Results showed that the inhibition efficiency of the PEG synergistically increased in the presence of halide ions. PEG became effective by adsorbing on the metallic surface that obeyed the Flory–Huggins and Temkin adsorption mechanism. Results showed the influence of halide ions on the anticorrosion effect of PEG, which obeyed the order: I > Br> Cl. Inhibition effectiveness of the PEG increases with PEG concentration and decreases with temperature.

Corrosion of aluminum (Al) in 5 M KOH solution with and without inorganics (ZnO), organics (PEG), and their blends (ZnO-PEG) was described elsewhere [48]. The formulation with 5000 ppm PEG and 16 gL^{-1} ZnO showed the highest protection efficiency. The surface morphological studies carried out using SEM, energy dispersive X-ray, and X-ray powder diffraction methods showed that surface morphology of the Al was greatly influenced by the formulations.

16.3 Conclusions

This chapter has described the corrosion inhibition effect of PEG and its derivatives for metals and alloys in different electrolytes. PEG and PEG derivatives are an environmentally sustainable class of corrosion inhibitors. The PEG and PEG derivatives are widely used for mild steel, carbon steel, aluminum, etc. corrosion in several electrolytic media, including H_2SO_4, HCl, NaCl, MOH, and many more. The environment-friendly behavior of PEG-based corrosion inhibitors are attributed to their nonvolatile and relatively higher vapor density. The PEG-based compounds are highly biodegradable with low flammability. Literature studies showed that PEG-based corrosion inhibitors are extensively used for steel alloys, however their use for other metals and alloys, such as zinc and aluminum, should be further explored. Because of their high resistivity for acid and basic solutions, they can be used as effective corrosion inhibitors at various pH ranges. The PEG-based compounds are also thermally stable, therefore they can be used as high-temperature corrosion inhibitors. They are relatively very stable against oxidative and reductive degradation. Because of their high solubility in aqueous media, the PEG-based compounds are widely used as aqueous phase corrosion inhibitors. Nevertheless, they are also used as corrosion inhibitors in coating phase. In view of the high demands for environment-friendly alternatives, the implementation of PEG-based compounds as corrosion inhibitors should be further explored.

Conflict of interest statement

The authors declare that there is no conflict of interest in any way.

Competing interest statement

The authors declare that there is no competing interest in any way.

Author's contributions

All authors collectedly contributed in the design and write-up of the book chapter.

Important links

https://www.medicinenet.com/polyethylene_glycol_peg_3350-oral/article.htm.
https://www.longdom.org/open-access/recent-applications-of-polyethylene-glycols-pegs-and-peg-derivatives-2329-6798.1000132.pdf.
https://www.oxiteno.us/what-is-polyethylene-glycol-peg-uses-side-effects/.

References

[1] G. Gece, Drugs: a review of promising novel corrosion inhibitors, Corros. Sci. 53 (2011) 3873−3898.

[2] C. Verma, E.E. Ebenso, I. Bahadur, M. Quraishi, An overview on plant extracts as environmental sustainable and green corrosion inhibitors for metals and alloys in aggressive corrosive media, J. Mol. Liq. 266 (2018) 577−590.

[3] R. Webster, V. Elliott, B.K. Park, D. Walker, M. Hankin, P. Taupin, PEG and PEG conjugates toxicity: towards an understanding of the toxicity of PEG and its relevance to PEGylated biologicals, PEGylated Protein Drugs: Basic Science and Clinical Applications, Springer, 2009, pp. 127−146.

[4] M. Kidwai, D. Bhatnagar, N.K. Mishra, Polyethylene glycol (PEG) mediated green synthesis of 2, 5-disubstituted 1, 3, 4-oxadiazoles catalyzed by ceric ammonium nitrate (CAN), Green Chem. Lett. Rev. 3 (2010) 55−59.

[5] A. Mero, G. Pasut, L. Dalla Via, M.W. Fijten, U.S. Schubert, R. Hoogenboom, et al., Synthesis and characterization of poly (2-ethyl 2-oxazoline) conjugates with proteins and drugs: suitable alternatives to PEG conjugates? J. Controlled Rel. 125 (2008) 87−95.

[6] J. Chen, S.K. Spear, J.G. Huddleston, R.D. Rogers, Polyethylene glycol and solutions of polyethylene glycol as green reaction media, Green Chem. 7 (2005) 64−82.

[7] S. Umoren, E. Ebenso, P. Okafor, O. Ogbobe, Water-soluble polymers as corrosion inhibitors, Pigment Resin Technol. 35 (2006) 346−352.

[8] V. Srivastava, D.S. Chauhan, P.G. Joshi, V. Maruthapandian, A.A. Sorour, M.A. Quraishi, PEG-functionalized chitosan: a biological macromolecule as a novel corrosion inhibitor, ChemistrySelect 3 (2018) 1990−1998.

[9] H. Ashassi-Sorkhabi, N. Ghalebsaz-Jeddi, F. Hashemzadeh, H. Jahani, Corrosion inhibition of carbon steel in hydrochloric acid by some polyethylene glycols, Electrochim. Acta 51 (2006) 3848−3854.

[10] N. Akiya, P.E. Savage, Roles of water for chemical reactions in high temperature water, Chem. Rev. 102 (2002) 2725−2750.

[11] U.M. Lindström, Stereoselective organic reactions in water, Chem. Rev. 102 (2002) 2751−2772.

[12] C. Verma, E.E. Ebenso, M. Quraishi, Ionic liquids as green and sustainable corrosion inhibitors for metals and alloys: an overview, J. Mol. Liq. 233 (2017) 403−414.

[13] C. Verma, J. Haque, M. Quraishi, E.E. Ebenso, Aqueous phase environmental friendly organic corrosion inhibitors derived from one step multicomponent reactions: a review, J. Mol. Liq. 275 (2019) 18−40.

[14] C. Verma, M. Quraishi, E.E. Ebenso, Microwave and ultrasound irradiations for the synthesis of environmentally sustainable corrosion inhibitors: an overview, Sustain. Chem. Pharm. 10 (2018) 134−147.

[15] F. Bailey Jr, J. Koleske, Poly (ethylene oxide), Academic Press, New York, 1976.

[16] D. Candy, J. Belsey, Macrogol (polyethylene glycol) laxatives in children with functional constipation and faecal impaction: a systematic review, Arch. Dis. Child. 94 (2009) 156−160.

[17] A.A. D'souza, R. Shegokar, Polyethylene glycol (PEG): a versatile polymer for pharmaceutical applications, Expert Opin. Drug Delivery 13 (2016) 1257−1275.

[18] M. Roberts, M. Bentley, J. Harris, Chemistry for peptide and protein PEGylation, Adv. Drug Delivery Rev. 54 (2002) 459−476.

[19] G. Totten, N. Clinton, Poly (ethylene glycol) and derivatives as phase transfer catalysts, J. Macromol. Sci. C, 38 (1998), 77−142.

[20] F.M. Kerton, R. Marriott, Alternative Solvents for Green Chemistry, Royal Society of Chemistry Publishing, 2013.

[21] N.F. Leininger, R. Clontz, J.L. Gainer, D.J. Kirwan, Aqueous Polyglycol Solutions as Alternative Solvents, ACS Publications, 2002.

[22] D.J. Heldebrant, P.G. Jessop, Liquid poly (ethylene glycol) and supercritical carbon dioxide: a benign biphasic solvent system for use and recycling of homogeneous catalysts, J. Am. Chem. Soc. 125 (2003) 5600–5601.

[23] P. Wang, D. Zhang, Y. Zhou, Y. Li, H. Fang, H. Wei, et al., A well-defined biodegradable 1, 2, 3-triazolium-functionalized PEG-b-PCL block copolymer: facile synthesis and its compatibilization for PLA/PCL blends, Ionics 24 (2018) 787–795.

[24] H.F. Naguib, M.S.A. Aziz, S.M. Sherif, G.R. Saad, Thermal properties of biodegradable poly (PHB/PCL-PEG-PCL) urethanes nanocomposites using clay/poly (ε-caprolactone) nanohybrid based masterbatch, Appl. Clay Sci. 57 (2012) 55–63.

[25] K. Kocheva, T. Kartseva, S. Landjeva, G. Georgiev, Parameters of cell membrane stability and levels of oxidative stress in leaves of wheat seedlings treated with PEG 6000, Gen. Appl. Plant Physiol. 35 (2009) 127–133.

[26] Y. Li, H. Zhang, Y. Yao, T. Gong, R. Dong, D. Li, et al., Promoted off-on recognition of H_2O_2 based on the fluorescence of silicon quantum dots assembled two-dimensional PEG-MnO_2 nanosheets hybrid nanoprobe, Microchim. Acta 187 (2020) 347.

[27] A.R. Kiasat, M. Zayadi, T.F. MOHAMMAD, M.M. Fallah, Simple, practical and eco-friendly reduction of nitroarenes with zinc in the presence of polyethylene glycol immobilized on silica gel as a new solid-liquid phase transfer catalyst in water, Iran. J. Chem. Chem. Eng. 30 (2011) 37–41.

[28] E. Rytting, K.A. Lentz, X.-Q. Chen, F. Qian, S. Venkatesh, A quantitative structure-property relationship for predicting drug solubility in PEG 400/water cosolvent systems, Pharm. Res. 21 (2004) 237–244.

[29] H. Yang, J.J. Morris, S.T. Lopina, Polyethylene glycol–polyamidoamine dendritic micelle as solubility enhancer and the effect of the length of polyethylene glycol arms on the solubility of pyrene in water, J. Colloid Interface Sci. 273 (2004) 148–154.

[30] H. Ashassi-Sorkhabi, N. Ghalebsaz-Jeddi, Inhibition effect of polyethylene glycol on the corrosion of carbon steel in sulphuric acid, Mater. Chem. Phys. 92 (2005) 480–486.

[31] H. Ashassi-Sorkhabi, N. Ghalebsaz-Jeddi, Effect of ultrasonically induced cavitation on inhibition behavior of polyethylene glycol on carbon steel corrosion, Ultrason. Sonochem. 13 (2006) 180–188.

[32] S. Umoren, O. Ogbobe, P. Okafor, E. Ebenso, Polyethylene glycol and polyvinyl alcohol as corrosion inhibitors for aluminium in acidic medium, J. Appl. Polym. Sci. 105 (2007) 3363–3370.

[33] S. Umoren, O. Ogbobe, I. Igwe, E. Ebenso, Inhibition of mild steel corrosion in acidic medium using synthetic and naturally occurring polymers and synergistic halide additives, Corros. Sci. 50 (2008) 1998–2006.

[34] S. Umoren, O. Ogbobe, E. Ebenso, The adsorption characteristics and synergistic inhibition between polyethylene glycol and halide ions for the corrosion of mild steel in acidic medium, Bull. Electrochem. 22 (2006) 155–168.

[35] S.A. Umoren, Synergistic inhibition effect of polyethylene glycol–polyvinyl pyrrolidone blends for mild steel corrosion in sulphuric acid medium, J. Appl. Polym. Sci. 119 (2011) 2072–2084.

[36] S. John, M. Kuruvilla, A. Joseph, Surface morphological and impedance spectroscopic studies on the interaction of polyethylene glycol (PEG) and polyvinyl pyrrolidone (PVP) with mild steel in acid solutions, Res. Chem. Intermed. 39 (2013) 1169–1182.

[37] M. Mobin, M.A. Khan, Adsorption and corrosion inhibition behavior of polyethylene glycol and surfactants additives on mild steel in H_2SO_4, J. Mater. Eng. Perform. 23 (2014) 222–229.

[38] A. Algaber, E.M. El-Nemma, M.M. Saleh, Effect of octylphenol polyethylene oxide on the corrosion inhibition of steel in 0.5 M H_2SO_4, Mater. Chem. Phys. 86 (2004) 26–32.

[39] S. Deng, X. Li, J. Sun, Adsorption and inhibition effect of polyethylene glycol 20000 on steel in H_2SO_4 solution, Clean. World 4 (2011).

[40] M.M. Fares, A. Maayta, J.A. Al-Mustafa, Synergistic corrosion inhibition of aluminum by polyethylene glycol and ciprofloxacin in acidic media, J. Adhes. Sci. Technol. 27 (2013) 2495–2506.

[41] M. Awad, M. Metwally, S. Soliman, A. El-Zomrawy, Experimental and quantum chemical studies of the effect of poly ethylene glycol as corrosion inhibitors of aluminum surface, J. Ind. Eng. Chem. 20 (2014) 796–808.

[42] H. Boudellioua, Y. Hamlaoui, L. Tifouti, F. Pedraza, Comparison between the inhibition efficiencies of two modification processes with PEG–ceria based layers against corrosion of mild steel in chloride and sulfate media, J. Mater. Eng. Perform. 26 (2017) 4402–4414.

[43] H. Boudellioua, Y. Hamlaoui, L. Tifouti, F. Pedraza, Effects of polyethylene glycol (PEG) on the corrosion inhibition of mild steel by cerium nitrate in chloride solution, Appl. Surf. Sci. 473 (2019) 449–460.

[44] S. Božović, S. Martinez, V. Grudić, A novel environmentally friendly synergistic mixture for steel corrosion inhibition in 0.51 M NaCl, Acta Chim. Slovenica 66 (2019) 112.

[45] S. Yahya, N. Othman, M. Ismail, Corrosion inhibition of steel in multiple flow loop under 3.5% NaCl in the presence of rice straw extracts, lignin and ethylene glycol, Eng. Fail. Anal. 100 (2019) 365–380.

[46] S.A. El Wanees, A.A. El Aal, E.A. El Aal, Effect of polyethylene glycol on pitting corrosion of cadmium in alkaline solution, Br. Corros. J. 28 (1993) 222–226.

[47] Y. Ein-Eli, M. Auinat, D. Starosvetsky, Electrochemical and surface studies of zinc in alkaline solutions containing organic corrosion inhibitors, J. Power Sources 114 (2003) 330–337.

[48] D. Gelman, I. Lasman, S. Elfimchev, D. Starosvetsky, Y. Ein-Eli, Aluminum corrosion mitigation in alkaline electrolytes containing hybrid inorganic/organic inhibitor system for power sources applications, J. Power Sources 285 (2015) 100–108.

17

Advances in the synthesis and use of 8-hydroxyquinoline derivatives as effective corrosion inhibitors for steel in acidic medium

Mohamed Rbaa[1], Younes El Kacimi[2], Brahim Lakhrissi[1], Abdelkader Zarrouk[3]

[1]LABORATORY OF ORGANIC CHEMISTRY, CATALYSIS AND ENVIRONMENT, FACULTY OF SCIENCES, IBN TOFAIL UNIVERSITY, KENITRA, MOROCCO [2]LABORATORY OF MATERIALS ENGINEERING AND ENVIRONMENT: MODELLING AND APPLICATION, FACULTY OF SCIENCE, UNIVERSITY IBN TOFAIL, KENITRA, MOROCCO [3]LABORATORY OF MATERIALS, NANOTECHNOLOGY AND ENVIRONMENT, FACULTY OF SCIENCES, MOHAMMED V UNIVERSITY, RABAT, MOROCCO

Chapter outline

17.1 Introduction

The concept of metal protection has received significant recognition and also been adapted into a variety of fields, such as the industrial field because a large amount of the production of industries is composite bodies of metal parts [1]. But a large amount of these metal bodies has been destroyed by corrosion, especially in acidic environments. The issue of the presence of protection methods has received considerable critical attention for bodies based on metal parts [2]. There are several methods of protection against corrosion, but the protection by organic inhibitors remains the most conventional because it is cheaper and more effective [3–5]. In recent years, there has been an increasing amount of literature on adaptive

Environmentally Sustainable Corrosion Inhibitors. DOI: https://doi.org/10.1016/B978-0-323-85405-4.00002-1

evolution of the use of 8-hydroxyquinoline derivatives for the security of metals against corrosion in a different acidic medium [6−8]. In the biological, pharmaceutical, chemical, and agrochemical, analytical, and electrochemical fields, 8-hydroxyquinoline and its derivatives have been used as potential agents. This is the reason that researchers are orienting their research toward the synthesis and the development of new heterocyclic compounds based on 8-hydroxyquinoline [9−12].

According to the Scopus database statistics it has been clear that 8-hydroxyquinoline and its derivatives behave as well performed corrosion inhibitors for metals in different environments (Fig. 17−1A). The 8-hydroxyquinoline and its derivatives are widely used as corrosion inhibitors of steel in acidic environments, like HCl, H_2SO_4, and H_3PO_4, in several countries, such as Morocco, China, Egypt, Palestine, and Germany (Fig. 17−1B). In addition, several centers have worked hard on the use of 8-hydroxyquinoline and its derivatives as inhibitors of the corrosion of steel in acidic media, for example, Ibn Tofail University (Morocco), Qatar University (Qatar), King Fahd University of Petroleum and Minerals (Arabie Saoudite), An-Najah National University (Palestine), and Instituto Superior Técnico (Portugal) (Fig. 17−1B). The Scopus database statistics show that there are several authors internationally working on the synthesis and applications of 8-hydroxyquinoline as an inhibitor of the corrosion of steel in acidic environments, for example, B. Lakhrissi, A. Zarrouk, M. Rbaa, and I.B. Obot (Fig. 17−1C).

This work is the summary of the synthesis, characterization, and applications of 8-hydroxyquinoline compounds and their derivatives for the inhibition of various metals corrosion in different acidic environments. The data presented in this review are information from recent years (2018−20), that have been published in scientific journals or by metal bodies companies.

17.2 Literature reviews

Rbaa et al. [12] have synthesized novel organic compounds based on 8-hydroxyquinoline derivatives [(QDO :5-((4,5-dihydrooxazol-2-yl)methyl)quinolin-8-ol and QIM: 5-((4,5-dihydro-1H-imidazol-2-yl)methyl)quinolin-8-ol)] using a conventional method consisting of the condensation of the compound 5-CNMHQ [2-(8-hydroxyquinolin-5-yl) acetonitrile] and the compounds binucleophilic in ethanol (EtOH) and in the attendance of NaOH at reflux for 8 hours. The obtained compounds QDO and QIM were purified by silica column chromatography followed by recrystallization in absolute ethanol (Fig. 17−2) and then the elemental data of the products were identified by nuclear magnetic resonance (NMR) spectroscopy, infrared (IR), and elemental analysis (EA). The corrosion inhibition performance of 8-hydroxyquinoline has been examined through the gravimetric, hydrogen evolution, potentiodynamic polarization (PDP), and electrochemical impedance spectroscopy (EIS) methods. The theoretical results have been calculated by the density functional theory method (DFT) method. We found that the corrosion inhibition performance is affected in a positive manner by increasing the concentration of both inhibitors, QIM and QDO. In contrast the inhibitors'

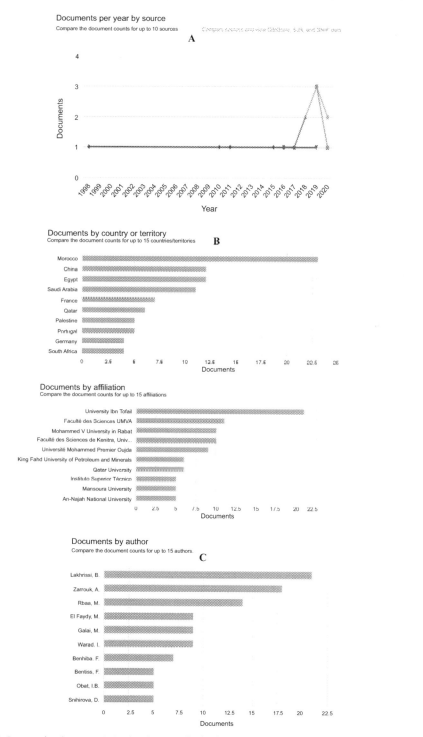

FIGURE 17–1 (A—C): Scopus database statistics for the use of 8-hydroxyquinoline in inhibiting corrosion of steel in acidic media.

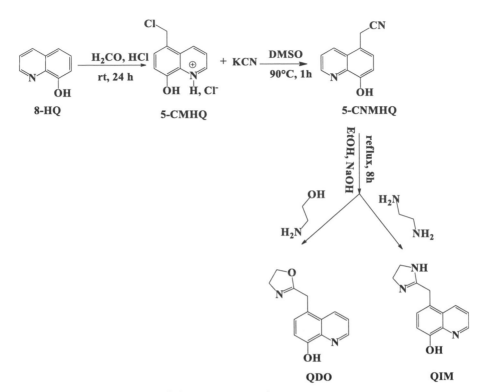

FIGURE 17–2 Route of organic compounds formation QDO and QIM.

performances decreased on increasing the temperature. Both of the inhibitors acted as mixed-type inhibitors and follow the Langmuir adsorption isotherm. Also, the results of the theoretical study corroborated with the results of the experimental study.

Rbaa et al. [13] synthesized new organic compound derivatives of 8-hydroxyquinoline (HQ-ZH, HQ-ZNO$_2$ and HQ-ZOH) and investigated their applications for the mild steel corrosion inhibition in 1.0 M HCl. These derivatives were synthesized by the condensation of quinazoline derivatives with 5-chloromethyl-8-hydroxyquinoline (5-CMHQ) in pure solvent tetrahydrofuran (THF) and in the presence of Et$_3$N (Fig. 17–3). The elemental data of these inhibitors was identified by IR and NMR spectroscopy, and further tested for mild steel corrosion inhibition in 1.0 M HCl by conventional and electrochemical (PDP and EIS) means. Theoretical calculations were done by the (DFT calculations and MC simulations) techniques. From the results it is clear that these inhibitors acted effectively and were mixed type for mild steel corrosion in acidic environment. And also, the thermodynamic parameters imply that the Langmuir adsorption isotherm is obeyed by these tested inhibitors. UV−Visible and scanning electron microscopy (SEM) analysis confirm the good adsorption of these inhibitors onto the mild steel surface. In addition, the results of the theoretical studies confirm the results of the experimental studies.

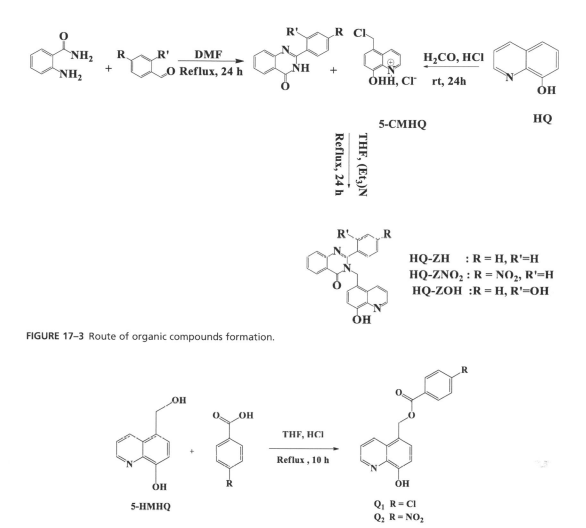

FIGURE 17–3 Route of organic compounds formation.

FIGURE 17–4 Route of organic compounds formation Q1 and Q2.

Rbaa et al. [14] prepared new 8-hydroxyquinoline derivatives by the condensation of 5-hydroxymethyl-8-hydroxyquinine (HMHQ) with derivatives of 4-alkylbezoic acid in THF and in the presence of HCl (37%) at reflux for 10 hours. The reaction was monitored by thin-layer chromatography and the obtained product was purified by silica column chromatography and then recrystallized in absolute ethanol to find two heterocyclic compounds derived from 8-hydroxyquinoline ((8-hydroxyquinolin-5-yl) methyl-4-chlorobenzoate (Q_1) and (8-hydroxyquinolin-5-yl) methyl-4-nitrobenzoate (Q_2)) (Fig. 17–4), and characterized by IR, NMR, and EA. On completion of the synthesis of the compounds the authors studied the corrosion-inhibiting effect of steel in 1 M HCl, by electrochemical (PDP and EIS), gravimetric (WL), and theoretical (DFT and MC) methods. The electrochemical study results revealed that the two inhibitors act effectively against mild steel corrosion in 1 M HCl, and the inhibition performance reached a maximum of 96% for Q_1 and 92% Q_2. In addition, the thermodynamic studies show that inhibitors adsorbed onto the metal surface with a chemical mode of adsorption (chemisorption) and according to the Langmuir isotherm. DFT calculations and Monte Carlo (MC) simulation were close to the obtained experimental results.

Rbaa et al. [15] prepared new benzimidazole derivatives based on 8-hydroxyquinoline. This synthesis was carried out by the condensation of the benzimidazole derivatives (A and B) with 5-CMHQ in dimethylformamide (DMF) and in the attendance of potassium bicarbonate (K_2CO_3) under reflux for 24 hours (Fig. 17–5). The purification of the obtained products (BQM) and (BQ) was done by silica column chromatography, and inspected as corrosion inhibitors for mild steel in 1 M HCl through conventional and electrochemical and theoretical means. The outcomes of the electrochemical studies reveals that both compounds behave nicely against mild steel corrosion in 1 M HCl and the inhibition effect likely depends upon the type of substitution attached on the molecule. Also, the electrochemical results reveal the mixed-type behavior of the studied inhibitors and efficiencies increased on an increase in the concentration of the inhibitors (BQM > BQ). The thermodynamic constant reveals its chemisorptions $\Delta G > -40$ kJ-mol^{-1} nature on to the mild steel surface. The Arrhenius coefficient preexponential factor (R^2) reveals the Langmuir mode of adsorption isotherm. The results of the theoretical studies (DFT and MD) are verified with the experimental studies.

El Faydy et al. [16] prepared a new organic compound based on 8-hydroxyquinoline by an effective and conventional method with a yield of 68%. This method consists of the

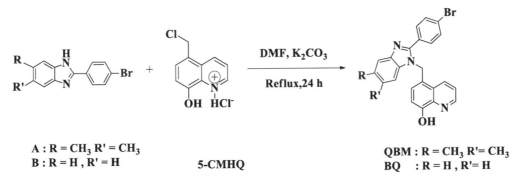

A : R = CH$_3$ R' = CH$_3$
B : R = H , R' = H   5-CMHQ   QBM : R = CH$_3$ R'= CH$_3$
                                                           BQ   : R = H , R'= H

FIGURE 17–5 Route of organic compounds formation BQ and BQM.

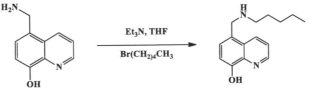

5-AAMHQ   P-QN

FIGURE 17–6 Route of organic compound formation P-QN. *P-QN*, 5-Penthylaminomethyl-8-hydroxyquinoline.

substitution of the 5-aminomethyl-8-hydroxyquinoline (5-AMHQ) by the 1-bromopentan in THF in the attendance of triethylamine (Et$_3$N) under reflux for 6 hours (Fig. 17–6). The product obtained [5-penthylaminomethyl-8-hydroxyquinoline (P-QN)] was characterized by proton and carbon NMR spectroscopy (^{1}H NMR and ^{13}C NMR). The obtained product has been tested as an inhibitor for carbon steel corrosion in 1.0 M HCl medium by weight loss (WL) method and electrochemical (PDP) and (EIS) methods. The electrochemical outcomes prove that P-QN is more effective against carbon steel corrosion and reaches an optimal value of 92% at the concentration of 10^{-3} M. The results of the EIS study show that there is charge transfer during the corrosion process. The thermodynamic constants (ΔHa^*, ΔSa^*, and ΔGa^*) show that the P-QN inhibitor was well adsorbed on the metal surface by chemisorption. In addition, the activation energy of P-NQ is larger than that of the blank specimen which proves that there is the formation of energies in-between the inhibitor and the metal surface. The SEM morphology shows that the carbon steel surface becomes smoother on addition of the inhibitor.

Dkhireche et al. [17] combined two heterocyclic compounds by the condensation of the benzimidazole derivatives (B$_1$ and B$_2$) and compound (5-CMHQ) (Fig. 17–7). The authors tested these products as a corrosion inhibitor of m-steel in sulfuric acid (0.5 M H$_2$SO$_4$) by electrochemical (PDP and EIS) and WL methods. The surface study of the tested steel was examined by SEM. As in the hydrochloric acid medium (1.0 M HCl), the 8-hydroxyquinoline derivatives are more effective in the sulfuric acid medium (0.5 M H$_2$SO$_4$). PDP results confirm that for both compounds, (Q-NO$_2$) and (Q-N(CH$_3$)$_2$), increasing the inhibition concentrations increases the inhibition efficiency and on increasing the temperature range there is a decrease in the inhibition efficiencies. The thermodynamic results clearly show that the two organic compounds tested are absorbed on the surface of the steel by chemical bonds following the Langmuir adsorption isotherm. The SEM images clearly indicate its positive adsorption on to the metal surface.

Rbaa et al. [18] obtained [1,4-bis ((8-hydroxyquinolin-5-yl) methyl)-6-methylquinoxalin-2,3-(1H, 4H)-dione Q-NH and 1,4-bis ((8-hydroxyquinolin-5-yl) methyl) quinoxalin-2,3-(1H, 4H)-dioneQ-NCH$_3$], by the reaction of the quinoxalinone derivatives (A and B) with 5-hydroxymethyl-8-hydroxyquinoline (5-HMHQ) in the presence of triethylamine (Et$_3$N) in

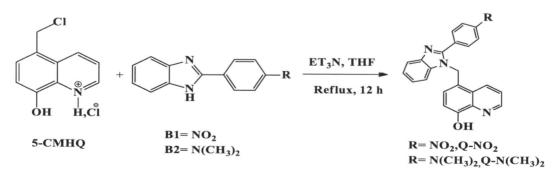

FIGURE 17–7 Route of organic compounds formation Q-NO$_2$ and Q-N(CH$_3$)$_2$.

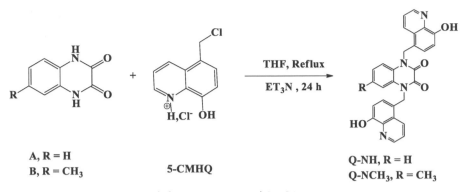

FIGURE 17–8 Route of organic compounds formation Q-NH and Q-NCH₃.

refluxing THF for 24 hours (Fig. 17–8). The obtained products were characterized by NMR spectroscopy, and tested as mild steel corrosion inhibitors in 1 M HCl by EIS and PDP and conventional WL methods. The mild steel surface was examined by SEM analysis. The calculated data confirmed a good correlation between the polarization, impedance spectroscopy, and gravimetric analysis. The quinoxalines based on 8-hydroxyquinolines act as mixed-type inhibitors with a cathodic predominance. The thermodynamic parameter reveals that the 8-hydroxyquinoline derivatives adsorb to the mild surface by chemical mode of adsorption, following the Langmuir isotherm. The SEM images show that the 8-hydroxyquinoline derivatives adsorb very well on to the mild steel surface. And the surface in the presence of the inhibitors has been well protected.

El Faydy et al. [19] synthesized a new type of 8-hydroxyquinoline compounds, namely, 5-propoxymethyl-8-hydroxyquinoline (PMHQ), 5-methoxymethyl-8 hydroxyquinoline (MMHQ), HMHQ. The synthesis of these compounds was carried out in two steps; the first step was the conversion of 8-hydroxyquinoline (1) to 5-chloromethyl-8-hydroxyquinoline hydrochloride (5-CMHQ). In the second step, the product 5-CMHQ was treated with water and ammonia to obtain HMHQ. The compounds (PMHQ and MMHQ) were obtained by condensing the product 5-CMHQ with alcohol in the presence of sodium hydrogen carbonate (NaHCO₃) at reflux for 2 hours (Fig. 17–9). After the synthesis the authors studied the inhibitory action of these compounds on corrosion of C-steel in 1.0 M HCl by experimental method Potentiodynamic Polarization (PDP) and Electrochemical Impedance Spectroscopy (EIS) (EIS and PDP) and theoretical calculations Density-functional Theory (DFT) and Molecular Dynamics (MD)(DFT and MC simulations) means. The surface of C-steel was examined by Scanning Electron Microscopy and X-ray Photoelectron Spectroscopy (SEM and XPS) analysis. The obtained results showed the excellent corrosion-inhibiting performance of C-steel in a 1.0 M HCl. The polarization curves indicate that the three 8-hydroxyquinoline compounds studied inhibit both partial corrosion reactions. They were classified as mixed-type inhibitors. The impedance results obtained also show an increase in protective power with the concentration of inhibitors, with the estimated inhibitory efficacy evolving in the order PMHQ (94, 89%) > MMHQ (89%) > HMHQ, (81%); surface analysis by SEM and XPS shows that a protective

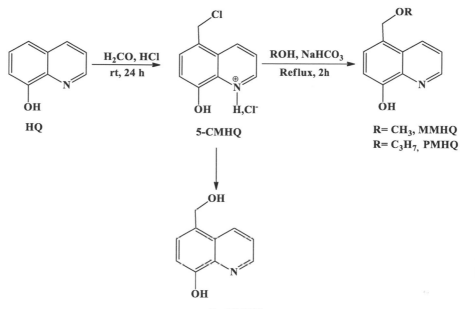

FIGURE 17–9 Route of organic compounds formation HMHQ, MMHQ and PMHQ. *HMHQ*, 5-hydroxymethyl-8-hydroxyquinine; *MMHQ*, 5-methoxymethyl-8 hydroxyquinoline; *PMHQ*, 5-propoxymethyl-8-hydroxyquinoline.

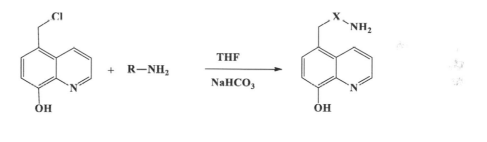

FIGURE 17–10 Route of organic compounds formation HMQN and AMQN. *AMQN*, 5-[(Aminooxy) methyl] quinolin-8-ol; *HMQN*, 5-(hydrazinylmethyl) quinolin-8-ol.

layer is formed on the metal surface to inhibit corrosion, The topology and adsorption energy obtained by MC simulation reveals that the three 8-hydroxyquinoline compounds take up a location just above the surface, reflecting the ability of these molecules to interact with the steel-based metal substrate.

Rouifi et al. [20] synthesized novel 8-hydroxyquinoline derivatives (Fig. 17–10) and tested their use in inhibiting the C-steel corrosion in a 1 M HCl. The derivatives were synthesized by the action of some binucleophiles on compound 5-CMHQ in pure solvent (THF) in the attendance of sodium hydrogen carbonate (NaHCO$_3$) to obtain the compounds

5-(hydrazinylmethyl) quinolin-8-ol (HMQN) and 5-[(aminooxy) methyl] quinolin-8-ol. Both the products were later characterized by NMR and IR spectroscopy, and the inhibition performance in 1 M HCl for carbon steel was tested by using electrochemical (PDP and EIS) and WL techniques. The carbon steel surface morphology has been examined by SEM. From DFT calculations and MD simulations the theoretical results have been calculated. The obtained results revealed that these molecules proved to be good corrosion inhibitors for carbon steel and showed 96.82% at 10^{-3} M inhibition efficiency for the compound HMQN. However, the inhibition efficiency decreases with the increase in the temperature. The adsorption of molecules on to the metal surface obeys the Langmuir adsorption isotherm. Theoretical calculations show that molecules with high donor and acceptor power and a small energy gap have good inhibitory efficacy.

Douche et al. [21] synthesized novel 8-hydroxyquinoline derivatives (HM1), (HM2), and (HM3). These products were obtained by a mixture of 5-alkylmethyl-8-hydroxyquinoline derivative (A and B), para formaldehyde, and piperidine in ethanol at reflux for 10 hours (Fig. 17–11). These compounds were characterized by NMR, IR, and mass spectroscopy (MS) and their inhibition actions for mild steel corrosion in 1 M HCl were judged through the EIS and PDP methods, gravimetric methods, and further theoretical calculations by DFT and MC simulations. The calculated results imply that these organic compounds behave effectively as good corrosion inhibitors for mild steel in 1 M HCl. The PDP result shows that there is a decrease in i_{corr} on increasing the concentration of the inhibitors. This implies that the higher the concentration, the better the surface coverage. These inhibitors exhibit cathodic inhibition in 1 M HCl. The percentage inhibition efficiency in terms of R_{ct} values follows the order: HM1 > HM2 > HM3. A quantum chemical study indicated that the electrons in the aromatic rings, oxygen and nitrogen heteroatom's were the main centers of absorption for donor–acceptor interactions with the unoccupied MS orbital.

Alamshany, Ganash [22] obtained 1,3-bis(quinoline-8-yloxy) propane from 8-hydroxyquinoline (HQ) and 1,3-dibromopropane in DMF in the presence of potassium hydroxide (KOH) (Fig. 17–12). The synthesized compound structure was confirmed by IR spectrum, RMN spectra

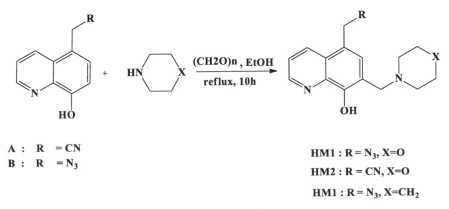

A : R = CN
B : R = N_3

HM1 : R = N_3, X=O
HM2 : R = CN, X=O
HM1 : R = N_3, X=CH$_2$

FIGURE 17–11 Route of organic compounds formation HM1–HM3.

FIGURE 17–12 Route of organic compound formation BQYP.

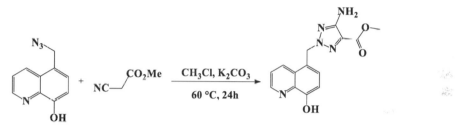

FIGURE 17–13 Route of organic compound formation methyl 5-amino-2-((8-hydroxyquinolin-5-yl)methyl)-2H-1,2,3-triazole-4-carboxylate (MHTC).

(^{13}C and ^{1}H), and MS. After the synthesis, the authors studied the compound as an inhibitor of mild steel corrosion in 1 M HCl using electrochemical methods (PDP and EIS), WL study, and theoretical study (DFT calculations). The compound showed best inhibition efficacy at 10^{-3} M. Impedance diagrams obtained at different concentrations showed a phase shift with respect to the real axis which was attributed to the frequency dispersion expressed by a constant phase element. A theoretical calculation at level B3LYP/6−311þG (d, p) was also carried out to correlate the experimental results with the structural parameters of the molecule studied. Analysis of structural parameters has shown a good correlation with their inhibitory capacities. Analysis of structural parameters has shown a good correlation with their inhibitory capacities.

Rouifi et al. [23] prepared new triazole derivatives based on 8-hydroxyquinoline (MHTC). The MHTC was obtained by condensation of 5-azidomethyl 8-hydroxyquinoline (5-AMHQ) and methyl 2-cyanoacetate in the presence of potassium bicarbonate (K_2CO_3) in chloroform (CHCl$_3$) (Fig. 17−13). Later MHTC was identified by ^{1}H and ^{13}C NMR spectroscopy and EA. The synthesized MHTC was tested as a carbon steel corrosion inhibitor in 1 M HCl through WL and electrochemical studies (EIS and PDP). The gravimetric solutions were analyzed by UV−Visible analysis and the surface of steel was examined by SEM and EDX analysis. A theoretical study was done by using the DFT method at (B3LYP)/6−31G (d, p) level and molecular dynamic (MD) simulation in order to make a significant contribution to the determination of the most likely mechanisms of interactions of the inhibitory molecules with the metal surface. The results show that MHTC behaves as an excellent corrosion inhibitor and the inhibition efficiency reached maximum values above 90% on increasing the inhibitor concentration. Electrochemical studies proclaimed that the studied corrosion inhibitor compounds acted as an inhibitor with mixed-type performance, meaning the inhibition both partial corrosion reactions. The UV−Visible spectrum shows the formation of a complex between Fe^{2+} and MHTC in 1.0 M HCl. SEM and EDX indicate the presence of an adsorbed layer that isolates the steel surface from the corrosive medium. The analysis of the atomic sites of the studied molecule shows highly reactive centers which are the origin of their inhibitory effectiveness. Simulation by MD shows the great interaction between the tested inhibitors and the metallic surface.

El Faydy et al. [24] synthesized benzimidazole derived from 8-hydroxyquinoline; these products were produced in several steps. The first step was to convert the compound 5-CMHQ (5-chloromethyl-8-hydroxyquinoline) to 5-cyanomethyl-8-hydroxyquinoline (5-CNMHQ) by the action of KCN in DMSO for 2 hours. The product (5-CMHQ) was followed by acid hydrolysis to give HQAA. The last step was to condense this intermediate (HQAA) with o-phenylenediamine derivatives to give 5-[(1H-benzimidazol-2-yl) methyl] quinolin-8-ol, 5-[(5-methyl-1H-benzimidazol-2-yl)methyl]quinolin-8-ol, 5-[(5-chloro-1H-benzimidazol-2-yl)methyl]quinolin-8-ol, and 5-[(5,6-dichloro-1H-benzimidazol-2-yl)methyl]quinolin-8-ol (Fig. 17−14). The inhibitory effect of the four compounds was assessed. Inhibitory potency was demonstrated by electrochemical methods as well as surface analysis (SEM and XPS), and theoretical studies were done by DFT. These four compounds were effective in preventing the depletion of carbon steel in 2.0 M H_3PO_4 phosphoric acid medium by forming an insoluble complex with ferrous species. The potency of inhibition increased with the increase in the inhibitor concentration. The

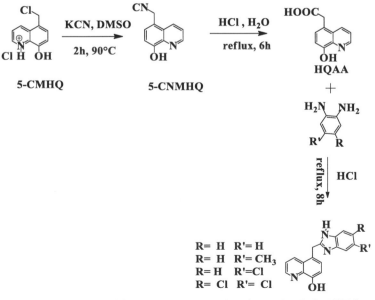

FIGURE 17−14 Route of organic compound formation HQAA, BIMQ and MBMQ. *BIMQ*, 5-[(1H-benzimidazol-2-yl) methyl] quinolin-8-ol; *MBMQ*, 5-[(5-methyl-1H-benzimidazol-2-yl)methyl]quinolin-8-ol.

effectiveness of inhibitor protection depends on inhibitor concentration, temperature, and phosphoric acid concentration. Thermodynamic parameters revealed that these inhibitors chemically adsorb on to the carbon steel surface and obey the Langmuir adsorption isotherm. SEM and XPS measurements showed that the inhibition properties of four inhibitors are due to the formation of a protective film on the carbon steel surface. The results of the theoretical studies are suitable for that of the experimental studies.

El Faydy et al. [25] synthesized new heterocyclic derivatives of 8-hydroxyquinoline (BHQC) and (EHQP). The synthesis of these compounds was carried out in two steps. In the first step: the 5-azidomethyl-8-hydroxyquinoline (5-AZHQ) was converted into 5-AMHQ. In the second step, the product 5-AMHQ was condensed with boc-anhydride in acetonitrile (CH$_3$CN) for 6 hours to obtained the final product BHQC, and the compound EHQP was obtained by condensing the product (5-AMHQ) with diethyl malonate in acetonitrile for 10 hours (Fig. 17−15). These two products (EHQP, BHQC) were further identified by IR, EA, and ^{1}H and ^{13}CNMR spectroscopy. Later on these two products tested as mild steel corrosion inhibitors in 1 M HCl by electrochemical (PDP and EIS), UV−Visible, and surface analysis by SEM and theoretically using DFT and MC calculations. The obtained results revealed that these compounds behave as excellent corrosion inhibitors against mild steel corrosion in 1 M HCl, and the inhibitory proficiency increases with increasing inhibitor concentration and reached maximum values above 95%. The Nyquist diagrams suggest that these inhibitors act by adsorption by protective layer formation on the mild steel surface. The adsorption of 5-substituted-8-hydroxyquinoline derivatives on the steel surface follows the Langmuir adsorption isotherm. The thermodynamic outcome confirms that these molecules adsorb to the metal surface by forming more stable chemical bonds as well as electrostatic bonds in nature. The increase in temperature does not affect the inhibitory efficiencies of the latter,

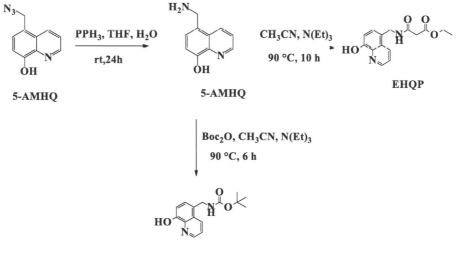

FIGURE 17–15 Route of organic compound formation EHQP and BHQC.

indicating high stability even at high temperatures. SEM study of steel surface has shown the formation of an inhibitory film on the metallic surface and UV−Visible spectroscopy shows that compounds of 5-substituted-8-hydroxyquinoline derivatives possess an iron-inhibitory complexing property. In addition, the results of theoretical studies agree with the results of experimental studies.

Lachiri et al. [26] prepared new azomethine derivatives of 8-hydroxyquinoline and trested their application in the acid corrosion (1 M HCl) inhibition of C38 steel. The synthesis of this product occurred by the action of 4-hydroxybenzaldehyde on 5-AMHQ in ethyl acetate in the presnece of triethylamine (Et₃N) to obtain the 5-[((4 hydroxybenzylidene) amino) methyl] quinolin-8-ol (HBHQ) at reflux for 2 hours (Fig. 17−16). The EA of the products has been done by NMR spectroscopy. The corrosion inhibition study of the tested compounds has been calculated by WL method and electrochemical (EIS and PDP) methods for C-steel in 1.0 M HCl. The theoretical study has been perceived through DFT calculations and MC simulations. The obtained results clearly show that HBHQ has excellent corrosion-inhibiting properties. The adsorption of HBHQ on C38 steel obeys the Langmuir adsorption isotherm with a correlation coefficient close to 1. In addition, the study of the influence of temperature on inhibition efficiency shows that it decreases with the increase in temperature. This indicates that the adsorption of HBHQ on the surface C38 steel takes place via an intermediate adsorption between physisorption and chemisorption. The analysis of structural parameters (i.e., E_{HOMO}, E_{LUMO}, ΔE) showed a good correlation with their inhibitory capacities.

Rbaa et al. [27] studied the effect of acid corrosion inhibition of biobased polymer chitosan (CH) alone and further modified (CH-HQ) by 5-CMHQ for m-steel. The modified chitosan has been realized by the reaction of 5-CMHQ with the biobased polymer in water (H₂O) at room temperature for 72 hours (Fig. 17−17). The compounds were characterized by ¹H and ¹³C NMR spectroscopy and FT-IR analysis. The corrosion inhibition efficiency of the obtained compounds was measured using electrochemical techniques (EIS and PDP). The surface morphology of the steel has been examined by SEM/EDX and the gravimetric solutions have been identified by UV−Visible spectroscopy. The experimental test results have been corroborated with the theoretical studies results (DFT and MD simulations). The outcome clearly reveals that modified chitosan (CH-HQ) is more effective (93%) than chitosan

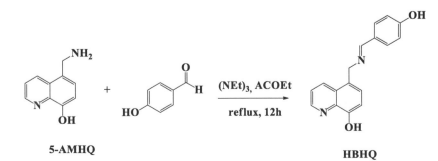

5-AMHQ

HBHQ

FIGURE 17–16 Route of organic compound formation of HBHQ.

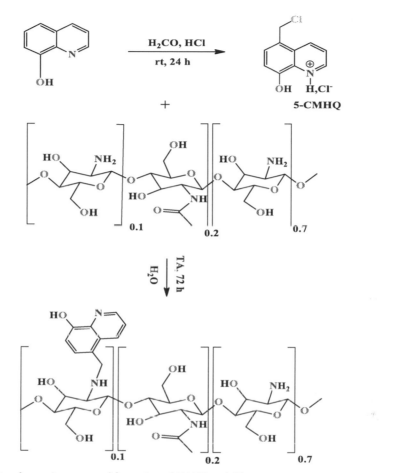

FIGURE 17–17 Route of organic compound formation of CH-HQ and CH.

alone (CH, 78%). Potentiodynamic polarization (PDP) results indicate that biobased polymer chitosan and modified chitosan behave as mixed-type corrosion inhibitors. Both the inhibitors adsorbed on the metal surface by mode of chemisorption followed by Langmuir adsorption isotherm. DFT calculations and MC simulations show a good correlation with the experimental study results.

Rbaa et al. [28] synthesized and characterized two new heterocyclic compounds, derivatives of 8-hydroxyquinoline (N-HQ) and (O-HQ) (Fig. 17–18), by conventional means in a single step reaction between 5-CMHQ and the nucleophiles in THF at reflux for 8 hours. The obtained products have been tested for mild steel acid corrosion (1.0 M HCl). The anticorrosion activity was carried out by PDP, impedance spectroscopy (EIS), DFT calculations, and MC simulations. The electrochemical studies revealed that these compounds act as efficient corrosion inhibitors (98% for N-HQ and 96% for O-HQ) and also behave as mixed-type corrosion inhibitors. In addition, they also acted nicely at high temperatures (328K).

The adsorption of the compounds on to the metal surface is by chemical mode of adsorption following the Langmuir adsorption isotherm. Infrared spectroscopy (FT-IR) and SEM-EDS confirm that the metal surface was very well shielded on the addition of inhibitors solution to the blank solution. In addition, the identification of UV−Visible gravimetric solutions showed that inhibitor have complexing properties with ferrous ions.

Rbaa et al. [29] investigated a new type of organic compound based on 8-hydroxyquinoline (Q-CH$_3$, Q-Br, and Q-H) as a corrosion inhibitor of steel in an acid medium (1 M HCl). The synthesis of these compounds was carried out by the condensation of 8-hydroxyquinoline derived (HMHQ) from the derivatives of para-substituted benzoic acid in THF at reflux for 10 hours (Fig. 17−19). The anticorrosion activity was evaluated by standard electrochemical (PDP and EIS) and gravimetric methods. The surface of the steel was examined by SEM/EDS spectroscopy. The gravimetric solutions were identified by the UV−Visible method. The PDP result revealed that the tested inhibitors act as mixed-type corrosion inhibitors and adsorb on the steel surface

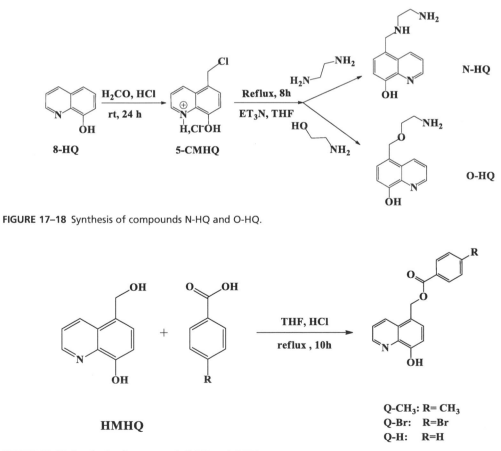

FIGURE 17−18 Synthesis of compounds N-HQ and O-HQ.

FIGURE 17−19 Synthesis of compounds N-HQ and O-HQ.

by chemisorption. The experimental results corroborated the theoretical results (DFT calculations and MC simulations).

Rbaa et al. [30] synthesized new copper and zinc complexes based on 8-hydroxyquinoline (CuQ₂ and ZnQ₂), via the reaction of 5-azidomethyl-8-hydroxyquinoline (5-AMHQ) with a solution of copper (CuCl₂, 2H₂O) and zinc chloride (ZnCl₂, 2H₂O) in ethanol (EtOH) at reflux for 3 hours (Fig. 17−20). The obtained products were characterized by IR and NMR. Both the products have been tested as steel corrosion inhibitors in 1.0 M HCl by PDP, impedance spectroscopy (EIS), DFT calculations, and MC simulations. The experimental and theoretical result indicates that both the two inhibitors acted as mixed-type inhibitors, and that the inhibition efficiency increases with the increase in inhibitor concentrations. The thermodynamics parameters prove that adsorption on a steel surface obeyed the Langmuir isotherm. The adsorption of the tested compounds on the steel surface is confirmed by SEM/EDS spectroscopy.

Rbaa et al. (2020h) synthesized novel oxathiolan and triazole based on 8-hydroxyquinoline derivatives (Q-Ox and Q-T) by a conventional method consisting of the condensation of 2-(8-hydroxyquinolin-5-yl) acetonitrile (5-CNMHQ) and binucleophilic compounds in ethanol (ETOH) in the presence of sodium hydroxide (NaOH) at reflux for 8 hours. The obtained products were purified by silica column chromatography (acetone/hexane 85:15) and recrystallization was done by absolute ethanol (Fig. 17−21). The corrosion inhibition effect was calculated by PDP, impedance spectroscopy (EIS), DFT calculations, and MC simulations. The obtained results showed that that the two inhibitors (Q-Ox and Q-T) act as mixed-type corrosion inhibitors and that the inhibition efficiency increases with the increase in the inhibitors concentration. The thermodynamic parameter proves that the two inhibitors (Q-Ox and Q-T) adsorbed onto the metal surface by Langmuir adsorption isotherm. The adsorption of inhibitors on to the metal surface was confirmed by SEM/EDS spectroscopy. The gravimetrical solutions

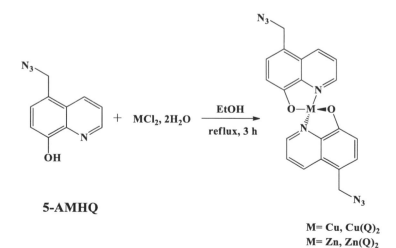

FIGURE 17–20 Synthesis of compounds CuQ2 and ZnQ2.

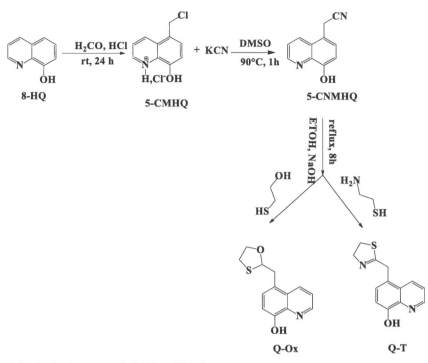

FIGURE 17–21 Synthesis of compounds N-HQ and O-HQ.

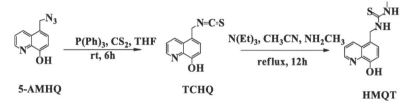

FIGURE 17–22 Preparation of N-HQ and O-HQ compounds.

analysis was done by UV–Visible spectroscopy and showed that the products are capable of making a bond with iron ions.

El Faydy et al. [31] carried out the synthesis of new organic derivatives based on 8-hydroxyquinoline (HQMT and TCHQ). These compounds were derived by the reaction of 5-azidomethyl-8-hydroxyquinoline (5-AMHQ) with carbon disulfide (CS_2) in the presence of triphenylphosphine ($P(Ph)_3$) in THF at room temperature for 6 hours to give 5-isothiocyantomethyl-8-hydroxyquinoline (TCHQ). From the compound TCHQ and methyl-amine (CH_3-NH_2) and in the presence of pure solvent (Et_3N) as base at reflux for 12 hours, the authors obtained the compound HMQT (Fig. 17–22). The steel surface corrosion was examined by SEM spectroscopy and the gravimetric solutions were characterized by UV–visible spectroscopy. The obtained experimental results were simulated with the theo-retical calculations results (DFT and MD). The electrochemical results showed that the

TCHQ compound (91%) is more effective than the HQMT compound (89%). and further explained this result by the presence of an electron donor methyl group ($-CH_3$) carried on the TCHQ molecule. The PDP study results revealed the inhibitors adsorption on to the steel surface by mixed mode, and acted as mixed-type inhibitors. In addition, the calculated thermodynamic results show that both the products are adsorbed on the steel surface by chemisorptions followed by Langmuir adsorption isotherms. The theoretical calculations (DFT and MD) have been corroborated by experimental tests.

17.3 Conclusion

This literature review deals with two main objectives, the first one is to show the importance of 8-hydroxyquinoline and its derivatives in the last year as an anticorrosion agent for mild steels in acidic domain by quoting from several recently published articles (2018−20). The second objective is the presentation of results concerning the corrosion test, showing the mechanism of action of inhibitors of 8-hydroxyquinoline, as well as the methods of synthesis, characterization, and purification. Finally, this review has several advantages in general for researchers in the same field of inhibition of corrosion by 8-hydroxyquinoline derivatives, and more precisely gives ideas to new researchers with regard to the results in the literature.

References

[1] B. El Ibrahimi, A. Jmlal, L. Bazzi, S. El Issami, Amino acids and their derivatives as corrosion inhibitors for metals and alloys, Arab. J. Chem. 13 (2020) 740−771.

[2] N. Asadi, M. Ramezanzadeh, G. Bahlakeh, B. Ramezanzadeh, Utilizing Lemon Balm extract as an effective green corrosion inhibitor for mild steel in 1M HCl solution: a detailed experimental, molecular dynamics, Monte Carlo and quantum mechanics study, J. Taiwan. Inst. Chem. Eng. 95 (2019) 252−272.

[3] A. Fateh, M. Aliofkhazraei, A.R. Rezvanian, Review of corrosive environments for copper and its corrosion inhibitors, Arab. J. Chem. 13 (2020) 481−544.

[4] K.E. Mouaden, D.S. Chauhan, M.A. Quraishi, L. Bazzi, Thiocarbohydrazide-crosslinked chitosan as a bioinspired corrosion inhibitor for protection of stainless steel in 3.5% NaCl, Sustain. Chem. Pharm. 15 (2020) 100213.

[5] D. Zhang, H.Q. Zhang, S. Zhao, Z.G. Li, S.X. Hou, Electrochemical impedance spectroscopy evaluation of corrosion protection of X65 carbon steel by halloysite nanotube-filled epoxy composite coatings in 3.5% NaCl solution, Int. J. Electrochem. Sci. 14 (2019) 4659−4667.

[6] W. Al Zoubi, Y.G. Ko, Freestanding anticorrosion hybrid materials based on coordination interaction between metal-quinoline compounds and TiO2-MgO film, J. Colloid Interface Sci. 565 (2020) 86−95.

[7] M. Rbaa, A.S. Abousalem, M. Galai, H. Lgaz, B. Lakhrissi, I. Warad, et al., New N-heterocyclic compounds based on 8-hydroxyquinoline as efficient corrosion inhibition for mild steel in HCl solution: experimental and theoretical assessments, Arab. J. Sci. Eng. (2020) 1−18.

[8] C.T. Ser, P. Žuvela, M.W. Wong, Prediction of corrosion inhibition efficiency of pyridines and quinolines on an iron surface using machine learning-powered quantitative structure-property relationships, Appl. Surf. Sci. 512 (2020) 145612.

[9] A.G. Erturk, Synthesis, structural identifications of bioactive two novel Schiff bases, J. Mol. Struct. 1202 (2020) 127299.

[10] M. Rbaa, A.S. Abousalem, Z. Rouifi, R. Benkaddour, P. Dohare, M. Lakhrissi, et al., Synthesis, antibacterial study and corrosion inhibition potential of newly synthesis oxathiolan and triazole derivatives of 8-hydroxyquinoline: Experimental and theoretical approach, Surf. Interfaces 19 (2020) 100468.

[11] M. Rbaa, A.S. Abousalem, Z. Rouifi, L. Lakhrissi, M. Galai, A. Zarrouk, et al., Selective synthesis of new sugars based on 8-hydroxyquinoline as corrosion inhibitors for mild steel in HCl solution-effect of the saturated hydrocarbon chain: Theoretical and experimental studies, Inorg. Chem. Commun. (2020) 108019.

[12] M. Rbaa, A.S. Abousalem, M. Ebn Touhami, I. Warad, F. Bentiss, B. Lakhrissi, et al., Novel Cu (II) and Zn (II) complexes of 8-hydroxyquinoline derivatives as effective corrosion inhibitors for mild steel in 1.0M HCl solution: computer modeling supported experimental studies, J. Mol. Liq. 290 (2019) 111243.

[13] M. Rbaa, F. Benhiba, I.B. Obot, H. Oudda, I. Warad, B. Lakhrissi, et al., Two new 8-hydroxyquinoline derivatives as an efficient corrosion inhibitors for mild steel in hydrochloric acid: Synthesis, electrochemical, surface morphological, UV—visible and theoretical studies, J. Mol. Liq. 276 (2018) 120—133.

[14] M. Rbaa, M. Galai, F. Benhiba, I.B. Obot, H. Oudda, M. Ebn Touhami, et al., Synthesis and investigation of quinazoline derivatives based on 8-hydroxyquinoline as corrosion inhibitors for mild steel in acidic environment: experimental and theoretical studies, Ionics 25 (2018) 1—19.

[15] M. Rbaa, M. Galai, M. El Faydy, Y. Lakhrissi, M. Ebn Touhami, A. Zarrouk, et al., Synthesis and characterization of new quinoxaline derivatives of 8- hydroxyquinoline as corrosion inhibitors for mild steel in 1.0M HCl medium, J. Mater. Environ. Sci. 1 (2018) 172—188.

[16] M. El Faydy, M. Galai, M. Rbaa, M. Ouakki, B. Lakhrissi, M. Ebn Touhami, et al., Synthesis and application of new quinoline as hydrochloric acid corrosion inhibitor of carbon steel, Anal. Bioanal. Electrochem. 7 (2018) 815—839.

[17] N. Dkhireche, M. Galai, Y. El Kacimi, M. Rbaa, M. Ouakki, B. Lakhrissi, et al., New quinoline derivatives as sulfuric acid inhibitor's for mild steel, Anal. Bioanal. Electrochem. 1 (2018) 111—135.

[18] M. Rbaa, H. Lgaz, Y. El Kacimi, B. Lakhrissi, F. Bentiss, A. Zarrouk, Synthesis, characterization and corrosion inhibition studies of Novel 8-Hydroxyquinoline derivatives on the acidic corrosion of mild steel: experimental and computational studies, Materials. Discovery 12 (2018) 43—54.

[19] M. El Faydy, R. Touir, M. Ebn Touhami, A. Zarrouk, C. Jama, B. Lakhrissi, et al., Corrosion inhibition performance of newly synthesized 5-alkoxymethyl-8-hydroxyquinoline derivatives for carbon steel in 1M HCl solution: experimental, DFT and Monte Carlo simulation studies, Phys. Chem. 30 (2018) 20167—20187.

[20] Z. Rouifi, M. Rbaa, F. Benhiba, T. Laabaissi, H. Oudda, B. Lakhrissi, et al., Preparation and anti-corrosion activity of novel 8-hydroxyquinoline derivative for carbon steel corrosion in HCl molar: computational and experimental analyses, J. Mol. Liq. (2020) 112923.

[21] D. Douche, H. Elmsellem, L. Guo, B. Hafez, B. Tüzün, A. El Louzi, et al., Anti-corrosion performance of 8-hydroxyquinoline derivatives for mild steel in acidic medium: gravimetric, electrochemical, DFT and molecular dynamics simulation investigations, J. Mol. Liq. (2020) 113042.

[22] Z.M. Alamshany, A.A. Ganash, Synthesis, characterization, and anti-corrosion properties of an 8-hydroxyquinoline derivative, Heliyon 11 (2019) e02895.

[23] Z. Rouifi, F. Benhiba, M.E. Faydy, T. Laabaissi, H. About, H. Oudda, et al., Performance and computational studies of new soluble triazole as corrosion inhibitor for carbon steel in HCl, Chem. Data. Collect. 22 (2019) 100242.

[24] M. El Faydy, B. Lakhrissi, C. Jama, A. Zarrouk, L.O. Olasunkanmi, E.E. Ebenso, et al., Electrochemical, surface and computational studies on the inhibition performance of some newly synthesized 8-hydroxyquinoline derivatives containing benzimidazole moiety against the corrosion of carbon steel in phosphoric acid environment, J. Mater. Res. Technol. 1 (2020) 727—748.

[25] M. El Faydy, F. Benhiba, B. Lakhrissi, M. Ebn Touhami, I. Warad, F. Bentiss, et al., The inhibitive impact of both kinds of 5-isothiocyanatomethyl-8-hydroxyquinoline derivatives on the corrosion of carbon steel in acidic electrolyte, J. Mol. Liq. 295 (2019) 111629.

[26] A. Lachiri, M. El Faydy, F. Benhiba, H. Zarrok, M. El Azzouzi, A. Zertoubi, et al., Inhibitor effect of new azomethine derivative containing an 8-hydroxyquinoline moiety on corrosion behavior of mild carbon steel in acidic media, Int. J. Corros. Scale, Inhib. 4 (2018) 632.

[27] M. Rbaa, F. Benhiba, A.S. Abousalem, M. Galai, Z. Rouifi, H. Oudda, et al., Sample synthesis, characterization, experimental and theoretical study of the inhibitory power of new 8-hydroxyquinoline derivatives for mild steel in 1.0M HCl, J. Mol. Struct. 1213 (2020) 128155.

[28] M. Rbaa, P. Dohare, A. Berisha, O. Dagdag, L. Lakhrissi, M. Galai, et al., New Epoxy sugar-based glucose derivatives as eco friendly corrosion inhibitors for the carbon steel in 1.0M HCl: experimental and theoretical investigations, J. Alloy. Compd. (2020) 154949.

[29] M. Rbaa, M. Fardioui, C. Verma, A.S. Abousalem, M. Galai, E.E. Ebenso, et al., 8-Hydroxyquinoline based chitosan derived carbohydrate polymer as biodegradable and sustainable acid corrosion inhibitor for mild steel. experimental and computational analyses, Int. J. Biol. Macromol. 155 (2020) 645−655.

[30] M. Rbaa, M. Galai, A.S. Abousalem, B. Lakhrissi, M. Ebn Touhami, I. Warad, et al., Synthetic, spectroscopic characterization, empirical and theoretical investigations on the corrosion inhibition characteristics of mild steel in molar hydrochloric acid by three novel 8-hydroxyquinoline derivatives, Ionics 8 (2020) 1−20.

[31] M. El Faydy, B. Lakhrissi, A. Guenbour, S. Kaya, F. Bentiss, I. Warad, et al., In situ synthesis, electrochemical, surface morphological, UV−visible, DFT and Monte Carlo simulations of novel 5-substituted-8-hydroxyquinoline for corrosion protection of carbon steel in a hydrochloric acid solution, J. Mol. Liq. 280 (2019) 341−359.

Sustainable corrosion inhibitors for electronics industry

18

Environmentally sustainable corrosion inhibitors used for electronics industry

Baimei Tan[1], Lei Guo[2], Da Yin[1], Tengda Ma[1], Shihao Zhang[1], Chenwei Wang[1,*]

[1]SCHOOL OF ELECTRONICS AND INFORMATION ENGINEERING, HEBEI UNIVERSITY OF TECHNOLOGY, TIANJIN, P.R. CHINA [2]SCHOOL OF MATERIAL AND CHEMICAL ENGINEERING, TONGREN UNIVERSITY, TONGREN, P.R. CHINA
*CORRESPONDING AUTHOR. E-MAIL ADDRESS: WANGCHENWEI@HEBUT.EDU.CN

Chapter outline

Environmentally Sustainable Corrosion Inhibitors. DOI: https://doi.org/10.1016/B978-0-323-85405-4.00007-0

18.1 Corrosion inhibitor used for copper chemical mechanical planarization

18.1.1 Introduction of copper corrosion inhibitor

During the process of multilayer copper wiring chemical mechanical planarization (CMP), the removal of copper is generally controlled by the synergistic effect of complexing agents, oxidants, and corrosion inhibitors [1]. The surface of the wafer is uneven, and the copper in the convex part needs to be removed faster. The oxidant can oxidize the copper into ions, and the complexing agent can complex copper ions, promoting the positive movement of the balance and increasing the removal rate (RR) of copper. Copper in concave needs to be protected. Although the presence of the oxidant will form an oxide film on the surface to protect the wafer, its passivation ability is far from reaching the requirements of global planarization. Corrosion inhibitors are added to polishing liquid to further inhibit the copper corrosion in concave to protect the copper and achieve global planarization [2].

Corrosion inhibitors are generally divided into two categories: inorganic corrosion inhibitors and organic corrosion inhibitors. Inorganic corrosion inhibitors mainly include polyphosphate, tungstate, silicate, nitrite, etc. Organic corrosion inhibitors include benzotriazole (BTA), mercaptobenzothiazole, and other heterocyclic compounds containing nitrogen and oxygen [3]. Inorganic corrosion inhibitors generally contain metal salts, which will introduce metal ions into the polishing solution. Therefore organic corrosion inhibitors are generally used in the CMP process of multilayer copper wiring. The unfilled outer electron orbitals of the surface copper atoms can provide electrons to form coordination bonds. The nitrogen, sulfur, and other atoms in the organic corrosion inhibitors contain lone pairs of electrons and paired electrons, which can form chemical bonds with metals by obtaining electrons or sharing lone pairs of electrons. They will adsorb on the metal surface and inhibit metal corrosion [4,5].

18.1.2 Benzotriazole

BTA is a good corrosion inhibitor. It is a bicyclic nitrogen-containing heterocyclic organic compound containing a benzene ring and a 1,2,3-triazole ring. In the process of multilayer copper interconnect CMP, BTA is widely used as an inhibitor. It can effectively reduce the corrosion of the copper surface and help flatten the wafer surface after CMP. However, it is flammable and poisonous. It can react with copper and produce Cu-BTA passivation, which requires a large mechanical force to remove and seriously affects the surface defects. On the other hand, under high pH conditions, its ability to correct the height difference is weakened [6,7].

In aqueous solution, different forms of BTA have two ionization equilibrium relationships [8,9]:

$$BTAH_2^+ = BTAH + H^+, pH < 1.0 \tag{18.1}$$

$$BTAH = BTA^- + H^+, pH > 8.0 \tag{18.2}$$

The existence of BTA in solution changes due to different pH values. Fig. 18−1 shows three forms of BTA. Under the near neutral condition, it mostly exists in the form of unionized

(A) BTAH   (B) BTAH^{2-} (C)   BTA$^-$

FIGURE 18–1 Three forms of BTA in aqueous solution. *BTA*, benzotriazole.

BTAH molecules, as shown in Fig. 18–1A; under the strongly acidic condition (pH < 1), protonated BTAH^{2+} will form, as shown in Fig. 18–1B; and under the alkaline condition (pH > 8), BTA exists in the form of BTA$^-$ (shown in Fig. 18–1C).

Under the acidic oxygen-containing condition, the copper surface is easily oxidized to form Cu$_2$O film, thereby inhibiting corrosion of the copper surface. When the acidity is strong, the Cu$_2$O passivation film is easily dissolved. The corrosion inhibition effect will be weakened. BTA will react with copper to form Cu-BTA after CMP, effectively protecting the surface from corrosion. There are two types of reactions for the formation of Cu BTA: BTA molecules can react directly with Cu$_2$O, as shown in Eq. (18.3). Under acidic conditions, BTA molecules will react with cuprous ions dissociated from Cu$_2$O, as shown in Eqs. (18.4) and (18.5).

$$2BTA + Cu_2O = 2Cu(I)BTA + H_2O \tag{18.3}$$

$$Cu_2O + 2H^+ = 2Cu^+ + H_2O \tag{18.4}$$

$$Cu^+ + BTA = Cu(I)BTA + e^+ \tag{18.5}$$

Under neutral condition, BTA is replaced to form Cu:BTA$_{(ads)}$ for adsorption with the action of water molecules on copper surface. The reaction equations are shown as follows [10]:

$$Cu:H_2O_{(ads)} + BTA_{(aq)} \rightleftharpoons Cu:BTA_{(ads)} + H_2O_{(aq)} \tag{18.6}$$

$$Cu:BTA_{(ads)} \rightleftharpoons Cu(I)BTA_{(s)} + H^+_{(aq)} + e^- \tag{18.7}$$

According to the above Eqs. (18.6) and (18.7), when the BTA concentration and pH value increase, the chemical equilibrium will shift to the right and finally reach a steady state.

Under the alkaline environment, CuO formed by oxidation is dissolved and Cu$_2$O is left on the copper surface. Under the action of electrostatic attraction, BTA$^-$ adsorbs easily on the positively charged Cu$_2$O surface, while BTAH chemically reacts with Cu$_2$O to produce Cu/Cu$_2$O/Cu(I)-BTA composite passivation protective film, forming chemical adsorption. Fig. 18–2 presents Cu(I)-BTA adsorption structure.

The adsorption of BTA on copper surface includes both physical adsorption and chemical adsorption [12]. As depicted in Fig. 18–3, in the process of approaching the copper surface, the BTA closest to copper chemically adsorbs with copper, and the BTA adsorbing on the

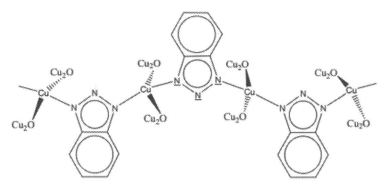

FIGURE 18–2 Structure of Cu(I)-BTA adsorption on Cu₂O [11].

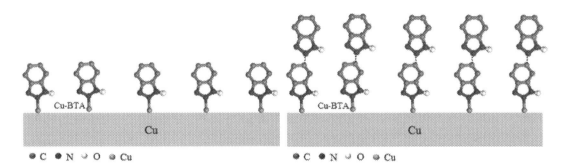

FIGURE 18–3 Schematic diagram of Cu-BTA adsorption model: (left) chemical adsorption state; (right) chemical and physical adsorption state.

first layer forms Cu-BTA polymer with copper ions. The subsequent BTA layer is attached to the Cu-BTA polymer by physical adsorption.

At present, the film-forming mechanism of BTA is generally accepted: BTA molecules can adsorb at the top position through the hybridization of N7 atoms with the *sp* orbitals of copper atoms on the metal surface, or adsorb on the copper surface through the bridge sites of N7 and N8 atoms. The nitrogen atom on the triazole ring of the BTA molecule contains a lone pair of electrons, and the benzene ring contains a π bond. Conjugation effects can occur between the nitrogen atom and the π bond. They are alternately combined with the metal in the form of coordinate bonds and covalent bonds, covering the copper surface with the multielement long-chain molecular structure. Al-Kharafi et al. [13] proposed that BTA will form a double-layer inhibitor film on the copper surface, which is mainly composed of the inner film Cu(I)BTA and the physically adsorbed outer film Cu(II)BTA. With the increase of BTAH concentration, and the extension of time, the thickness of the film will gradually increase.

Kokalj and Peljhan [14] confirmed that the adsorption of BTAH is affected by the properties of the copper oxide matrix (Cu_2O or CuO). The adsorption on Cu_2O is very fast, while the adsorption on CuO is very slow. Some scholars have proposed that at high temperature

(60°C), Cu(I)BTA will be gradually oxidized to Cu(II)BTA after forming on the Cu$_2$O matrix in aqueous solution [15]. However, subsequent studies have shown that the formation of Cu (II)BTA complex is caused by the atmospheric oxidation of the sample during the transfer process. It does not originate from the oxidation of the Cu(I)BTA film.

Modestov et al. [16] studied the surface morphology of Cu-BTAH films adsorbing on Cu$_2$O substrates with different thicknesses. Cu-BTAH presents a star-shaped structure composed of large needle-like crystallites on thin Cu$_2$O substrate, while a finer and denser film is formed on thick Cu$_2$O substrate; the latter is more protective.

18.1.3 1,2,4-Triazole and 2-mercaptobenzothiazole

1,2,4-triazole (TAZ) is widely used as a pesticide and pharmaceutical intermediate due to its solubility and low toxicity. As a potential corrosion inhibitor, it has been proven to achieve the same copper RR while achieving lower pitting and corrosion compared with BTA. Cu-TAZ passivation film is relatively weak and easy to remove. The copper surface becomes hydrophilic after cleaning [17].

Jiang et al. [18] proved that 1,2,4-triazole molecular passivation film results from the inhibitory behavior of TAZ on the copper surface. The formation of Cu-TAZ passivation film can be divided into two categories: one can directly form on the copper surface, while on the other hand, some copper ions react with TAZ to form Cu-TAZ complexes and precipitate in the form of particles. The Cu-TAZ complex is deposited on the copper oxide surface to form a weak passivation film. The dissociation constants pKa$_1$ and pKa$_2$ of TAZ are 2.3 and 10.2, respectively [19]. According to the dissociation equation, the equilibrium composition of TAZ (the degree of dissociation) in different pH solutions can be calculated. The calculation formula is as follows:

$$pH = pK_a + \lg(L^-/HL) \tag{18.8}$$

where L^- represents electron acceptor and HL represents electron donor.

Zhang et al. [20] proposed that the hydrogen bonds within and between the TAZ molecules could adsorb on the copper surface through the interaction of N4 sites and neutral copper atoms, thereby forming a dense H-bonded conjugate film composed of TAZ molecules. The findings of Muniz-Miranda et al. [21] also showed that N4 in TAZ molecules was more negative than N2. It is easier to bridge with metals. Therefore 1,2,4-TAH adsorbs on the copper surface by sharing electrons with N4. The 1,2,4-TAH molecules form a dense conjugate film through hydrogen bonding, which covers the copper surface to form a passivation film, as shown in Fig. 18−4. From the distribution of the three forms of TAZ in different pH solutions, we find that when the pH is 10, TAZ exists in the forms of 1,2,4-TAH and 1,2,4-TA$^-$. The content of 1,2,4-TAH is 61.31%, and the content of 1,2,4-TA$^-$ is 38.69%. The inhibition effect of TAZ is achieved by adsorption. Under the alkaline condition, TAZ has both physical and chemical adsorption on copper surface.

Since van der Waals forces exist between any two molecules, physical adsorption can occur on any solid surface [22]. The physical adsorption of TAZ on the copper surface is determined by the 1,2,4-TAH neutral molecule and the interaction of van der Waals forces

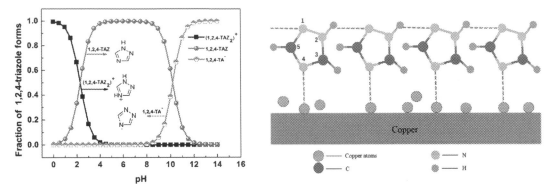

FIGURE 18–4 (left) Distribution of TAZ in different pH solutions. (right) Adsorption model of TAZ on the surface of neutral copper atoms [20]. *TAZ*, 1,2,4-triazole.

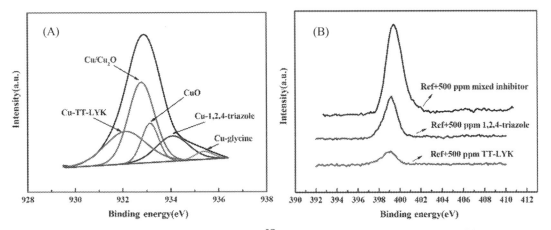

FIGURE 18–5 XPS spectrum of copper surface: (A) Cu 2p$^{3/2}$ spectrum treated with mixed inhibitors; (B) N 1s spectrum treated with different inhibitors [23].

between metal surfaces. The physical adsorption is reversible. It has a weak binding force and a small adsorption heat. It can adsorb and desorb quickly. The chemical adsorption of TAZ on the copper surface is mainly caused by 1,2,4-TAH molecules and 1,2,4-TA$^-$ anions.

In addition, Zhou et al. [23] found that the mixtures 2,2'-[[(methyl-1H-benzotriazol-1-yl) methyl]imino]diethanol (namely TT-LYK) and TAZ acted synergistically to form a dense double-layered passivation film, achieving better protection of the copper surface. When mixed inhibitors are used, the copper surface film is mainly composed of Cu/Cu$_2$O, CuO, Cu-glycine, Cu-TT-LYK, and Cu-1,2,4-triazole. The XPS Cu 2p$^{3/2}$ spectrum of copper surface treated with mixed inhibitors is shown in Fig. 18–5A. The XPS N1s spectrum of copper surface treated with different solutions with different inhibitors is shown in Fig. 18–5B. Since there are few

Cu-glycine complexes on the copper surface, a small amount of N1s can be observed in N1s spectrum. Compared with two single inhibitors, the amount of N1s significantly increases after using the mixed inhibitor. The reason is that the mixed inhibitors significantly enhance the adsorption effect. The corrosion inhibition effect is better during the copper CMP process. The surface quality is also improved.

2-Mercaptobenzothiazole (MBT) contains S and N elements. It is one of the most effective corrosion inhibitors for copper and alloys [24]. Electron transfer occurs between MBT and copper. MBT coordinates with Cu(I) ions in the form of thiol to form a polymer film on the surface. At a lower potential (less than 0 V, vs SCE), MBT adsorbs upright on the copper surface in the form of thiol ions through the S atoms outside the ring [25]. The surface film prevents the corrosion of the corrosive medium effectively.

18.1.4 Other green corrosion inhibitors

For generations, many scholars have researched and produced environment-friendly compounds as effective corrosion inhibitors to reduce the risk of environmental pollution. Researchers have found that molybdate is a promising green and environment-friendly corrosion inhibitor [26]. It has extremely low toxicity and no environmental pollution. However, if molybdate is used alone, its dosage will be very large. And the corrosion inhibition performance is not perfect. But researchers found that if molybdate is compounded as the main part with other inhibitors, the effect will become better. Rocca et al. [27] introduced a linear sodium heptanoate corrosion inhibitor, with a formula of $CH_3 - (CH_2)_5 - COONa$, the results demonstrated that it showed excellent corrosion inhibition performance in a slightly alkaline environment.

Another promising corrosion inhibitor is amino acid. It is an amphoteric compound with basic and acidic groups. It can be completely decomposed in a natural environment and is effective in inhibiting the local corrosion of copper [28]. In addition, we believe that high-molecular polymers are also good candidates. Different organic polymers are less toxic than monomers and will not cause negative effects on the environment. Polymers have better film-forming properties than monomers. They can be used to form a large protective film on the copper surface.

18.2 Corrosion inhibitors used for cobalt chemical mechanical polishing

18.2.1 Introduction of cobalt and its corrosion inhibitors

Copper ions are easily diffused into the medium and form moveable ions in the substrate, resulting in the failure of interconnect structure. So, it is necessary to deposit a barrier layer between the substrate and medium to prevent the ions' diffusion. The basic requirements of barrier materials include dense and low solubility with copper, good adhesion for both copper and the medium, good barrier properties to electromigration, and low deposition temperature to ensure that the medium is not destroyed. Cobalt has been considered and reported by industry and academia due to its good properties as a barrier. In addition, cobalt has been widely used as a wiring material. Whether it is used as barrier or wiring material, it is necessary to inhibit its galvanic corrosion and self-corrosion for integrated circuits [29].

18.2.2 Benzotriazole

BTA has been widely used in chemical fertilizer, iron and steel, and other industrial fields, as a basic corrosion inhibitor in integrated circuit manufacturing. Lu and coworkers [30] investigated the inhibition effect in hydrogen peroxide-free polishing slurry system at pH $= 3, 5, 9, 11$. The results showed that the polishing rate of cobalt is reduced obviously after 5 mM BTA added, indicating that BTA can inhibit cobalt corrosion effectively. The principle of its action can be similar to that of BTA suppress copper polishing: BTA adsorbs on the cobalt surface, inhibiting the corrosion of cobalt and reducing the chemical effect of cobalt CMP, thereby reducing the polishing rate. Gallant and coworkers reported that a compact polymeric network $[Co(II)(BTA)_2 \cdot H_2O]_n$ can be formed at the onset of the anodic dissolution reaction, the mechanism is as follow [31]:

$$nCo + 2n(BTAH)_{ads} + nH_2O \rightarrow [Co(II)(BTA)_2 \cdot H_2O]_n + 2nH^+_{(aq)} + 2ne^- \quad (18.9)$$

If a large number of insoluble products cover a wafer surface entirely, it would effectively prevent the corrosion from solution, block the mass transfer process, and play a role in inhibiting the continued decay of the candle, and it would appear as a current reduction in the dynamic potential polarization test. The insoluble products can not only reduce the chemical effect of the polishing slurry, but also reduce the mechanical friction effect from polishing pad, and reduced the polishing rate of cobalt. The greater the concentration of BTA, the lower the polishing rate of cobalt, which may be caused by the colloidal particles generated from ions reaction being more tightly connected and having a stronger isolation effect on corrosion from solution. It is worth noting that the inhibition from BTA is based on consuming some of the atoms. However, few products formed on the cobalt surface will generate cobalt ions. It can also accelerate the oxidation reaction rate, which showed a current increasing in the dynamic potential polarization test.

18.2.3 Other effective inhibitors

Similar to copper, 1,2,4-triazole (TAZ) can also avoid chemical dissolution and inhibit galvanic corrosion for cobalt. Under an alkaline environment, TAZ exists in two forms: 1,2, 4-TAH and 1,2,4-TA⁻, thus TAZ can adsorb on the cobalt surface with both physical and chemical adsorption. The physical adsorption is caused by the interaction of van der Waals forces between the neutral molecule and Co, which can be quickly adsorbed and desorbed. While chemical adsorption is caused by two aspects: 1,2,4-TAH and 1,2,4-TA⁻, there are polar groups with N as the central atom in the neutral molecules of 1,2,4-TAH. Moreover, 1,2,4-TA⁻ can chelate Co ions in solution and form $Co(1,2,4-TA)_2$ that adsorbs on the Co surface [32].

2-Mercaptothiazoline (2-MT) has been widely used as a steel corrosion inhibitor. Since Co is a ferromagnetic metal next to Fe in the periodic table, iron corrosion inhibitors are generally effective for cobalt. Lu et al. [33] reported that 2-MT can inhibit the oxidation of cobalt into Co^{2+}, thus achieving the purpose of inhibiting the corrosion on cobalt. Since the S atom and the N atom in 2-MT contain lone pairs of electrons and unpaired electrons, while Co contains the outermost electron orbital $3d^74s^2$, they can connect by gaining and losing electrons or share a lone pair of electrons. The experiment indicates that the corrosion inhibition efficiency of cobalt can reach 90% when the 2-MT concentration is 0.2 mM, the dosage is smaller than BTA, but the inhibition performance is better than BTA.

According to Popuri [34], potassium oleate (PO) was used as a passivating additive to H_2O_2 and citric acid-based silica slurry. The cobalt corrosion currents are approximately one or two orders of magnitude lower at pH 7 with PO, and the postpolish surface quality was excellent. Moreover, it can be concluded from the Langmuir adsorption isotherm model that the standard free energy of adsorption value of PO on Co is $-40\ kJ\ mol^{-1}$, indicating a chemical adsorption occurs. It can be seen from FTIR spectroscopy measurements that PO is bound to Co through bridge coordination between the negatively charged oxygen of the carboxylic acid.

Sagi et al. [35] investigated the use of a mixed aqueous solution containing H_2O_2, oxalic acid (OA), and nicotinic acid (NA) to determine the dissolution and corrosion of chemical vapor deposition Co films. The results show that under the condition of pH = 10, the solution consisting of 1 wt.% H_2O_2, 40 mM OA, and 80 mM NA can effectively reduce the dissolution and corrosion current of the Co film. When pH = 10, positively charged $Co(OH)_2$ species are formed by Co. The isoelectric point is 11.4. As a result, the negatively charged Nic^{2-} and Ox^{2-} species will be attracted. The adsorbed surface complexes are formed, thereby inhibiting the corrosion of the Co film, as shown in following reactions:

$$Co(OH)_2 + Nic^{2-}_{(aq)} \rightarrow Co(OH)_2(Nic^{2-})_{ads} \qquad (18.10)$$

$$Co(OH)_2 + Ox^{2-}_{(aq)} \rightarrow Co(OH)_2(Ox^{2-})_{ads} \qquad (18.11)$$

As illustrated in Fig. 18−6, the surface roughness (Sq) is 0.6 nm, indicating that the excellent inhibition performance requires a synergy between the components.

18.2.4 Cu/Co galvanic corrosion inhibition

During the Co chemical mechanical polishing process, there exists a large electrochemical corrosion potential difference at the Co/Cu interface since the chemical reaction activity of Co is stronger than Cu [36]. Hence it is necessary to reduce the galvanic corrosion between copper and cobalt by adding inhibitors. According to Zhang's report [20], galvanic corrosion between Cu and Co can be inhibited by TAZ, as shown in Fig. 18−7. The corrosion potential of Cu and Co are −172 mV and −253 mV with the TAZ concentration of 500 ppm. The corresponding corrosion potential difference ΔE is 81 mV.

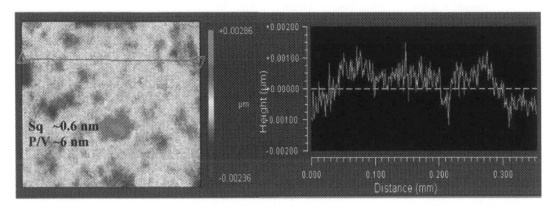

FIGURE 18–6 Surface image of Co films postpolish with (3 wt.% silica +) 1 wt.% H$_2$O$_2$ + 40 mM OA + 80 mM NA at pH = 10 [35].

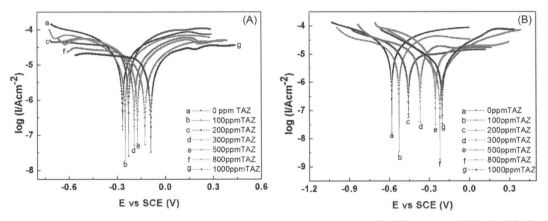

FIGURE 18–7 Potentiodynamic polarization curves as a function of 1,2,4-triazole for (A) copper and (B) cobalt [20]. *SCE*, saturated calomel electrode; *TAZ*, 1,2,4-triazole.

The corrosion inhibition of TAZ on Cu/Co surface mainly comes from the mixed adsorption film formed by TAZ on the surface of the two metals. On the one hand, the TAZ neutral molecule forms a physical adsorption film with the metal surface through van der Waals force; on the other hand, the TAZ molecule and the active atoms on the two metal surfaces share an electron pair with the N4 position on the azole ring, and the deprotonated TAZ forms a chemical adsorption film with copper and cobalt ions by chemical bonding, which hinders the reaction of the two metals and forms passivation protection.

Hu et al. [37] investigated the galvanic corrosion between Cu/Co by using TT-LYK containing slurry. As shown in Fig. 18−8, the results show that the addition of TT-LYK is powerful for inhibiting the corrosion on Cu and Co. The corrosion potential difference between Cu and Co is reduced by the formation of the Cu$_2$O and Cu(I)-TT-LYK bilayer structure on Cu surface.

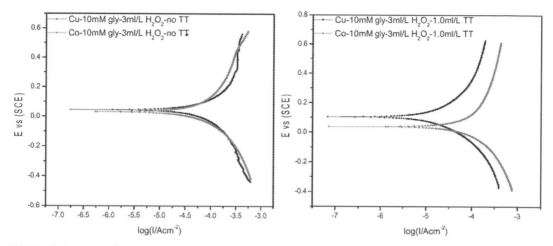

FIGURE 18–8 Potentiodynamic polarization curves for Cu and Co (left) without and (right) with 1.0 mL L^{-1} TT-LYK under static conditions [37].

18.3 Corrosion inhibitors used for tantalum/tantalum nitride barrier layer chemical mechanical planarization

In the IC manufacturing process beyond 10 nm, copper is the predominant metal used as a multilevel interconnection layer material. Meanwhile tantalum, tantalum nitride, ruthenium, or cobalt are used as barrier metal, which is deposited between the wiring metal and the dielectric to improve the adhesion of copper to the dielectric and prevent copper from diffusing into surrounding materials. The differences in the properties of the wiring metal and barrier metal lead to variations in RRs which result in the selectivity problems during CMP as well as galvanic corrosion occurring at the interface between the barrier metal and copper [38]. With the premise of reducing galvanic corrosion and improving selectivity of different materials, various inhibitors are introduced in the barrier CMP slurry to achieve an acceptable level of dishing or erosion toward a single-step CMP.

18.3.1 Corrosion inhibitors for acid slurry

Tantalum and its nitride have been identified as diffusion barrier layers for copper metallization because of their excellent chemical and thermal stability. However, the wide differences in properties between copper and tantalum layers result in selectivity problems during the CMP process. Consequently, defects such as metal dishing occur during the CMP process. So it is essential during Ta/TaN barrier CMP that the selectivity between Cu and Ta/TaN is ideally at a 1:1 ratio.

Due to these different surface chemistries of the two metals, it is necessary in single-slurry CMP of Cu and Ta, that the removal of Ta not be affected by the surface passivating action of inhibitor utilized in the Cu removal step. Sulyma and Roy [39] used electrochemical characterization to study the inhibition effect of ammonium dodecyl sulfate (ADS) during Cu/Ta CMP. It is found that ADS as an anodic inhibitor can suppress Cu dissolution without significantly affecting the surface chemistry of Ta using slurry in pH $2-5$, containing succinic acid as well as hydrogen peroxide, and hence can improve planarization efficiencies by adequately tuning the polishing rate. This selective behavior of ADS toward Cu is essential for single-slurry CMP of Cu and Ta. Surisetty et al. [40] utilized several simple oxyanions in slurry chemistries to meet the requirement for low pressure CMP of Ta barrier and Cu lines in nonalkaline dispersions. It was found that measured at a downpressure of 2 psi, slurries containing 5 wt.% SiO_2, 5 wt.% H_2O_2, and 0.13 M potassium salts of chlorate or nitrate with a pH of 4 can provide a ratio at 1:1 of the removal selectivity between Ta and Cu. Rock et al. [41] examined the selective CMP mechanisms of Ta and Cu using guanidine carbonate (GC)-based slurry consisting of H_2O_2 as an oxidizer and dodecyl benzene sulfonic acid (DBSA) as an inhibitor. Electrochemical investigations suggest that DBSA strongly affects the CMP chemistry of Cu, but exhibits relatively weaker effects on the surface activity of Ta, and thus plays a vital role in dictating the selectivity of Ta:Cu polish rates. Vijayakumar et al. [42] reported that in silica slurry at $pH = 4$, with 5% H_2O_2, Cu RR is almost three times higher than that of Ta, however, with the addition of 0.1 wt.% BTA inhibitor, the best Cu/Ta selectivity of 1 was obtained.

18.3.2 Corrosion inhibitors for basic slurry

Hydrogen peroxide (H_2O_2) is one of the materials that are commonly used as an oxidizing agent in copper CMP. However, it usually needs some stabilizers to prevent decomposition. Kim and coworkers [43] evaluated phosphoric acid (H_3PO_4) as a stabilizer of hydrogen peroxide for TaN barrier slurry. The stability of the hydrogen peroxide in slurry including the phosphoric acid increases, indicating that the decomposition rate of the active peroxide is slowed down by the inhibition effect of H_3PO_4. In the absence of stabilizer, the decomposition rate of hydrogen peroxide is -0.040 wt.% day^{-1}, while it is only -0.002 wt.% day^{-1} for slurry containing 0.5 wt.% H_3PO_4. However, taking zeta potential into consideration, the addition of H_3PO_4 with a proper dosage is preferred, because the dispersion stability of alumina and colloidal silica present negatively with the increase of H_3PO_4 (shown in Fig. 18−9), which should be avoided.

Xiao et al. [44] studied the function of FA/O II chelating agent (based on a derivative of ethylenediamine tetraacetic acid and containing multiple functional groups of tertiary amine, hydroxyl and ethanol) and BRIJ30 nonionic surfactant on Cu/TaN CMP in weakly alkaline slurry ($pH = 10$). It turns out that the RR of TaN exhibits a positive variation with the increase of the concentration of FA/O II, despite there also existing severe corrosion on the wafer surface due to the formation of a Ta-FA/O II complex (see reactions 18.12−18.15). On the other hand, the addition of BRIJ30 slightly reduces the RR of TaN, but the corrosion potential as well as corrosion current density drastically decreases by half compared to the value without BRIJ30, indicating that the reactions occurring on TaN surface are restrained and the corrosion on TaN surface is being well controlled. The investigators proposed that the nonionic surfactant molecular preferentially adsorbed on the TaN surface formed

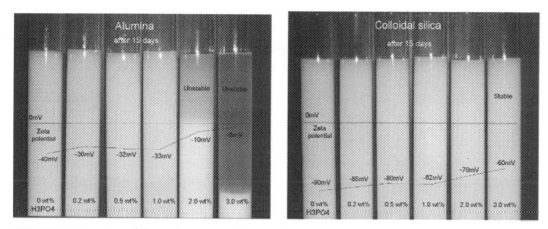

FIGURE 18–9 Dispersion stability and zeta potential test with the contents of phosphoric acid (slurry pH = 8) [43].

a protective layer which hindered the reactants from passing through it, accounting for the inhibition effect of corrosion.

$$4TaN_{x \leq 0.5} + 2.5O_2 = 2TaN_{2x} + Ta_2O_5 \tag{18.12}$$

$$O_2 + 2e^- \rightarrow O_2^{2-} \tag{18.13}$$

$$Ta_2O_5 + 4O_2^{2-} + H_2O \rightarrow 2\left[Ta(O_2)_4\right]^{3-} + 2OH \tag{18.14}$$

$$3\left[R(NH_2)_2\right]^{2+} + 2\left[Ta(O_2)_4\right]^{3-} \rightarrow \left[R(NH_2)_2\right]_3\left[Ta(O_2)_4\right]_2 \tag{18.15}$$

BIT (1,2-benzisothiazolin-3-one) is a stable and highly soluble inhibitor. In addition, BIT can effectively inhibit fungi, syphilis, bacteria, and algae in organic media. Thus it is generally used for inhibiting bacteria in colloidal silica. In our previous work [12], BIT was used for inhibiting corrosion in the polishing slurry (pH = 10). Compared with BTA, BIT can inhibit contamination on the pattern wafer surface better than that of BTA in the process of barrier layer CMP. In addition, the ligand benzothiazole of BIT can form a two-dimensional layered coordination polymer with N and S elements and copper salt, which can effectively protect the copper wiring from corrosion.

18.4 Corrosion inhibitors used for ruthenium barrier layer chemical mechanical planarization

Compared to Ta and TaN, the resistivity of Ru is lower. It can electroplate Cu directly without using seed layers. Thus Ru is used for diffusion barriers in Cu-interconnects [45]. However, galvanic corrosion often occurs during the process of multilayer CMP owing to the direct contact between Ru and Cu [46]. In addition, the RR of Ru will be improved by using strong oxidants. It would also cause spot corrosion of Ru, making the RR of Cu higher than that of Ru. Researchers are engaged in searching for competent inhibitors providing anodic or cathodic inhibition, or both, to mitigate the galvanic corrosion and to achieve a respectable flatness. Whereas the dominated slurry for Ru CMP is neutral or basic, the introduction of inhibitors into the slurry and the relevant effect will be briefly summarized in the subsequent part.

18.4.1 Corrosion inhibitors for neutral slurry

The chemical backbone structures and attached functional groups affect the inhibitor efficiency. Under near-neutral conditions (pH = 6), Cui et al. [47] investigated several effects of corrosion inhibition behavior, surface roughness, and CMP performance of Ru film of polymer and monomer organic inhibitors with amine or amide functional groups. It is found that the order of corrosion inhibition efficiency is DETA (diethylene triamine) > BTA > EDTA (ethylenediamine tetraacetic acid) > PAM 10 k (polyacrylamide, Mw 10k).

However, as shown in Fig. 18−10, the surface roughness is not directly correlated with corrosion behaviors. The roughness of wafer treated by DETA decreases to 13.1 Å, while the roughness of wafer treated by EDTA decreases to 2.2 Å. The reason is that both chemical corrosion and mechanical abrasion affect the surface roughness after CMP. As the effect of corrosion inhibition is low, the surface content of RuO$_3$ is not enough. And EDTA can effectively chelate with Ru film. The surface roughness can be reduced significantly by EDTA. Furthermore, the X-ray photoelectron spectroscopy verified the surface chemical composition difference with and without inhibitors.

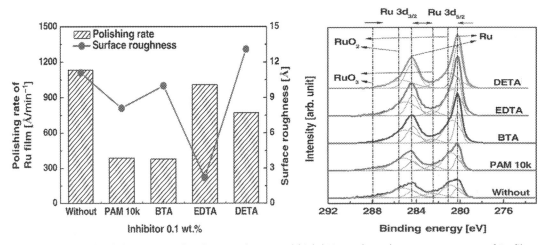

FIGURE 18–10 (left) Polishing rate and surface roughness, and (right) X-ray photoelectron spectroscopy of Ru films treated with different inhibitors [47]. *BTA*, benzotriazole; *DETA*, diethylene triamine; *EDTA*, ethylenediamine tetraacetic acid; *PAM*, polyacrylamide.

18.4.2 Corrosion inhibitors for basic slurry

BTA is commonly used as a corrosion inhibitor as it forms a protective film by adsorption or by complex formation with the metal, which can provide a RR selectivity for Cu/Ru CMP of ~1 as well as reduce the possibility of corrosion. Lots of experiments were made to study the effect of BTA for galvanic corrosion inhibition resulting from Cu/Ru coupled direct contact [48]. A comparison was conducted between the NaOCl-based slurry with and without BTA, utilizing potentiodynamic polarization measurements and electrochemical impedance spectroscopy. Yadav et al. [49] obtained a desirable selectivity of 1:1 for Ru and Cu RR at pH 9 when the slurry formulation was 1 wt.% titania, 0.5 wt.% NaOCl, and 5 mM BTA. The values of cathodic and anodic Tafel slopes for both Cu and Ru systems indicated that addition of BTA affects both the cathodic hydrogen evolution as well as anodic dissolution processes, and a tendency is obviously that the more BTA is added, the higher the inhibition efficiency obtained. However, with the presence of BTA, the adsorption of BTA molecules makes the surface relatively resistant to dissolution, resulting in a decreased RR, hence a compromise should be reached between the inhibition effect and the RR about the dosage of BTA introduced in the slurry.

Nontoxic molybdate salts, which are stable in both strongly acidic and alkaline solutions, have been extensively used in combination with BTA to prevent Cu corrosion. In the research of Cheng et al. [50], the synergetic effect on the passivation of BTA and K_2MoO_4 in the KIO_4-based slurry with a pH of 9 was investigated via electrochemical methods. The results confirm that the galvanic corrosion of Cu is mitigated by the addition of 5 mM BTA and 20 mM K_2MoO_4 from 250.9 $\mu A\ cm^{-2}$ (without inhibitor) to 7.2 $\mu A\ cm^{-2}$. MoO_4^{2-} is an active oxygen acid ion, and it may easily be adsorbed on the surface of metals causing an increase of the activation energy of the electrode reaction. With the absorption of MoO_4^{2-}, the ion − dipole would support the adsorption stability of BTAH, thus forming a three-dimensional network complex film on the protected metal surface, while K_2MoO_4 does not directly participate in passivation of the metal.

Peethala et al. [51] proposed that in the process of Ru CMP, 5 mM BTA and 7 mM L-ascorbic acid (AA) could effectively minimize the galvanic corrosion of Cu and Ru in alkaline (pH = 9) KIO_4 slurries. The difference of E_{oc} (open circuit potential) (ΔE_{oc}) between Ru and Cu is $\Delta E_{oc} = E_{oc}(Ru) - E_{oc}(Cu) \approx 0.6$ V without the inhibitors. which causes galvanic corrosion easily between Cu and Ru in CMP slurry. But with the addition of inhibitor, the ΔE_{oc} diminishes from 540 mV to the minimum value of 20 mV. Furthermore, it can be seen from the potentiodynamic polarization curves that i_{corr} of BTA and AA is lower and E_{corr} is higher compared to without these inhibitors, as shown in Fig. 18−11.

Besides, it should be considered that the other compositions of slurry, such as oxidizer and chelate agent, can act as influencing factors on inhibition efficiency. Turk et al. [52] reported that the specific slurry (pH = 10) containing sodium percarbonate, AA and BTA could effectively inhibit galvanic corrosion between Cu and Ru but has almost no effects on the polishing rate of Ru and Cu. Since the reagents used in slurry differ from Peethala's, the corresponding optimal concentration of the same inhibitors—AA and BTA—for the best inhibition performance can be disparate. AA is confirmed to effectively alleviate galvanic corrosion during the

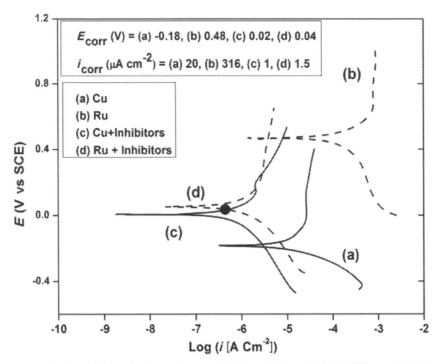

FIGURE 18–11 Potentiodynamic plots for Cu and Ru recorded in a solution of 0.015 M KIO$_4$ at pH 9 with and without mixed corrosion inhibitors of 5 mM BTA + 7 mM AA [51].

Ru CMP mixed with BTA. It can be used as a cathodic inhibitor, and AA may also act as an anodic inhibition, which is consistent with other reported research [53]. As a result, effective inhibition will not be obtained using AA alone but it can be obtained by combining AA with other inhibitors such as BTA. Considering that BTA is toxic and hydrophobic which can cause severe problems for post-CMP cleaning, there is still a lot to do with regard to the synergistic inhibition effect of AA as a green and eco-friendly corrosion inhibitor.

Ruthenium indeed acts as a good barrier material for Cu interconnection. However, the columnar structure of atomic layer deposited (ALD) Ru can provide a diffusion path for copper and has poor adhesion to underlying dielectric. Thus Ru is not a good diffusion barrier [54]. Hence, a thin Ru liner ($\sim$2 nm) over a thin TiN barrier layer ($\sim$2 nm), both deposited by ALD, has been proposed as a promising barrier liner for future Cu interconnects [55]. As is shown in Fig. 18−12A, Ru contacts with TiN and Cu. And the differences in the standard reduction potentials of these three materials are different, galvanic corrosion may occur. To reduce the galvanic corrosion of the Cu/Ru/TiN films during CMP in the BEOL, Sagi et al. [56] proposed an alkaline slurry (pH = 10) with 5 wt.% silica abrasive, 10 mM KMnO$_4$, 1 wt.% GC and 1 mM BTA to mitigate the galvanic corrosion of Cu/Ru. It turns out that the potential gap between Cu and Ru is 10 mV and that for Ru and TiN <20 mV. In addition, the RR ratio of Ru to Cu has increased to $\sim$0.8 minimizing the possibility of dishing. Moreover, this candidate slurry was

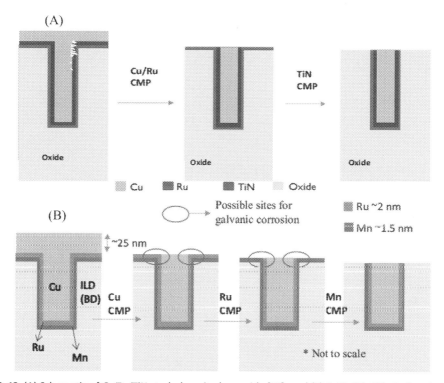

FIGURE 18–12 (A) Schematic of Cu/Ru/TiN stack deposited on oxide [56], and (B) Cu/Ru/Mn/BD stack to be polished [57]. *CMP*, chemical mechanical planarization; *ILD*, interlayer dielectric.

also applied to polish Cu/Ru/Mn/SiCOH patterned (showed in Fig. 18–12B) wafers. There was no corrosion and no post-CMP loss, where the Cu line dishing is limited to ~3 nm [57].

Generally, a low RR of Cu by barrier layer (and dielectrics) leads to dishing, while under different pattern densities, the nonuniform mechanical abrasion will lead to corrosion. Wang et al. [58] used GC and 1,2,4-triazole (TAZ) to research the RR selectivity of Ru to Cu in hydrogen peroxide based slurry. The RR selectivity rises from 0.3 to 1.6 through adjusting the GC and TAZ contents. The dishing and erosion of Cu patterned wafer can also be corrected. However, when the concentration of BTA is low, the copper RR of Cu is higher than that without inhibitor. It is due to this that BTA is used as an anodic inhibitor. Pitting corrosion may appear in this area [59]. According to the research above, TAZ can function as an excellent inhibitor for preventing Ru/Cu galvanic corrosion, as well as Cu pitting, if the dosage of is appropriate. In other words, insufficient or excessive TAZ could not provide an efficient passivation.

Different oxidizers such as pypocholoride, periodate, percarbonate, ammonium sulfate, potassium bromate, oxone, and potassium molybdate were used to achieve the planarized surface for Ru along with the desired RR [60,61], but employing hydrogen peroxide as an oxidizer for Ru CMP has not been widely reported. Unlike the strong oxidizers mentioned above, hydrogen peroxide can provide oxidation without introducing metal ions into slurry, which is considered as a potential green oxidizer. Nonetheless, the inertness of Ru is extremely high and using hydrogen peroxide alone is not strong enough to convert Ru into the corresponding oxidation state. As a result, auxiliary reagents such as complexing agents would be introduced in Ru barrier slurry [62]. As a result, the RR of Cu could be much higher. Thus the application of inhibitors in hydrogen peroxide-based slurry has become the study focus of research.

Recently, removal selectivity of Ru to Cu was researched when complexing agent ammonium sulfate and inhibitor TT-LYK were introduced into the H_2O_2-based slurry [63]. As demonstrated in Fig. 18−13, the RR of Cu and Ru are almost positively correlated with $(NH_4)_2SO_4$ concentration, which can significantly improve the RR of Ru and Cu. While the addition of TT-LYK dramatically reduces the RR of Cu, on the contrary, it rarely introduces an impact on that of Ru and the effect of TT-LYK concentration on the RR selectivity of Ru and Cu. This is because TT-LYK can inhibit corrosion by forming a passivation film on the the copper surface while it does not occur on ruthenium surface. Additionally, when the concentrations of $(NH_4)_2SO_4$ and TT-LYK are 40 mM and 2000 ppm, respectively (pH = 8.0), the optimal removal rate selectivity of Ru to Cu will be 2.7.

Though azole compounds showed great inhibition effect on account of consisting of an aromatic heterocyclic structure that is similar to BTA, the cost of complicated synthesis processes as well as to some extent toxic properties restrict their application to CMP slurry. To avoid environmental pollution as much as possible, green additives are widely used to develop novel CMP slurries used in semiconductor and microelectronics industries [64]. Compared to the traditional toxic and corrosive ones employed in machining and manufacturing, the environment-friendly CMP slurries tremendously reduce the pollution and contamination to the environment, which is a breakthrough and milestone contribution. Among them, chitosan (CTS) has active hydroxyl groups and amino groups, which have strong chemical reaction ability. It can not only complex with metal ions, but also form adsorption on the metal surface, while whether it inhibits or promotes the removal of metal materials depends on the amount of CTS added in slurry and the material itself. Zhou et al. [65] investigated the adsorption mechanism of CTS on Ru employing Freundlich adsorption isotherm and EIS test. The results show that physical and chemical adsorption coexist on the adsorption behavior of CTS on Ru surface. The dense and continuous double passivation

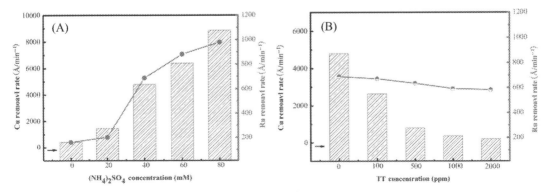

FIGURE 18–13 The effect of (A) $(NH_4)_2SO_4$ and (B) TT-LYK concentration on the removal rates of Ru and Cu [63].

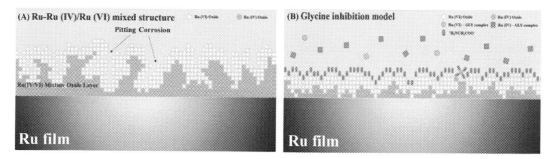

FIGURE 18–14 The schematic illustrations of (A) the formed Ru-Ru(IV)/Ru(VI) mixed passivation layer structure and (B) glycine inhibition and complexation model [66].

structure is owing to the cover of complex products comprising CTS and Ru, which can effectively avoid corrosion.

Glycine can accelerate the RR of Cu, which is achieved by the formation of Cu-glycine complex during Cu CMP, while it can also prevent oxidant reacting with Ru by absorbing on Ru surface. Thus the formation rate of oxides is reduced. Shao et al. [66] investigated the inhibition effect of glycine on Ru in NaClO-based slurry. Under the condition of alkaline NaClO solution (pH = 10), glycine can inhibit corrosion well for Ru. The adsorption isotherm of glycine on Ru surface follows TEMKIN's model. The passivation/corrosion model is shown in Fig. 18–14. In conclusion, there are two ways for glycine to inhibit corrosion: (1) glycine can prevent oxidant from reacting with Ru, thus reducing the formation rate of oxides; (2) glycine will accelerate the oxidation layer dissolution by forming the soluble complex (Ru(IV) and Ru (VI) oxide). Finally, in order to study Ru and Cu properties in NaClO slurries, polishing experiments were conducted. It can be concluded that when the concentration of NaClO is 0.5 wt.%, the concentration of glycine is 1 mM, the selectivity between Cu and Ru RR is 1.09.

18.5 Conclusions

Metal interconnection is the key material in the process of integrated circuit. Chemical mechanical polishing is the key step to ensure the reliability of the integrated circuit. In order to obtain high flatness, the corrosion inhibition of wiring metal and barrier metal is the first problem to be solved. The selection of the corrosion inhibitor will be based on the premise of green environmental protection. It is still the focus of research to find new inhibitors with high inhibition efficiency. The compatibility of inhibitors, pH regulators, complexing agents, and oxidants should be considered comprehensively. Although nontoxic and environment-friendly corrosion inhibitors have made great progress in recent decades, there are some problems that need to be resolved: (1) the amount is large and the use-cost is high; (2) the evaluation standards for corrosion inhibitors are not uniform; and (3) more further theoretical studies are required on the inhibition mechanism.

Useful links

https://www.sciencedirect.com/search?qs = Copper%20corrosion%20inhibitor.
https://link.springer.com/search?query = BTA.
https://link.springer.com/search?query = azole + inhibitor.
https://www.sciencedirect.com/search?qs = adsorption%20mechanism.

References

[1] M. Krishnan, J.W. Nalaskowski, L.M. Cook, Chemical mechanical planarization: slurry chemistry, materials, and mechanisms, Chem. Rev. 110 (2010) 178−204.

[2] C. Yao, X. Niu, C. Wang, et al., Study on the weakly alkaline slurry of Copper chemical mechanical planarization for GLSI, ECS J. Solid State Sci. Technol. 6 (2017) 499−506.

[3] L.T. Popoola, Organic green corrosion inhibitors (OGCIs): a critical review, Corros. Rev. 37 (2019) 71−102.

[4] K. Zhang, X. Niu, C. Wang, et al., Effect of chelating agent and ammonium dodecyl sulfate on the interfacial behavior of copper CMP for GLSI, ECS J. Solid State Sci. Technol. 7 (2018) 509−517.

[5] D. Tromans, J.C. Silva, Anodic behavior of copper in iodide solutions: comparison with chloride and effect of benzotriazole-type inhibitors, J. Electrochem. Soc. 143 (1996) 458−465.

[6] L. Jiang, Y. Lan, Y. He, et al., 1,2,4-Triazole as a corrosion inhibitor in copper chemical mechanical polishing, Thin Solid Films 556 (2014) 395−404.

[7] J. Zhou, J. Wang, X. Niu, et al., Chemical interactions and mechanisms of different pH regulators on copper and cobalt removal rate of copper film CMP for GLSI, ECS J. Solid State Sci. Technol. 8 (2019) 99−105.

[8] D. Tromans, Aqueous potential-pH equilibria in copper-benzotriazole systems, J. Electrochem. Soc. 145 (1998) 42−45.

[9] M. Finšgar, I. Milošev, Corrosion study of copper in the presence of benzotriazole and its hydroxy derivative, Mater. Corros. 62 (2011) 956−966.

[10] B.J. Cho, S. Shima, S. Hamada, et al., Investigation of Cu-BTA complex formation during Cu chemical mechanical planarization process, Appl. Surf. Sci. 384 (2016) 505−510.

[11] L. Yang, B. Tan, Y. Liu, et al., Effect of organic amine alkali and inorganic alkali on benzotriazole removal during post Cu-CMP cleaning, J. Semicond. 39 (2018) 126003.

[12] T. Ma, B. Tan, Y. Liu, et al., Role of 1,2-benzisothiazolin-3-one (BIT) in the Improvement of barrier CMP performance with alkaline slurry, ECS J. Solid State Sci. Technol. 8 (2019) 449−456.

[13] F.M. Al Kharafi, A.M. Abdullah, I.M. Ghayad, et al., Effect of sulfide pollution on the stability of the protective film of benzotriazole on copper, Appl. Surf. Sci. 253 (2007) 8986−8991.

[14] A. Kokalj, S. Peljhan, Density functional theory study of adsorption of benzotriazole on Cu_2O surfaces, J. Phys. Chem. C 119 (2015) 11625−11635.

[15] N.K. Allam, A.A. Nazeer, E.A. Ashour, A review of the effects of benzotriazole on the corrosion of copper and copper alloys in clean and polluted environments, J. Appl. Electrochem. 39 (2009) 961−969.

[16] A.D. Modestov, G.D. Zhou, Y.P. Wu, et al., A study of the electrochemical formation of Cu(I)-BTA films on copper electrodes and the mechanism of copper corrosion inhibition in aqueous chloride/benzotriazole solutions, Corros. Sci. 36 (1994) 1931−1946.

[17] S.M. Lai, Y.Y. Chen, C.P. Liu, et al., Degradation of inhibitor in alkaline cleaning solution for post-Cu CMP cleaning, Surf. Coat. Technol. 350 (2018) 1080−1084.

[18] L. Jiang, Y. He, J. Li, et al., Passivation kinetics of 1,2,4-triazole in Copper chemical mechanical polishing, ECS J. Solid State Sci. Technol. 5 (2016) 272−279.

[19] X. Yang, V.B. Birman, Acyl transfer catalysis with 1,2,4-triazole anion, Org. Lett. 11 (2009) 1499.

[20] W. Zhang, Y. Liu, C. Wang, et al., Role of 1,2,4-triazole in Co/Cu removal rate selectivity and galvanic corrosion during barrier CMP, ECS J. Solid State Sci. Technol. 6 (2017) 786−793.

[21] M. Muniz-Miranda, F. Muniz-Miranda, S. Caporali, SERS and DFT study of copper surfaces coated with corrosion inhibitor, Beilstein J. Nanotechnol. 5 (2014) 2489−2497.

[22] K.L. Stewart, J.J. Keleher, A.A. Gewirth, Relationship between molecular structure and removal rates during chemical mechanical planarization: comparison of benzotriazole and 1,2,4-triazole, J. Electrochem. Soc. 155 (2008) 625−631.

[23] J. Zhou, X. Niu, Y. Cui, et al., Study on the film forming mechanism, corrosion inhibition effect and synergistic action of two different inhibitors on copper surface chemical mechanical polishing for GLSI, Appl. Surf. Sci. 505 (2020) 144507.

[24] L.P. Kazansky, I.A. Selyaninov, Y.I. Kuznetsov, Adsorption of 2-mercaptobenzothiazole on copper surface from phosphate solutions, Appl. Surf. Sci. 258 (2012) 6807−6813.

[25] C.W. Yan, H.C. Lin, C.N. Cao, Investigation of inhibition of 2-mercaptobenzoxazole for copper corrosion, Electrochim. Acta 45 (2000) 2815−2821.

[26] F.J. Presuel-Moreno, M.A. Jakab, J.R. Scully, Inhibition of the oxygen reduction reaction on copper with cobalt, cerium, and molybdate ions, J. Electrochem. Soc. 152 (2005) 376−387.

[27] E. Rocca, G. Bertrand, C. Rapin, et al., Inhibition of copper aqueous corrosion by non-toxic linear sodium heptanoate: mechanism and ECAFM study, J. Electroanal. Chem. 503 (2001) 133−140.

[28] D.Q. Zhang, L.X. Gao, G.D. Zhou, Inhibition of copper corrosion in aerated hydrochloric acid solution by amino-acid compounds, J. Appl. Electrochem. 35 (2005) 1081−1085.

[29] X. Zhang, G. Pan, L. Hu, et al., Effects of nitrilotriacetic acid and corrosion inhibitor on cobalt barrier chemical−mechanical polishing: experimental and density functional theory analysis, Colloids Surf. A Physicochem. Eng. Asp. 605 (2020).

[30] H.S. Lu, X. Zeng, J.X. Wang, et al., The effect of glycine and benzotriazole on corrosion and polishing properties of cobalt in acid slurry, J. Electrochem. Soc. 159 (2012) 383−387.

[31] D. Gallant, M. Pézolet, S. Simard, Inhibition of cobalt active dissolution by benzotriazole in slightly alkaline bicarbonate aqueous media, Electrochim. Acta 52 (2007) 4927−4941.

[32] M. Zhong, S.S. Venkataraman, Y. Lan, et al., Role of 1,2,4-triazole as a passivating agent for cobalt during post-chemical mechanical planarization cleaning, J. Electrochem. Soc. 161 (2014) 138−144.

[33] H.S. Lu, J.X. Wang, X. Zeng, et al., The effect of H_2O_2 and 2-MT on the chemical mechanical polishing of cobalt adhesion layer in acid slurry, Electrochem. Solid State Lett. 15 (2012) 97−100.

[34] R. Popuri, H. Amanapu, C.K. Ranaweera, et al., Potassium oleate as a dissolution and corrosion Inhibitor during chemical mechanical planarization of chemical vapor deposited Co films for interconnect applications, ECS J. Solid State Sci. Technol. 6 (2017) 845−852.

[35] K.V. Sagi, L.G. Teugels, M.H. van der Veen, et al., Chemical mechanical polishing of chemical vapor deposited Co films with minimal corrosion in the Cu/Co/Mn/SiCOH patterned structures, ECS J. Solid State Sci. Technol. 6 (2017) 276−283.

[36] S. Shima, A. Fukunaga, M. Tsujimura, Effects of liner metal and CMP slurry oxidizer on copper galvanic corrosion, ECS Trans. 11 (2007) 285−295.

[37] L. Hu, G. Pan, X. Zhang, et al., Inhibition effect of TT-LYK on Cu corrosion and galvanic corrosion between Cu and Co during CMP in alkaline slurry, ECS J. Solid State Sci. Technol. 8 (2019) 437−447.

[38] C.C. Hung, Y.S. Wang, W.H. Lee, et al., Galvanic corrosion between TaNx barriers and copper seed, Electrochem. Solid State Lett. 10 (2007) 127−130.

[39] C.M. Sulyma, D. Roy, Electrochemical characterization of surface complexes formed on Cu and Ta in succinic acid based solutions used for chemical mechanical planarization, Appl. Surf. Sci. 256 (2010) 2583–2595.

[40] C.V.V.S. Surisetty, B.C. Peethala, D. Roy, et al., Utility of oxy-anions for selective low pressure polishing of Cu and Ta in chemical mechanical planarization, Electrochem. Solid State Lett. 13 (2010) 244–247.

[41] S.E. Rock, D.J. Crain, J.P. Zheng, et al., Electrochemical investigation of the surface-modifying roles of guanidine carbonate in chemical mechanical planarization of tantalum, Mater. Chem. Phys. 129 (2011) 1159–1170.

[42] A. Vijayakumar, T. Du, K.B. Sundaram, et al., Selectivity studies on tantalum barrier layer In copper CMP, MRS Proc. 767 (2011) 631–636.

[43] N.H. Kim, J.H. Lim, S.Y. Kim, et al., Effects of phosphoric acid stabilizer on copper and tantalum nitride CMP, Mater. Lett. 57 (2003) 4601–4604.

[44] Y. Xiao, G. Pan, Q. Tian, et al., Effect of chelating agent and surfactant on TaN CMP in weakly alkaline slurry, ECS J. Solid State Sci. Technol. 7 (2018) 608–614.

[45] S.C. Seo, C.C. Yang, C.K. Hu, et al., Thermal stability of copper contact metallization using Ru-containing liner, Electrochem. Solid State Lett. 14 (2011) 187–190.

[46] D. Tamboli, O. Osso, T. Mcevoy, et al., Investigating the compatibility of ruthenium liners with copper interconnects, ECS Trans. 33 (2010) 181–187.

[47] H. Cui, J.H. Park, J.G. Park, Corrosion inhibitors in sodium periodate slurry for chemical mechanical planarization of ruthenium film, ECS J. Solid State Sci. Technol. 2 (2013) 71–75.

[48] M.C. Turk, M.J. Walters, D. Roy, Tribo-electrochemical investigation of a slurry composition to reduce dissolution and galvanic corrosion during chemical mechanical planarization of Cu-Ru interconnects, Mater. Chem. Phys. 201 (2017) 271–288.

[49] K. Yadav, R. Manivannan, S.N. Victoria, Chemical mechanical planarization of ruthenium using sodium hypochlorite based titania slurry, ECS J. Solid State Sci. Technol. 6 (2017) 879–885.

[50] J. Cheng, T. Wang, H. Mei, et al., Synergetic effect of potassium molybdate and benzotriazole on the CMP of ruthenium and copper in KIO4-based slurry, Appl. Surf. Sci. 320 (2014) 531–537.

[51] B.C. Peethala, D. Roy, S.V. Babu, Controlling the galvanic corrosion of copper during chemical mechanical planarization of ruthenium barrier films, Electrochem. Solid State Lett. 14 (2011) 306–310.

[52] M.C. Turk, S.E. Rock, H.P. Amanapu, et al., Investigation of percarbonate based slurry chemistry for controlling galvanic corrosion during CMP of ruthenium, ECS J. Solid State Sci. Technol. 2 (2013) 205–213.

[53] H. Akrout, S. Maximovitch, L. Bousselmi, et al., Evaluation of corrosion non toxic inhibitor adsorption for steel in near neutral solution: L-ascorbic acid, Mater. Corros. 58 (2007) 202–206.

[54] W. Sari, T.K. Eom, C.W. Jeon, et al., Improvement of the diffusion barrier performance of Ru by incorporating a WNx thin film for direct-plateable Cu interconnects, Electrochem. Solid State Lett. 12 (2009) 248–251.

[55] H.P. Amanapu, K.V. Sagi, L.G. Teugels, et al., Role of guanidine carbonate and crystal orientation on chemical mechanical polishing of ruthenium films, ECS J. Solid State Sci. Technol. 2 (2013) 445–451.

[56] K.V. Sagi, H.P. Amanapu, S.R. Alety, et al., Potassium permanganate-based slurry to reduce the galvanic corrosion of the Cu/Ru/TiN barrier liner stack during CMP in the BEOL interconnects, ECS J. Solid State Sci. Technol. 5 (2016) 256–263.

[57] K.V. Sagi, L.G. Teugels, M.H. van der Veen, et al., Chemical mechanical polishing and planarization of Mn-based barrier/Ru liner films in Cu interconnects for advanced metallization nodes, ECS J. Solid State Sci. Technol. 6 (2017) 259–264.

[58] Q. Wang, J. Zhou, C. Wang, et al., Controlling the removal rate selectivity of ruthenium to copper during CMP by using guanidine carbonate and 1, 2, 4-triazole, ECS J. Solid State Sci. Technol. 7 (2018) 567–574.

[59] S.F.L.A. da Costa, S.M.L. Agostinho, J.C. Rubim, Spectroelectrochemical study of passive films formed on brass electrodes in 0.5 M H_2SO_4 aqueous solutions containing benzotriazole (BTAH), J. Electroanal. Chem. Interfacial Electrochem. 295 (1990) 203–214.

[60] Y.S. Chou, S.C. Yen, K.T. Jeng, Fabrication of ruthenium thin film and characterization of its chemical mechanical polishing process, Mater. Chem. Phys. 162 (2015) 477–486.

[61] S.N. Victoria, J. Jebaraj, I.I. Suni, et al., Chemical mechanical planarization of ruthenium with oxone as oxidizer, Electrochem. Solid State Lett. 15 (2012) 55–58.

[62] Z. Wang, J. Zhou, C. Wang, et al., Role of ammonium ions in colloidal silica slurries for Ru CMP, ECS J. Solid State Sci. Technol. 8 (2019) 285–292.

[63] Z. Wang, J. Zhou, C. Wang, et al., Controlling of Ru/Cu removal rate selectivity during CMP by using ammonium sulfate and inhibitor, ECS J. Solid State Sci. Technol. 8 (2019) 509–515.

[64] Z. Zhang, B. Wang, P. Zhou, et al., A novel approach of chemical mechanical polishing using environment-friendly slurry for mercury cadmium telluride semiconductors, Sci. Rep. 6 (2016) 22466.

[65] J. Zhou, X. Niu, Z. Wang, et al., Roles and mechanism analysis of chitosan as a green additive in low tech node copper film chemical mechanical polishing, Colloids Surf. A Physicochem. Eng. Asp. 586 (2020) 124293.

[66] S. Shao, B. Wu, P. Wang, et al., Investigation on inhibition of ruthenium corrosion by glycine in alkaline sodium hypochlorite based solution, Appl. Surf. Sci. 506 (2020) 144976.

Sustainable corrosion inhibitors for oil and gas industry

19

Corrosion inhibitors for refinery industries

Ruby Aslam[1], Mohammad Mobin[1], Jeenat Aslam[2]

[1]CORROSION RESEARCH LABORATORY, DEPARTMENT OF APPLIED CHEMISTRY, FACULTY OF ENGINEERING AND TECHNOLOGY, ALIGARH MUSLIM UNIVERSITY, ALIGARH, INDIA [2]DEPARTMENT OF CHEMISTRY, COLLEGE OF SCIENCE, YANBU, TAIBAH UNIVERSITY, AL-MADINA, SAUDI ARABIA

Chapter outline

Environmentally Sustainable Corrosion Inhibitors. DOI: https://doi.org/10.1016/B978-0-323-85405-4.00004-5

Abbreviations

AFM	Atomic force microscopy
EIS	Electrochemical impedance spectroscopy
EN	Electrochemical noise
LPR	Linear polarization resistance
PDP	Potentiodynamic polarization
PM-IRRAS	Polarization modulation-infrared reflection-adsorption spectroscopy
QCM	Quartz crystal microbalance
WL	Weight loss

19.1 Introduction

The petroleum industry started with the successful drilling of the first commercial oil well in 1859 and the establishment of the very first refinery two years later to convert crude oil into kerosene. Refinery processes typically consist either processing or treatment of crude oil. Process unit composition is unique for any refinery [1]. While certain process steps such as distillation, cracking and, removal of impurities and by-products are crucial, the technology adapted to the same processes can differ considerably across refineries. Petroleum refining [2] starts with the desalination of feedstock followed by the distillation or fractionation into different hydrocarbon groups of crude oils. The resulting products are specifically correlated to the processed crude oil characteristics. By changing the size and structure of the hydrocarbon molecules by cracking, reforming, and other transformation processes, most distillation products are further converted into more usable products. To eliminate unnecessary constituents and enhance product quality, these converted products are then subjected to various treatment and separation processes, such as extraction, hydrotreatment, and sweetening. Integrated refineries provide fractionation, refining, treatment and, blending operations, and may also include petrochemical processing. The refinery industry relies heavily on certain critical materials, including carbon and alloy steel, copper, aluminum, and nonferrous alloys, for the construction of petroleum refinery static equipment such as pressure vessels, columns, boilers, heat exchangers, condensers, and pipelines. Corrosion of these metallic components is, however, a major cause of inefficiency in the refining process as it leads to failure of equipment. Direct corrosion-related costs in the US oil sector were estimated at US$ 3.7 billion as of 1996 [3].

19.2 Sources of corrosion in refinery

The physicochemical properties of crude oils and their corrosivity [4] need to be clarified to better understand corrosion problems and solutions in the refineries. Crude oil is a complex mixture of many different hydrocarbons and a very small portion of nonhydrocarbons. Besides hydrocarbons, crude oils may include sulfur, nitrogen and, oxygen-containing compounds [5]. All such contaminants can occur as dissolved gases, liquids and, solid phases in crude oils. Microorganisms may also be found in the active or inactive state in the crude, water, and fuels. There are almost the same types of compounds in each crude but different ratios. As a result, crude oils differs by its corrosiveness.

The crude oil assays are graded by two methods [6]. The first one is based on the proportions of organic compounds such as paraffinic, naphthenic, aromatic, or mixed type. Another method of classifying crude oils is dependent on their American Petroleum Institute (API) gravity which is inversely proportional to the density of crude oil. The greater the API gravity, the lighter the crude is. Low carbon, high hydrogen, and, high API gravity crude oils are usually rich in paraffin and tend to produce higher proportions of petrol and light petroleum products; lower carbon, low hydrogen and, low API gravity oils are generally rich in aromatics [3]. It is widely agreed that heavier oils are more corrosion-protective than lighter ones.

Crude oils with large quantities of hydrogen sulfide (H_2S) are called "sour" crudes, and those with fewer H_2S are known as "sweet" crudes [7]. The problem of corrosion due to constituents of crude oil is predominantly associated with hydrogen chloride, organic and inorganic chlorides, hydrogen sulfide, mercaptans and, organic sulfur compounds, carbon dioxide, dissolved oxygen and water, organic acids, and nitrogen compounds.

19.3 Types of corrosion

Corrosion occurs in different ways in the refining process, such as water droplet pitting corrosion, hydrogen embrittlement, and sulfide stress corrosion cracking. This can be subdivided roughly into two categories [8]:

1. Low-temperature corrosion occurs below 260°C. The most common forms of corrosion are stress corrosion cracking and pitting. The main sources of this type of corrosion are contaminants present in crude oil process streams such as air, water, H_2S, CO_2, etc., and chemicals used like catalysts, caustic, neutralizers, and solvents.
2. High-temperature corrosion occurs above 260°C. The most common forms of corrosion are uniform thinning, local attack, and erosion-corrosion. This type of corrosion is primarily caused by sulfur compounds present in crude oil.

Details of the contaminants causing corrosion and the locations where deterioration occurs are discussed below in detail:

19.3.1 Naphthenic acid

Because of the presence of naphthenic acid, corrosion was reported in the refineries vacuum units at 220°C–370°C. Naphthenic acid is the main organic acid present in crude oils [9]. The vulnerability of carbon steel and stainless steel to naphthenic acid is high [10]. Coupled with high temperature and high velocity, even quite low doses of naphthenic acid can trigger extremely increased rates of corrosion. Corrosion is greatly increased by the presence of naphthenic acid and, sulfur compounds in the high-temperature areas of distillation systems. Formic, acetic, and propionic acids include many other organic acids that can also induce corrosion at low temperatures [11].

19.3.2 Salts

Inorganic salts such as sodium chloride, magnesium chloride, and calcium chloride are often found in crude oils in suspension or dissolved in entrained water (brine). To avoid catalyst contamination, equipment corrosion, and fouling, these salts must be removed or neutralized before processing. In refineries, hydrogen chloride is produced in raw distillation units by hydrolysis of chloride salts. HCl can combine to form ammonium chloride with ammonia, causing fouling and corrosion. To avoid this, the concentration of salt is kept low in the oil, and caustic injections are used to neutralize HCl. Often, amine neutralization and film-forming amine-based corrosion inhibitors may be injected to monitor overhead corrosion of crude units. Cleaning out of water may be important to prevent fouling.

19.3.3 Sulfur compounds

For heavier crude oil, the sulfur content of crude oil may be in the range of 2.0%–3.5% [12]. Sulfur compounds contain mercaptans, thiophenes, and elemental sulfur, in addition to H_2S. It is not just sulfur that is responsible for corrosion. It relies on the degree to which the sulfur compounds in the oil are converted into more corrosive compounds, including H_2S and HCl. H_2S can cause hydrogen attack, general corrosion, and pitting corrosion. The latter occurs as blisters or cracks. Hydrogen-induced cracking, sulfide stress cracking, and stress-oriented hydrogen-induced cracking are the most common types of cracking induced by wet H_2S [13]. H_2S can be released into the atmosphere through a crack, causing human death and adverse environmental impacts.

19.3.4 Carbon dioxide (CO_2)

Carbon dioxide can be present in trace quantities in crude oil and also in condensate and produced water. It is released from crudes produced usually in CO_2-flooded fields and crudes that contain a high naphthenic acid content. Dry CO_2 is noncorrosive, but when CO_2

is combined with water, carbonic acid (H_2CO_3) is formed which causes corrosion of carbon steel (sweet corrosion):

$$Fe + CO_2 + H_2O \rightarrow FeCO_3 + H_2 \qquad (19.1)$$

Corrosion by CO_2 is very severe, causing metal dissolution, intergranular effects, and provides the environment for stress corrosion cracks. Noticeable corrosion occurs in carbonic acid solution at a pH of 6.0.

19.3.5 Dissolved oxygen (O_2) and water (H_2O)

Though this volume of water is usually small, its corrosion impact can be high, as it normally contains a high percentage of dissolved corrosive salts, mainly sodium, calcium, and magnesium chloride. Oxygen, on the other hand, also plays a significant role in corrosion [14]. For corrosion to occur when combined with certain other conditions, such as deposits on the metal surfaces and dissolved oxygen, merely small quantities of O_2 are needed. This will trigger a pitting attack. The existence of O_2 enhances the corrosive effects of acid gases (CO_2 and H_2S) on iron and carbon steel.

19.3.6 Nitrogen compounds

Unless converted to ammonia or hydrogen cyanide, nitrogen compounds alone in crude oil do not lead to a corrosion problem. This often happens in operations involving catalytic cracking, hydrotreating, and hydrocracking.

19.3.7 Corrosion due to chemicals used

Sulfuric acid, caustic acid, ammonia, and chlorine are often used in the process of refining crude oil and can cause corrosion. Sulfuric acid used at lower amounts in water treatment plants (below 85%) triggers prevalent carbon steel corrosion in the form of pits or general metal loss with pits. Caustic and ammonia are usually used to neutralize acid components in manufacturing operations but overdosing them causes metal failure or stress corrosion cracking of carbon steel and copper-base alloys.

19.3.8 Impact of flow regime and rate of flow

The flow regime and flow rate have a direct impact on corrosion. By increasing flow rate, the introduction of aggressive substances to the metal surface, the removal of corrosion products, and protective layers (such as corrosion inhibitors and $FeCO_3$) may occur. Consequently, the rate of corrosion increases. Increasing the flow rate from 1 to 10 m s^{-1}, for example, induces a rise in the corrosion rate from 1 to 3 mm y^{-1} in CO_2-containing aqueous solution (1 bar and 20°C) in the absence of a $FeCO_3$ protective layer [15]. The type of wetting of the metal surface is determined by the flow regime, resulting in the occurrence of corrosion at the top of the pipe. Fast flow rates can lead to erosion corrosion. Aggressive materials have enough time to be in contact with the metal surface at low flow rates. Therefore various deposits may accumulate and cause localized corrosion.

19.4 Corrosion control schemes

The following preventive steps are taken to mitigate the impact of corrosion in the process plant for refining crude oil:

1. the use of corrosion-resistant structural materials;
2. desalting of the crude oil; and
3. chemical dosing.

1. The use of corrosion-resistant structural materials

Most building materials for petrochemical refineries fall into the following general categories: carbon steels, carbon—molybdenum and chromium—molybdenum alloy steels, austenitic chromium—nickel stainless steels, ferritic and martensitic chromium stainless steels, copper alloys, i.e., brass, bronze, cupronickel, nickel alloys, i.e., nickel, Monel, Inconel, Incoloy, titanium, Hastelloys, and aluminum alloys. A brief overview of widely used materials [8] and preventive cladding for corrosion mitigation is given below with respect to corrosion causing compounds:

a. Corrosion by HCl, organic, and inorganic chloride—Monel is usually resistant to weak HCl corrosion and is widely used in the upper section of crude towers, condensers, and distillate drum liners, and pipe still overhead systems. Therefore safe Monel cladding is used in the top portion of the column of crude distillation. The typical corrosion rate for Monel is 0.002 inchs y^{-1}.

b. Corrosion by H_2S, organic sulfur compounds—the chromium-steel, 12% Cr and those with 0.5%—1% molybdenum have good resistance to this type of corrosion. Additionally, the austenitic stainless steels such as 18Cr-8 nickel steel have excellent corrosion resistance.

c. Corrosion by CO_2—Monel, aluminum, stainless steel, and cupronickel is good against the CO_2 corrosion. In addition, the attack of wet CO_2 on steel is attenuated by the addition of Cr (12%), although the 304 stainless steel is the most widely used and considered to be most effective.

d. Corrosion by organic acid—Cu alloys can be used to lessen the corrosion due to the attack of organic acid which has more resistance than steel. Cr-Ni SS (type 316) is very resistant to naphthenic acid corrosion.

e. Corrosion by nitrogen compounds—SS or plastic laying may be needed for corrosion protection due to N compounds.

2. Desalting of the crude oil

Crude oil contains many unwanted impurities, such as sand, inorganic salts, drilling mud, rubber, by-products of corrosion, etc. As a consequence, impurities have detrimental effects such as rust, fouling, and upsets of the components.

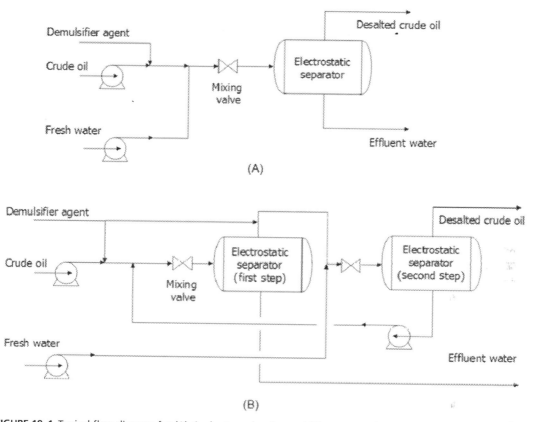

FIGURE 19–1 Typical flow diagram for (A) single-stage desalter and (B) two-stage desalter [16].

The aim of desalination is therefore to reduce the salt content of the treated crude oil, particularly calcium, magnesium, and sodium, by water-washing before distillation to appropriate levels. Depending on the required salt content in the desalted crude oil, a single- or two-step method may be implemented. A maximum salt concentration of 1.5 PTB (a pound of salt measured as NaCl per 1000 barrels) is required for refining purposes. Fig. 19–1 displays a general process flow diagram for desalting processes of one and two phases.

The desalted crude oil is heated to 100°C–150°C temperature and mixed with 4%–10% freshwater that dilutes the salt. The mixture is then injected into a settling tank, separating the salt water and draining the oil. Electrodes in the settling tank apply an electrostatic effect that allows the water droplets suspended in the larger volume of liquid to polarize. This causes the droplets of water to clump together and fall to the bottom of the tank [17].

It is possible to achieve 90% removal of salt in a single-step configuration shown in Fig. 19–2A. For higher percentages of salt removal, a two-step configuration is needed, as

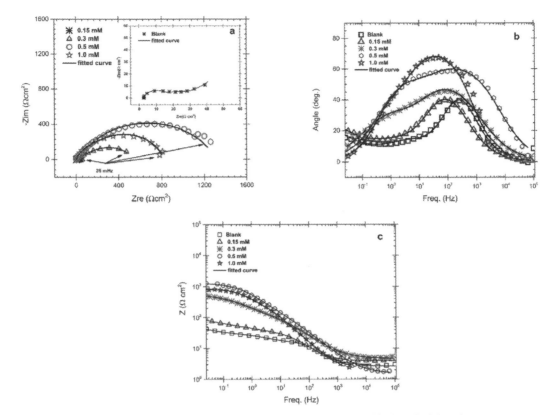

FIGURE 19–2 Electrochemical impedance spectroscopy (EIS) measurements of blank and inhibited system at 298K: (A) Nyquist plots; (B) Bode phase angle; (C) absolute impedance values against the frequency [28].

shown in Fig. 19−2B. Two demulsifier injection points are used with respect to the last process configuration, both in the first and second stages before the mixing valve. Also, wastewater is pumped to the second level and from this level, wastewater is returned to the first stage. This design will reduce salt by 99% [15].

3. Chemical dosing

This mainly depends on the crude type, equipment, and operating parameters. The following methods are reported for chemical dosing:

a. Demulsifiers

When collected from a reservoir, crude oil tends to be blended either with the forming water or with mixed formation and injection water. It is necessary to efficiently and quickly extract the crude oil from the water for processing operations. This allows the removal of oil and the discharge of clean water, thus increasing the crude oil value and reducing operational expenses. Demulsifiers are molecules that usually help to isolate low concentrations of oil from water. Demulsification is the division of a crude oil emulsion into oil and water phases. It is a cycle with two stages [18]:

i. Flocculation or aggregation: water droplets during this process clump together to form aggregates. The droplets are similar to each other, touching at some points, but still do not lose their identity.

ii. Coalescence: in this process, droplets of water fuse or coalesce together producing a broader drop. This is an inevitable mechanism that leads to a reduction in the number of droplets of water and inevitably to complete demulsification.

Commercial demulsifiers are polymeric surfactants with high molecular weight, such as polyoxyethylene and propylene copolymers or alkylphenol-formaldehyde resins or mixtures with various surfactant substances. The suggested maximum concentration is between only a few ppm and around 100 ppm. Overdosing should be avoided because it risks restoring the emulsion.

b. Corrosion inhibitors

Areas/units of refineries where inhibitors are in need

Areas and units in the refinery where the chemical inhibitors are used to combat corrosion include [19] (1) atmospheric and vacuum crude distillation, (2) fluid catalytic cracking, (3) coker, (4) hydroprocessing, (5) catalytic reforming, (6) amine regenerator, (7) boiler, (8) sour water stripper, and (9) hydrogen manufacturing.

19.4.1 Common forms of inhibitors used in refinery

1. Neutralizing corrosion inhibitors

Neutralizers are the chemicals used in refinery distillation units to neutralize the acids present in the condensing water. Acid reduction or neutralization is an easy solution to the problem of corrosion, as corrosion is known to result from acid attack on metals. It should be sufficient for any substance strong enough to neutralize the acid and raise the pH to the desired level. To preserve the pH between 5.5 and 7, solid HCl and dissolved acid gas such as H_2S is neutralized. Corrosion is extremely destructive at pH values below 5.5, whereas pH values above 7 may lead to fouling. Additionally, at pH values above 7, ammonia and amines can attack copper alloys such as admiralty brass and Monel. To provide high solubility in water, effective neutralizers are also developed. When the first droplet of water condenses, it will avoid HCl from damaging steel immediately. Neutralizers should produce one or more molecular amines to counteract HCl while preventing the formation of corrosive salts. Typical categories of neutralizing amines include ethylamine, methoxypropylamine, ethylenediamine, morpholine, dimethylethanolamine, dimethylisopropanolamine, ethylenediamine, methoxypropylamine, monoethanolamine, morpholine, picolines, and trimethylamine [17]. Polyamines like ethylenediamine provide greater regulation of pH than ammonia or morpholine. While there are potentially a huge number of amines that could be used to prevent strong acid corrosion, the industry is limited to those that are economically viable, i.e., those amines that generally cost less than about US$4.5/kg [20].

Ammonia is the most commonly used material because of its high neutralizing strength, low unit cost, easy accessibility, and ease of handling. Under pressure, it can be injected into the cylinder as a liquid and flashed into the crude vapor process. Ammonia can dissolve into the condensate water to increase its pH when the vapors are diluted. Despite the above-mentioned benefits, there are many disadvantages of using ammonia. In the condensing method, if copper alloys are found, the incorporation of ammonia beyond neutralization, i.e., pH >7, is a risky operation. This is due to the formation of soluble cuprammonium complex by copper at pH levels >7–8.5, which can be expected to deteriorate the materials like CDA 443-445 (Admiralty). Moreover, higher molecular weight amines having strong buffering ability compared to ammonia and morpholine are also recommended as they do not form deposits of chloride from either the hydrocarbon or water process. Such content allows for better pH regulation and effectively eliminates the possibility of high pH i.e., >7.5 copper corrosion.

2. Film-forming inhibitors

To minimize the risk of corrosion under mildly acidic conditions i.e., pH 5–7, filming inhibitors are applied to the refinery distillation-column overheads. Filming inhibitors are typically organic compounds with an alkyl chain (tail) bound to a polar group (head). The polar group usually contains nitrogen and oxygen, and also sulfur or phosphorus. The polar group binds to the surface of the metal, while a hydrophobic layer is created by the hydrocarbon skeleton that repels water molecules and then retards corrosion reactions. From this thin hydrophobic layer, which may be thicker than a few molecules, the word filming inhibitor is derived. Generally, it is dosed to the overhead of the crude distillation column just before the aerocondenser to prevent corrosion of aerocondenser tubes. An average rate of injection is of the order of 3–5 vppm for operation [17]. Injection rates can be temporarily increased to a level of 12 vppm during startup or unit disruption to regenerate the protective film. Some of the widespread film-forming amine inhibitors are imidazoline derivative, alkyl quaternary amines, alkyl morpholine, aminoamides, amides, alkylamines, ethoxylates, pyridine and their salts, and quaternary ammonium salts. As described above, inhibitors differ in solubility, etc., and must also be selected in accord with the pH range and other fluid properties. Water-soluble filming amines offer excellent corrosion protection properties. Molecular structures for some of the most commonly used filming corrosion inhibitors in the refinery system [8] are given in Table 19–1.

19.4.2 Imidazoline

Imidazoline is a heterocycle organic compound with two nitrogen atoms [21,22]. They are among the most efficient film-forming combinations that are considered in several patent applications as a starting point for formulations [23]. Usually, it is expected that such categories of corrosion inhibitors would form a self-assembled layer and protect the surface from corrosion [24]. Some research has provided head and functional group modeling associated with the adhesion of corrosion inhibitor molecules to the surface of the metal, while the hydrocarbon tail limits the migration of corrosion products from the bulk of the solution to the surface of the metal and forms a

Table 19–1   Basic molecular structures of refinery corrosion inhibitors.

Amide	
Imidazoline	
Quaternary ammonium ion	
Aminoamide	
Alkyl pyridine	
Primary amine	

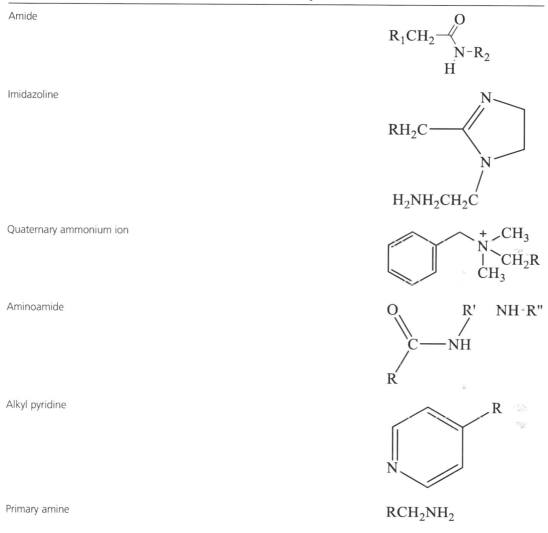

—R, R', R" corresponds to different alkyl group.

monolayer. In natural circumstances, however, they have poor solubility in water [25]. The recommended injection points are at the overhead vapor line and the reflux line. 2–5 ppm of the composition is typically injected as a spray [26]. In addition to its corrosion inhibition properties, imidazole also possesses the ability to minimize water and crude petroleum emulsion formation.

The few imidazoles tested as corrosion inhibitors, the aggressive media employed, the substrate, the concentration of the inhibitor, test conditions, testing techniques, and the efficacy of the inhibition are described in Table 19–1.

Recently Sotelo-Mazon et al. [27], studied corrosion resistance of 1018 carbon steel in the presence of hydroxyethyl-imidazoline derivate based on avocado oil in the mixture of brine–diesel saturated with CO_2 corrosive medium at 70°C. The inhibitor acted efficiently up to 25 ppm. Above this concentration corrosion rates increased due to the desorption of inhibitor molecules from the metal surface. This may be due to electrostatic repulsion forces between the charges of the hydrocarbon tails when the inhibitor is added in excess, thus leaving unprotected areas on the metal surface. The PDP plots show the tested compound acted as a cathodic type inhibitor. SEM images further confirmed the results, showing that the localized corrosion occurs in the form of pits at 50 and 100 ppm.

Studies on the anticorrosive properties of the newly synthesized inhibitor 2-(2-heptadecyl-4, 5-dihydro-1H-imidazole-1-yl) ethanol (HDIE) for Q-235 steel were undertaken by Munis et al. [28] in a 7.5% NH_4Cl solution. The findings demonstrate that at 298K and 0.5 mM L^{-1}, inhibition efficiency above 90% is achievable. From Fig. 19–2, the enhancement of corrosion inhibition is visualized by the diameter of the high frequencies capacitive loop that significantly increases inhibitor concentration. Quantum chemical analysis conducted by the author demonstrated that the active sites of the studied inhibitor are the imidazoline ring and heteroatoms. Through these active sites, the HDIE molecule can bind to the Fe surface. The electron lone pair of imidazoline ring heteroatoms and π-electrons produces coordination bonding with the empty Fe surface d-orbital.

Onyeachu et al. [29] examined the efficacy of 2-(2-pyridyl)benzimidazole as a CO_2 corrosion inhibitor for X60 steel in NACE brine ID196 (with a composition of 3.5% NaCl, 0.305% $CaCl_2 \cdot 2H_2O$ and 0.186% $MgCl_2 \cdot 6H_2O$ prepared with double distilled water) by electrochemical measurement. The findings have shown that the inhibitor behaves very well under static and hydrodynamic conditions. But not so good at high rotation speed, where the efficiency of inhibition is reduced. By the formation of the pyridinium ion after protonation of 2-pyridyl nitrogen, the strongest adsorption of the inhibitor is recognized. For this interaction, the underlying mechanisms can be seen in Fig. 19–3.

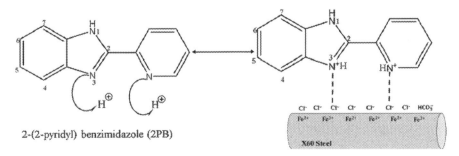

2-(2-pyridyl) benzimidazole (2PB)

FIGURE 19–3 Schematic representation of the interaction between 2PB and X60 steel surface during CO_2 corrosion inhibition in the NACE brine solution [29].

19.4.3 Aliphatic amines

The other filming inhibitor is aliphatic amines. High protective efficacy can be due to the existence of functional nitrogen atoms in the amino ($-NH_2$) group in aliphatic amines. Owing to its high electron density, the nitrogen atom in aliphatic amines serves as a binding center. It is a less costly class of film forming inhibitors than imidazolines and amides.

The inhibition efficacy of five low-molecular-weight straight-chain diamines, namely ethylenediamine, 1,3-propanediamine, 1,4-butanediamine, 1,8-octanediamine, and 1,12-dodecanediamine [30] in 1 M HCl and methylamine, ethylamine, propylamine, and butylamine [31], in 1 M H_2SO_4 for carbon steel was studied by Fouda et al. With an increased chain length of up to eight carbon atoms, the authors reported an improvement in inhibition effectiveness. Higher chain length amines with 12 carbon atoms, however, led to a decline in the inhibition efficacy. Xhanari et al. [32] tested corrosion inhibition effectiveness of 0.1 and 1.0 wt.% of eight amines, namely, butylamine, ethylamine, isopropylamine, octylamine, 2-ethylhexyl amine, aniline, benzylamine, and triethanolamine, for mild steel (C15 grade) in 3 wt.% NaCl solution at 25°C and 70°C. Immersion tests revealed that 2-ethylhexyl amine was the most effective corrosion inhibitor for a 0.1 wt.% amine concentration added at 25°C, while triethanolamine gave the lowest corrosion rate at 70°C.

19.4.4 Ethoxylates

Ethoxylates also play the role of film-former and surfactant in the formulation of corrosion inhibitors because of their excellent water solubility. Commonly such films are being used in conjunction with other film-forming entities as auxiliary film-formers and surfactants. Comparable to other fatty amines, such compounds also have alkaline characteristics and therefore should be neutralized to attain their cationic state by the organic acids [33].

19.4.5 Quaternary ammonium salts

The quaternary ammonium salts are another part of the organic corrosion inhibitors, which have found widespread applications in various industries because of their cost-effectiveness. These are applied to prevent corrosion of iron and steel in acid media.

There are various patent reports available which report anticorrosive properties of various quaternary pyridinium salts. A common example is N-(hydrophobe aromatic)pyridinium, N-aryl or N-alkyl substituted pyridinium halides, etc. Borgard et al. in their patent (US patent No. 5,368,774) [34] disclosed that quaternary ammonium salts in combination with organosulfur compounds can be used as refinery inhibitor. In another patent [35,36], cationic surfactants are reported to enhance the anticorrosive properties of quaternary pyridinium salts.

Zhao et al. [37] tested quaternary quinolinium ammonium salt (QB) and Gemini surfactant 1,3-bis(dodecyldimethylammonium chloride)-2-propanol (12-3OH-12) corrosion inhibition for mild steel in the solution of H_2S and CO_2 saturated brine. Local corrosion on the steel surface occurs in the presence of QB alone. The coadsorption of 12-3OH-12 keeps the inhibitor film more compact and ideal by covering the corrosion reactive sites when the

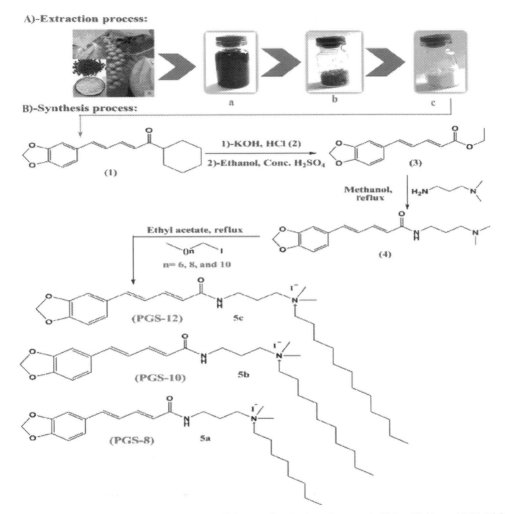

FIGURE 19–4 Synthetic route for the preparation of the novel cationic surfactants (PGS-8, PGS-10, and PGS-12) [39].

12-3OH-12 concentration is lower than 50 mg L^{-1}. A synergistic relationship between the compounds tested was documented by the authors.

In the simulated concrete pore solution containing 3.5 wt.% NaCl, Fei et al. [38] recorded electrochemical measurements and surface analysis results of a tailored cationic form of imidazoline quaternary ammonium salt to check its anticorrosive properties. The findings suggested that the inhibitor substantially boosted the polarization resistance and pitting potential. Most likely, the main inhibition mechanism was dependent on the adsorption of the molecules of corrosion inhibitor on the steel surface, leading to a reduction in the corrosion current density of the steel reinforcement. Tantawy et al. studied [39] the inhibition capacities of the series of cationic surfactants namely N-(3-((2E,4E)-5-(benzo[d][1,3]dioxol-5-yl)penta-2,4-dienamido)propyl)-N,N-dimethyloctan-1-aminium iodide (PGS-8), N-(3-((2E,4E)-5-(benzo

Table 19–2 Some organic film-forming corrosion inhibitors used for metals in different aggressive media.

Inhibitor	C	Material	Medium	T (°C)/pH	Techniques	IE (%)	Reference
Hydroxyethyl imidazoline	100 ppm	Mild steel	3 wt.% NaCl + H_2S	50	LPR EIS EN	95	[40]
Carboxyethyl-imidazoline	50 ppm	API X120	3 wt.% NaCl + H_2S with/without 10% diesel	50	PDP LPR EIS EN	98	[41]
Hydroxyethyl-imidazoline	20 ppm	1018 carbon steel	3 wt.% NaCl + H_2S	50	PDP LPR EIS EN	84.54	[42]
Aminoethyl-amine imidazoline		1018 carbon steel	3 wt.% NaCl + H_2S	50	PDP LPR EIS EN	90	[42]
Tall oil diethylenetriamine imidazoline	70 vppm	Mild steel	3 wt.% NaCl + CO_2	70/5	LPR EIS PDP	90	[43]
TOFA/DETA imidazoline	70 vppm	API X65	3 wt.% NaCl + CO_2	70/5	WL QCM AFM	95	[44]
N-[2-[(2-aminoethyl)amino] ethyl] 9 octadecenamide (AAOA)	5.44×10^{-5} mol dm^{-1}	Mild steel	5 wt.% NaCl + CO_2	25	PDS EIS PM-IRRAS	99	[45]
Amphiphilic amido-amine	5.44×10^{-5} mol dm^{-1}	Mild steel (AISI 1018)	5 wt.% NaCl + CO_2	25	PDS EIS PM-IRRAS	99	[46]
Amido-imidazoline	150 ppm	APIX52	5 wt.% NaCl + CO_2	25/4	PDP, EIS	95	[47]
Diethylenetetramine-derived imidazoline	100 ppm	Mild steel	0.5 M NaCl + CO_2	40/5–5.5	PDS LPR	84	[48]
Tetraethylenepentamine-derived imidazoline	100 ppm	Mild steel	0.5 M NaCl + CO_2	40/5–5.5	PDP LPR	95	[49]
Bis-imidazoline	100 ppm	Mild steel	0.5 M NaCl + CO_2	40/5–5.5	PDP LPR	96	[50]
Zwitterionic compound LZW-B12	75	AISI- 1018 steel	1.0 M NaCl + CO_2	70/3.5	EIS, PDP	91.6	[51]
Zwitterionic compound LZW-B16	25	AISI- 1018 steel	1.0 M NaCl + CO_2	70/3.5	EIS, PDP	99.94	[51]
Zwitterion compound LZW-B18	10	AISI- 1018 steel	1.0 M NaCl + CO_2	70/3.5	EIS, PDP	99.68	[51]
Imidazoline	100 ppm	Mild steel	0.1 M HCl	50	WL	97	[52]

AFM, Atomic force microscopy; *EIS*, Electrochemical impedance spectroscopy; *EN*, Electrochemical noise; *LPR*, linear polarization resistance; *PDP*, Potentiodynamic polarization; *PM-IRRAS*, Polarization modulation-infrared reflection-adsorption spectroscopy; *QCM*, Quartz crystal microbalance; *WL*, Weight loss.

[d][1,3]dioxol-5-yl)penta-2,4-dienamido)propyl)-N,N-dimethyldecan-1-aminium iodide (PGS-10), and N-(3-((2E,4E)-5-(benzo[d][1,3]dioxol-5-yl)penta-2,4-dienamido)propyl)-N,N-dimethyldodecan-1-aminium iodide (PGS-12) (Fig. 19—4), for C1018-steel corrosion in CO2-3.5%NaCl solution. If the length of the hydrophobic chain increases, inhibition potential increases, with the maximum

values reaching 91.4%, 93.7%, and 95.1% for PGS-8, PGS-10, and PGS-12, respectively, at 1.0 mM. The layer-forming ability of these surfactant species over the C1018-steel surface is confirmed by FE-SEM/EDX studies.

19.4.6 Others

Most of the sulfur components have even less possible impact on corrosion lowering, although they are able to strengthen their efficiency in combination with the film-forming amines. One of these materials used to develop such commercial products is mercaptoacetic acid.

Table 19−2 lists a few examples of the above-mentioned film-forming corrosion inhibitors used for various metals in the different aggressive media [40−52].

19.4.7 Main criteria of using corrosion inhibitors in refineries

In general, the following criteria shall be deemed and defined for the selection of the appropriate corrosion inhibitor:

19.4.8 Temperature

As higher temperature leads to the decomposition of organic molecules, film-forming inhibitors are recommended only for low temperatures. It can be said that there is limited use of film-forming inhibitors beyond 232°C−260°C. Furthermore, film-forming inhibitors operate by an adsorption mechanism that typically becomes less efficient at elevated temperatures, requiring higher treatment doses in order to maintain efficient films on the metal surfaces. This raises the cost of manufacturing chemicals.

19.4.9 Inadequate inhibitor concentration

Compared to untreated systems, film-forming type inhibitors can cause increased localized corrosion and pitting when used in quantities that are insufficient to create a viable corrosion resistant film. For that reason, it is not recommended to try to reduce the cost of inhibitors by decreasing the dose below healthy, effective levels. The high concentration reduces the time taken for old deposits to be scraped out and thus speeds up the formation of a film on the cleaned metal. The concentration then decreases steadily until the higher level of inhibition (as shown by coupons, resistance probes, and water analysis) is reached at an economic expense.

19.4.10 Surfactant properties of inhibitors

The inhibitors detergent activity should be useful in regulating the degradation of metals. The documented "detergent" effects, i.e., the ability to retain clean surfaces, are assumed to be the result of two processes: first, corrosion inhibition to such an extent that less corrosion product is formed; and second, the actual detergent action induced by those inhibitors' surface-active properties.

19.5 Conclusion

The chapter has furnished the corrosion problems and the preventive methods that occur in the refineries. The main corrosive species encountered in refinery corrosion are salts like sodium chloride, magnesium chloride, and calcium chloride; organic acids like naphthenic acid; sulfur compounds like mercaptans, thiophenes, elemental sulfur, H_2S; and CO_2. Therefore, various preventive methods like chemical dosing, inhibitor dosing, suitable desalting, and proper material selection are recommended to minimize the corrosion. The two most popular types of inhibitors employed in refinery systems are neutralizing and film-forming inhibitors. The content of this chapter shows great opportunity for economically ensuring the greater performance, effectiveness, and durability of the corrosion management programs. In addition, greater flexibility with minimal effort can be achieved by customizing the corrosion protection.

Acknowledgments

Ruby Aslam acknowledges the Council of Scientific & Industrial Research, New Delhi, India for providing financial aid under Research Associate fellowship (09/112(0616)2K19 EMR-I).

Conflict of interest statement

The authors declare that there is no conflict of interest in any way.

Competing interest statement

The authors declare that there is no competing interest in any way.

Author's contributions

All authors collectedly contributed in the design and write-up of the book chapter.

References

[1] Occupational Safety and Health Administration (OSHA), OSHA Instruction TED 1.15 CH-1, US Department of Labor, Washington, DC, 1996.

[2] W.L. Leffler, Petroleum Refining in Nontechnical Language, 4th Ed., 2008.

[3] R.D. Kane, Corrosion in petroleum refining and petrochemical operations, corrosion: environments and industries, ASM Handbook, 13C, ASM International, 2006, pp. 967−1014.

[4] A. Groysman, Corrosion problems and solutions in oil, gas, refining and petrochemical industry, Koroze a ochrana materialu 61 (2017) 100−117.

[5] B. Chambers, S. Srinivasan, K.M. Yap, M. Yunovich, Corrosion in crude distillation unit overhead operations: a comprehensive review, NACE Paper No.11360, 2011.

[6]   D.S.J. Jones, Introduction to crude oil and petroleum processing, Handbook of Petroleum Processing, 2014.

[7]   A.J. Kidnay, W.R. Parrish, D.G. McCartney, Fundamentals of Natural Gas Processing, 2nd Ed., CRC Press Taylor & Francis Group, Boca Raton, FL, USA, 2011, p. 552.

[8]   P.K. Bhowmik, M.E. Hossain, J.A. Shamim, Corrosion and its control in crude oil refining process, Proceedings of the 6th International Mechanical Engineering Conference & 14th Annual Paper Meet (6IMEC&14APM), 2012, IMEC&APM-ABS-000.

[9]   Z.M. Wang, J. Zhang, Corrosion of multiphase flow pipelines: the impact of crude oil, Corros. Rev. 34 (2016) 17−40.

[10] G.C. Laredo, C.R. López, R.E. Álvarez, J.J. Castillo, J.L. Cano, Identification of naphthenic acids and other corrosivity-related characteristics in crude oil and vacuum gas oils from a Mexican refinery, Energy Fuels 18 (2004) 1687−1694.

[11] S. Papavinasam, Corrosion Control in the Oil and Gas Industry, 1st ed., 2014.

[12] E.C. Greco, H.T. Griffin, Laboratory studies for determination of organic acids as related to internal corrosion of high pressure condensate wells, Corrosion 2 (1946) 138−152.

[13] NORSOK M-001, Materials Selection, fifth ed., 2014, p. 32.

[14] C.L. Eastonm, Corrosion control in petroleum refineries processing western Canadian crude oils, Corrosion 16 (1960) 275t−280t.

[15] D. Brondel, R. Edwards, A. Hayman, D. Hill, S. Mehta, T. Semerad, Corrosion in the oil industry, Oilf. Rev. 6 (1994) 4−18.

[16] S. Nesic, G.T. Solvi, J. Enerhaug, Comparison of the rotating cylinder and pipe flow tests for flow-sensitive carbon dioxide corrosion, Corrosion 51 (1995) 773−787.

[17] J. Gary, G. Handwerck (Eds.), Petroleum Refining: Technology and Economics, 4th ed., Marcel Dekker, Inc, New York. USA, 2001, p. 49. ISBN: 0-8247-0482-7.

[18] C. Noik, J. Chen, C. Dalmazzone, Electrostatic demulsification on crude oil: a state-of-the-art review. Society of Petroleum Engineers. SPE 103808, 2006.

[19] API Recommended Practice, Damage Mechanisms Affecting Fixed Equipment in the Refining Industry, 571, American Petroleum Institute (API), Washington, DC, 2003.

[20] S. Rennie, Corrosion and materials selection for amine service, Mater. Forum 30 (2006) 126−130.

[21] G. Schmitt, T. Simon, R.H. Hausler, Proceedings of NACE Corrosion 90, Paper No. 22, NACE International, Houston, 1990.

[22] G. Bereket, C. Ogretir, A. Yurt, Quantum mechanical calculations on some 4-methyl-5-substituted imidazole derivatives as acidic corrosion inhibitor for zinc, J. Mol. Struct. Theochem 571 (2001) 139−145.

[23] M. Askari, M. Aliofkhazraei, S. Ghaffari, A. Hajizadeh, Film former corrosion inhibitors for oil and gas pipelines—a technical review, J. Nat. Gas Sci. Eng. 58 (2018) 92−114.

[24] X. Guan, Y. Hu, Imidazoline derivatives: a patent review (2006−present), Expert Opin. Therapeutic Pat. 22 (2012) 1353−1365.

[25] K. Bílková, E. Gulbrandsen, Kinetic and mechanistic study of CO_2 corrosion inhibition by cetyltrimethyl ammonium bromide, Electrochim. Acta 53 (2008) 5423−5433.

[26] R.G. Bistline, J.W. Hampson, W.M. LinField, Synthesis and properties of fatty imidazolines and their N-(2-aminoethyl) derivatives, J. Am. Oil Chem. Soc. 60 (1983) 823−828.

[27] O. Sotelo-Mazon, S. Valdez, J. Porcayo-Calderon, J. Henao, C. Cuevas-Arteaga, C.A. Poblano-Salas, et al., Evaluation of corrosion inhibition of 1018 carbon steel using an Avocado oil-based green corrosion inhibitor, Prot. Met. Phys. Chem. Surf. 56 (2020) 427−437.

[28] A. Munis, T. Zhao, M. Zheng, A. Rehman, F. Wang, A newly synthesized green corrosion inhibitor imidazoline derivative for carbon steel in 7.5% NH_4Cl solution, Sustain. Chem. Pharm. 16 (2020) 100258.

[29] I.B. Onyeachu, I.B. Obot, A.A. Sorour, M.T. Abdul-Rashid, Green corrosion inhibitor for oilfield application I: electrochemical assessment of 2-(2-pyridyl) benzimidazole for API X60 steel under sweet environment in NACE brine ID196, Corros. Sci. 150 (2019) 183–193.

[30] A.S. Fouda, H.A. Mostafa, G.Y. Elewady, M.A. El-Hashemy, Low molecular weight straight-chain diamines as corrosion inhibitors for SS type 304 in HCl solution, Chem. Eng. Commun. 195 (2008) 934–947.

[31] A.S. Fouda, H.A. Mostafa, F. El-Taib, G.Y. Elewady, Synergistic influence of iodide ions on the inhibition of corrosion of C-steel in sulphuric acid by some aliphatic amines, Corros. Sci. 47 (2005) 1988–2004.

[32] K. Xhanari, N. Grah, M. Finsgar, R.F. Godec, U. Maver, Corrosion inhibition and surface analysis of amines on mild steel in chloride medium, Chem. Pap. (2016). Available from: https://doi.org/10.1007/s11696-016-0046-y.

[33] https://patentimages.storage.googleapis.com/f5/15/51/0c54f36f401e71/US3766053.pdf.

[34] J.A. Dougherty, B.T. Outlaw, B.A.O. Alink, Corrosion inhibition by ethoxylated fatty amine salts of maleated unsaturated acids, patent # US5582792A, 1996.

[35] B.G. Borgard, J.B. Harrell, Jr., J. Link, Water soluble corrosion inhibitor effective corrosion by carbon dioxide, US5368774A United States, 1994.

[36] S.S. Shah, W.F. Fahey, B.A.O. Alink, Corrosion inhibition in highly acidic environments by use of pyridine salts in combination with certain cationic surfactants Patent, Patent # 5,336,441, 1991.

[37] J. Zhao, H. Duan, R. Jiang, Synergistic corrosion inhibition effect of quinoline quaternary ammonium salt and Gemini surfactant in H_2S and CO_2 saturated brine solution, Corros. Sci. (2014). xxx–xxx, xxx.

[38] F. Fei, J. Hu, J. Wei, Q. Yu, Z. Chen, Corrosion performance of steel reinforcement in simulated concrete pore solutions in the presence of imidazoline quaternary ammonium salt corrosion inhibitor, Constr. Build. Mater. 70 (2014) 43–53.

[39] A.H. Tantawy, K.A. Soliman, H. M. Abd El-Lateef, Novel synthesized cationic surfactants based on natural piper nigrum as sustainable-green inhibitors for steel pipeline corrosion in CO2-3.5% NaCl: DFT, Monte Carlo simulations and experimental approaches, J. Clean. Prod. (2019). Available from: https://doi.org/10.1016/j.jclepro.2019.119510.

[40] E.F. Diaz, J.G. Gonzalez-Rodriguez, A. Martinez-Villafañe, C. Gaona-Tiburcio, H2S corrosion inhibition of an ultra-high strength pipeline by carboxyethyl-imidazoline, J. Appl. Electrochem. 40 (2010) 1633–1640.

[41] D.M. Ortega-Toledo, J.G. Gonzalez-Rodriguez, M. Casales, L. Martinez, A. Martinez-Villafañe, CO_2 corrosion inhibition of X-120 pipeline steel by a modified imidazoline under flow conditions, Corros. Sci. 53 (2011) 3780–3787.

[42] L.M. Rivera-Grau, M. Casales, I. Regla, D.M. Ortega-Toledo, J.A. Ascencio-Gutierrez, J.G. Gonzalez-Rodriguez, et al., H2S corrosion inhibition of carbon steel by a coconut-modified imidazoline, Int. J. Electrochem. Sci. 7 (2012) 12391–12403.

[43] I. Jevremovic, M. Singer, M. Achour, D. Blumer, T. Baugh, V. Miskovic-Stankovic, et al., A novel method to mitigate the top-of-the-line corrosion in wet gas pipelines by corrosion inhibitor within a foam matrix, Corrosion 69 (2012) 186–192.

[44] I. Jevremović, M. Singer, S. Nesic, V. Miskovic-Stankovic, Inhibition properties of self-assembled corrosion inhibitor talloil diethylenetriamine imidazoline for mild steel corrosion in chloride solution saturated with carbon dioxide, Corros. Sci. 77 (2013) 265–272.

[45] S. Mohanan, S. Maruthamuthu, R. Kalaiselvi, R. Palaniappan, G. Venkatachari, N. Palaniswamy, et al., Role of quaternary ammonium compounds and ATMP on biocidal effect and corrosion inhibition of mild steel and copper, Corros. Rev. 23 (2005) 425–444.

[46] J. Zhang, X. Sun, Y. Ren, M. Du, The synergistic effect between imidazoline-based dissymmetric bis-quaternary ammonium salts and thiourea against CO_2 corrosion at high temperature, J. Surfactants Deterg. 18 (2015) 981–987.

[47] M. Heydari, M. Javidi, Corrosion inhibition and adsorption behaviour of an amido-imidazoline derivative on API 5L X52 steel in CO_2-saturated solution and synergistic effect of iodide ions, Corros. Sci. 61 (2012) 148–155.

[48] A. de Oliveira Wanderley Neto, E.F. Moura, H.S. Júnior, T. N. d C. Dantas, A.A.D. Neto, A. Gurgel, Preparation and application of self-assembled systems containing dodecylammonium bromide and chloride as corrosion inhibitors of carbon-steel, Colloids Surf. 398 (2012) 76–83.

[49] A. Ulman, Formation and structure of self-assembled monolayers, Chem. Rev. 96 (1996) 1533–1554.

[50] M.M. Osman, A.M.A. Omar, A.M. Al-Sabagh, Corrosion inhibition of benzyl triethanol ammonium chloride and its ethoxylate on steel in sulphuric acid solution, Mater. Chem. Phys. 50 (1997) 271–274.

[51] E.G. Juarez, V.Y.M. Cervantes, J. Vazquez-Arenas, G.P. Flores, R. Hernandez-Altamirano, The inhibition of CO_2 corrosion via sustainable geminal zwitterionic compounds: effect of the length of the hydrocarbon chain from amines, ACS Sustain. Chem. Eng. 12 (2018) 17230–17238.

[52] Y. Wang, D. Han, D. Li, Z. Cao, A complex imidazoline corrosion inhibitor in hydrochloric acid solutions for refinery and petrochemical plant equipment, Petrol. Sci. Technol. 27 (2009) 1836–1844.

20

Environmentally sustainable corrosion inhibitors in the oil and gas industry

Megha Basik, Mohammad Mobin

CORROSION RESEARCH LABORATORY, DEPARTMENT OF APPLIED CHEMISTRY, FACULTY OF ENGINEERING AND TECHNOLOGY, ALIGARH MUSLIM UNIVERSITY, ALIGARH, INDIA

Chapter outline

20.1 Introduction

The oil and gas industries play a vital role in the world economy. But due to corrosion, these industries suffer a lot not only economically but also catastrophically. Generally, corrosion takes place in oil and gas pipelines, tanks, and processing equipment. The majority of equipment used in oil and field industries is low-carbon steel which has excellent mechanical properties to withstand the operating conditions required. Pipelines play an important role in transporting the oil and gas from wellheads, hence they are continuously exposed to a corrosion threat. Various reports have confirmed that pipelines are ruptured due to corrosion and experience oil spillages which create environmental pollution, loss of resources, and large-scale ecological damage [1]. A systematic diagram of pipeline rupture is shown in Fig. 20−1 [2]. Despite the defects in infrastructural component, crude oil is also responsible

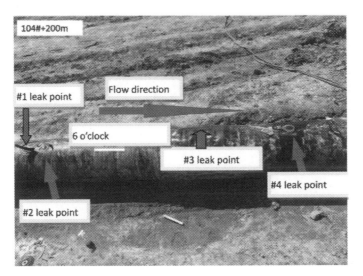

FIGURE 20–1 Pipe segment with pinholes.

for promoting corrosion due to the existence of impurities like acids, sulfur, etc. [3]. So to overcome this fact, it is really crucial to study the nature of the acids and the amount of the sulfurous components that exist in the crude oil to understand the root of corrosion and the performance needed to withstand it. Therefore the dependency of corrosion rate is on the quality of crude oil, its acidic constituents, and the transportation environment [4].

According to the latest report, the annual cost of corrosion in the oil and gas industries is about $1.372 billion, including surface pipeline and facility costs ($589 million), downhole tubing costs (463 million), and capital expenditures ($320 million) [5]. To avoid or minimize this problem, there are various alternatives, such as cathodic protection [6], application of coatings [7], corrosion-resistant alloys [8], anodic protection, and use of inhibitors [9]. Amongst them, the use of corrosion inhibitors has been the method of the choice to retard the corrosion rate in the petroleum and oil−gas industries. Inhibitors have been found to be successful in minimizing the corrosion in harsh and aggressive environments, such as CO_2, H_2S, polysulfides, organic acids, elemental sulfur, and oxygen (O_2). Generally, higher H_2S and CO_2 content together with high-temperature operating conditions are the major factors for corrosion to occur. By applying batch and continuous treatment of corrosion inhibitors the internal corrosion of piping and other equipment in oil−gas industries has been minimized.

A huge variety of corrosion inhibitors are commercially available and are being continuously developed by researchers or manufacturers. Traditionally, corrosion inhibitors are toxic in behavior and thus injurious to health as well as the environment. Governments have thus restricted their usage for safety considerations and have implemented the standards for using environment-friendly corrosion inhibitors. The rise in the era of "green chemistry" has introduced the concept in different fields of science, technology, engineering, etc., [10] to restrict the application of harmful corrosion inhibitors by implementing new ideas to reduce the contamination into the environment from being discharged and to use the chemicals which are less or nontoxic (eco-friendly) [11]. From this initiative, the use of green-based corrosion inhibitors such as natural gums [12,13], amino acids [14,15], plant extracts [16], chemical drugs [17], ionic liquids [18], and biocomposites [19], is being practiced.

Natural gums and polymers have large complex-type structures with different functional moieties attached to their molecular entity. They cover a huge surface area of the metal to be protected. Amino acids have both an amine group as well as an acidic group in its structure. So the adsorption process has been conducted between these groups and the surface to be protected. Plants and fruits-based corrosion inhibitors are rich in sources of minerals, vitamins, essential acids and oils, alkaloids, phenols, esters, etc. These all-natural and green inhibitors prevent the metal corrosion by their active functional groups through adsorption. Although these are not the only factors responsible for an inhibitor to act effectively, there are some factors that play a huge role in the path of minimizing the corrosion process. Some of these factors are the concentration of the medium, temperature, dispersion rate, pressure, the effect of ions and gases, pH, presence of microorganisms, velocity, etc. [20,21]. These factors along with the molecular structure of the inhibitor (S, N, O, P, and π-system) are responsible for minimizing the corrosion effect.

In the oil and gas industries, crude oil is the essential substance to be processed. It contains sulfur, nitrogen, and oxygen in its composition, hence corrosion is not an issue over there. But water containing substances or impurities can damage the structural domain of the industry. The presence of chlorides, sulfides, carbon dioxide, and oxygen acts as accelerating agents in the corrosion process, whereas ions like chromates and phosphates act as inhibiting agents [22]. Inorganic acids used for cleaning and degreasing are also responsible for the materialistic loss. Generally, there are eight major forms of corrosion that are observable in the oil and gas field. They are sweet corrosion (CO_2), sour corrosion (H_2S), oxygen corrosion, galvanic corrosion, erosion-corrosion, crevice corrosion, microbial corrosion, and stress corrosion cracking.

20.2 Mechanism of corrosion in oil and gas industries

A general mechanism of a corrosion process consists of an anode (oxidation), which releases the metals ions into the solution, and a cathode, which acts as an acceptor to complete the cathodic process (reduction). The general chemical anodic and cathodic reactions of corrosion in the presence of oxygen and moisture are as follows:

$$Fe \rightarrow Fe^{+2} + 2e^- \quad \text{(Oxidation)} \tag{20.1}$$

$$O_2 + 2H_2O + 4e^- \rightarrow 4OH^- \quad \text{(Reduction)} \tag{20.2}$$

In oil production the anode produces the Fe^{2+} ions from the steel. These ions combine with carbon dioxide, hydrogen sulfide, and oxygen to complete a cathodic hydrogen reaction. The presence of chlorides, sulfides, carbon dioxide, and oxygen in the medium alters the route of the corrosion process.

20.2.1 H₂S corrosion

The term sour corrosion refers to the corrosion of metal in a H_2S dominant environment. Due to the abundance of hydrogen sulfide in oil and gas processing plants, metals or alloys corrode severely and lose their mechanical strength, which leads to severe shutdown or tragic unwanted conditions. The presence of hydrogen sulfide is the root cause of pipeline failures due to sulfide stress corrosion cracking. This problem is becoming more significant because of the growing proportion of sulfur in crude oil [23]. Eq. (20.3) represents the corrosion reaction in the presence of a hydrogen sulfide medium.

$$Fe + H_2S \rightarrow FeS + H_2 \qquad (20.3)$$

Internal stresses are caused by the molecular hydrogen which creates hydrogen cracking. The outcomes of stress corrosion cracking are very dangerous and unexpected since no noticeable signs are shown during the early phase of this type of corrosion. The sulfide stress cracking is initiated by the simultaneous action of a tensile stress and a corrosive medium, hence it is very important to select proper material to reduce the equipment loss.

20.2.2 CO₂ corrosion

Corrosion due to dissolved carbon dioxide in defined conditions creates a drastic form of corrosion, termed sweet corrosion [24]. Carbon dioxide is produced and has been utilized in many ways in oilfields. CO_2 gas lowers oil viscosity and thus allows the oil to flow to obtain the desired recovery [25]. In the aqueous mix, CO_2 reacts with water to form carbonic acid (H_2CO_3) which leads to the lowering of the pH of the medium. CO_2 corrosion gets influenced by the temperature, composition of the aqueous system, flow condition, pH value, and metal composition [26].

$$Fe + CO_2 + H_2O \rightarrow FeCO_3 + H_2 \qquad (20.4)$$

During transportation and storage, CO_2 initiates internal corrosion which influences the structural integrities of the steel-based systems. The characteristic features in CO_2 based corrosion are deep sharp-edged pits imaged on the surface of the material.

20.2.3 Oxygen corrosion

Oxygen plays a crucial role in corrosion as it acts as a corrosive agent that initiates corrosion and thus damages various processing systems. It causes a detrimental effect on water injection equipment such as storage tanks, pipelines, etc. Dissolved oxygen generates pits in the pipelines which cause fatigue cracks and thus damage the overall structure.

20.2.4 Chloride corrosion

Chloride is another corrosive substance that initiates corrosion at high temperatures. Chlorides are generally present in mineralized water at the bottom of the well. The corrosion due to the presence of chloride is intergranular corrosion and chloride stress corrosion cracking, which leads to breakdown and failure of the metal structure.

20.3 Green-based corrosion inhibitors

Environmentally sustainable corrosion inhibitors from nontoxic and biodegradable sources are becoming more in demand compared to the commercially available toxic organic and inorganic inhibitors [27,28]. Green-based corrosion inhibitors are further divided into organic and inorganic classes. Organic inhibitors are nontoxic in nature; they are flavonoids, phenols, alkaloids, and by-products of waste plants; while inorganic inhibitors are toxic in nature up to a certain concentration and thus need to be functionalized before their usage [29].

The generally used natural and green corrosion inhibitors are plant extracts, polymers, and their composites, natural gums, drugs, plant wastes, etc. Plant extracts, being environment-friendly, have various active functional groups, good biological properties, and have a good ability to inhibit metals and alloys from corrosion. It should be noted that not only the leaf portion of the plant inhibits corrosion, the root, stem, flower, seed, fruit, bark, and peel may also act as excellent corrosion inhibitors [30−32]. However, leaves have been given the most preference because of their abundant phytochemicals and other active components [33,34]. These inhibitors are used in relatively low concentrations to inhibit the surface corrosion of metal in aggressive environments. The inhibitory action of *Plectranthus amboinicus* leaf extract was studied using weight loss and electrochemical techniques for mild steel in acidic medium. Electrochemical results show good inhibition efficiency (IE) and revealed that the inhibitor acts as a mixed-type inhibitor. The theoretical calculation reveals that thymol and cineole are mainly involved in adsorption to enhance IE [35]. *Lagerstroemia speciosa* leaf extract was examined as a corrosion inhibitor for mild steel corrosion. The corrosion mitigating efficiency was studied by weight loss (WL), potentiodynamic polarization (PDP), and electrochemical impedance spectroscopy (EIS). The inhibitor was found to be 94% effective at 500 ppm concentration at 333K. EIS and PDP data confirmed that an inhibitive film was formed over the mild steel surface and the inhibitive action was mixed type. Density functional theory (DFT) and molecular dynamics (MD) studies successfully correlated the IE with the molecular structure of active species present in the extract [34]. Plant extract with inorganic zirconium acetate has been studied as a novel and effective corrosion inhibitor for mild steel corrosion by weight loss, PDP, EIS, and DFT calculations. Studies revealed that a *Cissus quadrangularis*−zirconium acetate complex acts as a more effective corrosion inhibitor compared to *C. quadrangularis* alone. Fig. 20−2 shows the pictorial representation of the complex formation [16].

They function by altering the anodic and cathodic reaction rates and thus reduce the damage that can be generated without the inhibitor. Generally, most of the inhibitors act as a mixed type but sometimes one of either cathodic and anodic reactions may dominate. They form a protective film over the surface of the metal and thus do not allow the ions of the corrosive medium to interact with the metal being protected [36,37]. Although plant extracts are easily available there are certain limitations associated with them during their processing.

FIGURE 20–2 Pictorial representation of the solution of ZrAc (whitish color), CQ extract (light gray color) and CQ-ZrAc (dark gray color) complex.

Temperature as well as solubility plays an important role in the extraction process. Higher temperature decomposes the phytochemicals present in the extract, so the recommended temperature for the ideal extraction range is $60°C-80°C$ [38,39]. Similarly solvent also plays a crucial role in the extraction process. The choice of solvent is on the basis of their diffusion into the plant's tissue. The solvent penetrates and solubilizes the available phytochemicals easily and thus more chemical moieties become available for the adsorption process [40]. Generally, water, ethanol, and methanol are in high demand for selective plant extraction [41,42].

Natural polymers also gained attention for retarding metal corrosion. These polymers work by blanketing the metal surface via the heteroatoms present in their structure. They are chemically stable, inexpensive, and renewable sources [43,44]. Modified natural inhibitors are gaining interest due to their better performance as compared to natural inhibitors alone. Generally, natural inhibitors give better protection at higher concentrations, which creates drawbacks to their usage, so to enhance the performance and lower the optimum concentration they need to be modified by using certain active compounds or species. Modification is done in several ways, such as adding metal ions like Ag, Au, Zn, Zr, Ti, Fe, etc. to the matrix of the native species so that it is transformed into nanocomposite and thus its surface coverage get increased [19,45]. The incorporation of silver nanoparticles to the *carboxymethyl cellulose* matrix enhances the performance of the studied polymer. Corrosion studies, such as weight loss, DEIS, and PDP, were conducted to measure the inhibitive effect. The maximum IE of 92% was observed at 1000 ppm concentration. *Urtica dioica*—zinc acetate was synthesized and its inhibitory performance was studied using EIS and PDP technique. It was observed that *U. dioica* alone provided maximum IE of 45%, but this tended to increase in presence of the zinc acetate complex, providing 92% IE [46].

Some of the natural polymers are grafted to increase the side chains and stability of the natural polymer so that it does not degrade at high temperatures. Synergistic effect (surfactant, halide ions, rare earth metals, etc.) was also considered to enhance the IE at low concentration [47,48]. The performance of gum acacia alone and with SDS and CTAB surfactants on mild steel in an acidic medium was studied by weight loss, estimation of Fe^{+2} ions in the electrolyte, SEM, and AFM techniques. A small concentration of surfactant synergistically improved the inhibition performance of gum acacia [47]. Guar gum was studied as a green water-soluble corrosion inhibitor

Table 20–1 Summary of environment-friendly corrosion inhibitors.

Details		Reference
a. *Musa acuminata* (banana peels). b. Immersion test, gasometric, electrochemical polarization, and impedance spectroscopy (EIS) techniques. c. 0.1–4.0 g L^{-1} concentration.	The maximum IE of 90% at 4 g L^{-1} at 298.15K in 5.0 M HCl solution.	[49]
a. *Ziziphora* leaves extract. b. EIS, PDP and surface studies. c. 200–800 ppm concentration.	The maximum IE of 93% at 800 ppm in 1 M HCl	[50]
a. Aqueous extract of *Artemisia Herba Alba* (AHA). b. Weight loss, electrochemical impedance spectroscopy (EIS), and potentiodynamic polarization. c. 0.025–0.4 g L^{-1} concentration.	The highest IE of 92% observed at 0.4 g L^{-1} in 1 M HCl at 303 K.	[51]
a. Cationic surfactants based on natural *Piper nigrum*. b. Electrochemical (open-circuit potential versus time, Tafel, and EIS) measurements, surface morphology. c. 5 × 10^{-6}–1 × 10^{-3} M concentration.	The maximum IE of 96.7%–98.9% was observed at 50°C at 1 × 10^{-3} M concentration.	[52]
a. *Tunbergia fragrans* extract. b. Weight loss method, electrochemical studies and Surface morphological study. c. 100–500 ppm concentration.	Maximum IE of 81% observed at 500 ppm in 1 M HCl	[53]
a. *Myrobalan* aqueous extract. b. Electrochemical and surface studies. c. 200–800 ppm concentration.	Maximum IE of 91% observed at 800 ppm concentration.	[54]
a. *Magnolia grandiflora* leaves extract. b. Open circuit potential curves, Potentiodynamic polarization and electrochemical impedance spectroscopy studies. c. 50–500 ppm concentration.	Inhibition performance of 88% observed at 500 mg L^{-1}.	[55]
a. Garlic extract. b. Weight loss measurement, potentiodynamic polarization, and electrochemical noise analysis. c. 2–8 cc.L^{-1} concentration.	Maximum IE of 88% was observed at 8 cc. L^{-1}.	[56]
a. *Portulaca grandiflora* leaf (PGL) extract. b. Weight loss, open circuit potential, electrochemical impedance spectroscopy, and polarization measurement. c. 5–20 mL L^{-1} concentration.	Maximum IE of 95% observed at 20 mL L^{-1}.	[57]
a. *Lavandula mairei Humbert* extract. b. Weight loss and electrochemical measurements c. 0.05–0.40 g L^{-1} concentration.	92% IE obtained at 0.4 g L^{-1} concentration.	[58]

(Continued)

Table 20–1   (Continued)

Details		Reference
a. *Hyalomma tick* extract. b. Polarization and electrochemical impedance spectroscopy (EIS). c. 1–3 g L^{-1} concentration.	95% IE obtained at 3 g L^{-1} concentration.	[59]
a. *Aloysia citrodora* leaves extract. b. EIS and PDP studies. c. 200–800 ppm concentration.	94% IE observed at 800 ppm concentration.	[60]
a. *Luffa cylindrica* leaf extract. b. Gravimetric, depth of attack, and surface analysis techniques. c. 0.50–1.00 g L^{-1} concentration.	The optimum IE of 87.89% was obtained at 1 g L^{-1}.	[61]
a. *Garcinia* fruit rind extract (GIW). b. Weight loss and electrochemical methods. c. 1%–4% (V/V) GIW concentration.	93% efficiency in 0.5 M HCl and 87% in at 4% GIW extract	[62]
a. Lemon Balm extract, b. Potentiodynamic polarization and electrochemical impedance spectroscopy. c. 200–800 ppm concentration.	95% was obtained at 800 ppm concentration in 1 M HCl.	[63]
a. *Peganum harmala* seed extract. b. EIS and polarization studies. c. 200–800 ppm concentration.	Maximum IE of 95% observed at 800 ppm.	[64]
a. Mustard seed extract. b. Weight loss, impedance spectroscopy, and potentiodynamic polarization. c. 50–200 mg L^{-1} concentration.	Maximum IE of 97% obtained at 200 mg L^{-1}.	[65]
a. Parsley *(Petroselium sativum)* leaves extract. b. Weight loss, potentiodynamic polarization, and impedance spectroscopy. c. 1–5 g L^{-1} concentration.	95.28% obtained at 25°C for 5 g L^{-1} concentration.	[66]
a. *Eucalyptus* leaves extract (ELE). b. EIS and polarization test. c. 200–800 ppm concentration.	An IE of 88% was obtained at 800 ppm.	[67]
a. *Ircinia strobilina* crude extract. b. Weight loss method, potentiodynamic polarization, linear polarization resistance, and electrochemical impedance spectroscopy. c. 0.5–2.0 g L^{-1} concentration.	Maximum IE of 92% at 2 g L^{-1}.	[68]
a. *Eriobotrya japonica Lindl* (EJL) extract. b. EIS and potentiodynamic polarization. c. 200–800 ppm concentration.	The highest resistance (92%) was achieved at 800 ppm concentration.	[69]

(Continued)

Table 20–1   (Continued)

Details		Reference
a. *Glycyrrhiza glabra* extract. b. Potentiodynamic polarization and electrochemical impedance spectroscopy (EIS). c. 200–800 ppm concentration.	The maximum corrosion IE of 88% at 800 ppm.	[70]
a. Sunflower seed hull extract. b. Electrochemical impedance spectroscopy and potentiodynamic polarization studies. c. 50–400 ppm concentration.	98% IE in presence of 400 ppm concentration.	[71]
a. *Xanthium strumarium* leaves *(XSL)* extracts. b. Weight loss method and surface studies. c. 2–10 mL L^{-1} concentration.	Maximum efficiency was found to be 94.82% at the optimum concentration of 10 mL L^{-1}.	[72]
a. *Lannea coromandelica* leaf extract. b. Weight loss and electrochemical techniques. c. 50–250 mg L^{-1} concentration.	Maximum IE of 93.8% observed at 250 mg L^{-1} concentration.	[73]
a. *Griffonia simplicifolia* extract. b. Weight loss measurement and *electrochemical* monitoring techniques. c. 100–1000 ppm concentration.	91% IE observed at 1000 ppm in 1 M HCl.	[74]
a. *Longan* seed and peel extract. b. Weight loss method and electrochemical measurements. c. 300–600 ppm concentration.	92.9% IE observed at 600 ppm.	[75]
a. *Pennisetum purpureum* extract. b. Weight loss measurement and scanning electron microscopy. c. 0.1–0.5 g L^{-1} concentration.	The maximum IE of 95% observed at 0.5 g L^{-1} concentration.	[76]
a. *Morus alba pendula* leaves extract (MAPLE). b. Electrochemical impedance spectroscopy (EIS) and polarization test. c. 0.1–0.4 g L^{-1} concentration.	An IE of 93% achieved at 0.4 g L^{-1} MAPLE at 25°C.	[77]
a. *Ligularia fischeri* extract. b. Weight loss, potentiodynamic polarization and electrochemical impedance spectroscopy(EIS) studies. c. 100–500 ppm concentration.	A maximum IE of 92% was achieved using 500 ppm concentration.	[78]
a. Alkaloids extract of *Retama monosperma* (L.) *Boiss.* seeds. b. Electrochemical and surface characterization techniques. c. 50–400 ppm concentration.	The maximum IE of 94.4% was observed at 400 mg L^{-1} at 30°C.	[79]

(Continued)

Table 20—1   (Continued)

Details		Reference
a. Alkaloids extract (AE) from *Geissospermum laeve*. b. *Electrochemical* and surface studies. c. 5–100 mg L^{-1} concentration.	The IE of 92% is reached with 100 mg L^{-1} of AE at 25°C.	[80]
a. *Watermelon* rind extract (WMRE). b. Electrochemical impedance spectroscopy (EIS) and potentiodynamic polarization measurements. c. 0.10–2 g L^{-1} concentration.	The IE of 83% observed at 2 g L^{-1}.	[81]
a. *Borage* flower. b. Weight loss, EIS and surface analysis. c. 200–800 ppm concentration.	Maximum IE of 91% observed at 800 ppm concentration.	[82]
a. *Ficus religiose* (leaf, bodhi tree) b. EIS and gravimetric measurements. c. 100–500 ppm concentration.	Maximum IE of 88.29% observed at 500 ppm concentration.	[83]
a. *Myristica fragrans* (nutmeg fruit). b. Weight loss and surface studies. c. 100–500 ppm concentration.	Maximum IE of 87.81% observed at 500 ppm concentration.	[84]
a. *Gongronema latifolium* (utazi, herb). b. Gasometric method. c. 50–1000ppm concentration.	Maximum IE of 96.5% observed at 1000 ppm concentration at 50°C.	[85]
a. *Cuscuta reflexa* (morning glory family, fruit extract). b. Weight loss and electrochemical measurement. c. 100–500 ppm concentration.	95.47% IE is obtained at 500 ppm concentration.	[86]
a. *Rosa canina* fruit. b. Electrochemical and surface analysis. c. 200–800 ppm concentration.	Maximum IE of 85.7% observed at 600 ppm concentration.	[87]
a. *Capsicum annuum* fruit paste. b. Weight loss, contact angle measurements and surface analysis. c. 50–250 ppm concentration.	96.48% IE observed at 250 ppm concentration.	[88]
a. Salicylaldeyde-Chitosan Schiff base. b. Weight loss, electrochemical impedance spectroscopy (EIS) and potentiodynamic polarization (PDP) methods, c. 30–150 mg L^{-1} concentration.	95.2% IE observed at 150 mg L^{-1} concentration.	[89]
a. Metformin drug. b. Weight loss and electrochemical measurement techniques. c. 100–1000 ppm concentration.	The IE of 92% was observed at 1000 ppm at 60°C.	[90]
a. *Chitosan-cinnamaldehyde* Schiff base. b. Weight loss test, electrochemical impedance measurements, and polarization studies c. 50–400 mg L^{-1} concentration.	The maximum IE of 87.92% observed at 400 mg L^{-1}.	[91]

in an acidic medium over the concentration range of $0.1-1.0$ g L^{-1} at 298K$-$328K by weight loss and electrochemical methods. The maximum IE of 95% is observed at 1 g L^{-1} at 298K. FTIR, SEM, and XRD confirmed the adsorption of guar gum on the metal surface. Theoretical studies (DFT and MD simulation) gave a better overview of the reactivity of guar gum and provided a good correlation with the studied techniques [44]. Table 20$-$1 demonstrates detailed information about green inhibitors.

20.4 Conclusions

The corrosion issue is curable by following some safety measures. This problem is not possible to remove completely but its disastrous and harmful effect can be stopped by taking some safety measures. In oil and gas industries various types of corrosion and their active corroding agents have been examined alongside the different ways of mitigating them. However, it is important to understand the principle and mechanism of corrosion in order to select the proper material, and to design to withstand and to manage/observe the operating conditions in a timely manner to enhance the economic life of metal and safety conditions in oil and gas operations. For decades conventional corrosion inhibitors have been in practice but due to their toxic behavior toward the environment, their usage has been restricted. Thus the use of environment-friendly natural inhibitors is in practice as they are effective, easily available, and economically feasible. Hence, in this chapter, the aim has been to provide knowledge concerning green corrosion inhibitors for protecting pipelines and other structural materials.

Useful links

https://www.mdpi.com/2076-3417/10/10/3389
https://onlinelibrary.wiley.com/doi/abs/10.1002/9783527822140.ch7
https://onlinelibrary.wiley.com/doi/abs/10.1002/9783527822140.ch8
https://www.sciencedirect.com/science/article/abs/pii/S016773221835373X

References

[1] B.O. Oyelami, A.A. Asere, Mathematical modeling: an application to corrosion in a petroleum industry, NMC Proceeding Work. Environ., 2005. http://www.industchem.com/content/4/1/35.

[2] Q. Feng, B. Yan, P. Chen, S.A. Shirazi, Failure analysis and simulation model of pinhole corrosion of the refined oil pipeline, Eng. Fail. Anal. (2019). Available from: https://doi.org/10.1016/j.engfailanal.2019.104177.

[3] W.C. Wang, R.H. Natelson, L.F. Stikeleather, W.L. Roberts, CFD simulation of transient stage of continuous countercurrent hydrolysis of canola oil, Comput. Chem. Eng. (2012). Available from: https://doi.org/10.1016/j.compchemeng.2012.04.008.

[4] M. Mahmoodian, C.Q. Li, Failure assessment and safe life prediction of corroded oil and gas pipelines, J. Pet. Sci. Eng. (2017). Available from: https://doi.org/10.1016/j.petrol.2016.12.029.

[5] M.R. Simmons, Report of offshore technology conference (OTC) presentation, 2008.

[6]   H.H. Zhang, Y. Chen, Z. Zhang, Comparative studies of two benzaldehyde thiosemicarbazone deriva-
      tives as corrosion inhibitors for mild steel in 1.0 M HCl, Results Phys. (2018). Available from: https://doi.
      org/10.1016/j.rinp.2018.09.038.

[7]   K.F. Al-Azawi, I.M. Mohammed, S.B. Al-Baghdadi, T.A. Salman, H.A. Issa, A.A. Al-Amiery, et al.,
      Experimental and quantum chemical simulations on the corrosion inhibition of mild steel by 3-((5-(3,5-
      dinitrophenyl)-1,3,4-thiadiazol-2-yl)imino)indolin-2-one, Results Phys. (2018). Available from: https://
      doi.org/10.1016/j.rinp.2018.02.055.

[8]   K. Zakaria, N.A. Negm, E.A. Khamis, E.A. Badr, Electrochemical and quantum chemical studies on car-
      bon steel corrosion protection in 1 M H_2SO_4 using new eco-friendly Schiff base metal complexes,
      J. Taiwan Inst. Chem. Eng. (2016). Available from: https://doi.org/10.1016/j.jtice.2015.12.021.

[9]   A. Ismail, H.M. Irshad, A. Zeino, I.H. Toor, Electrochemical corrosion performance of aromatic functio-
      nalized imidazole inhibitor under hydrodynamic conditions on API X65 carbon steel in 1 M HCl solu-
      tion, Arab. J. Sci. Eng. (2019). Available from: https://doi.org/10.1007/s13369-019-03745-6.

[10] B. Ramezanzadeh, H. Vakili, R. Amini, The effects of addition of poly(vinyl) alcohol (PVA) as a green
      corrosion inhibitor to the phosphate conversion coating on the anticorrosion and adhesion properties of
      the epoxy coating on the steel substrate, Appl. Surf. Sci. (2015). Available from: https://doi.org/10.1016/
      j.apsusc.2014.11.167.

[11] M. Srivastava, P. Tiwari, S.K. Srivastava, R. Prakash, G. Ji, Electrochemical investigation of Irbesartan
      drug molecules as an inhibitor of mild steel corrosion in 1 M HCl and 0.5 M H_2SO_4 solutions, J. Mol.
      Liq. (2017). Available from: https://doi.org/10.1016/j.molliq.2017.04.017.

[12] M. Mobin, M. Rizvi, Inhibitory effect of xanthan gum and synergistic surfactant additives for mild steel
      corrosion in 1 M HCl, Carbohydr. Polym. (2016). Available from: https://doi.org/10.1016/j.
      carbpol.2015.09.027.

[13] M. Mobin, M. Basik, J. Aslam, Boswellia serrata gum as highly efficient and sustainable corrosion inhibi-
      tor for low carbon steel in 1 M HCl solution: experimental and DFT studies, J. Mol. Liq. 263 (2018)
      174−186. Available from: https://doi.org/10.1016/j.molliq.2018.04.150.

[14] M. Basik, M. Mobin, M. Shoeb, Cysteine-silver-gold nanocomposite as potential stable green corrosion
      inhibitor for mild steel under acidic condition, Sci. Rep. 10 (2020) 1−12. Available from: https://doi.org/
      10.1038/s41598-019-57181-5.

[15] M. Parveen, M. Mobin, S. Zehra, R. Aslam, L-proline mixed with sodium benzoate as sustainable inhibi-
      tor for mild steel corrosion in 1M HCl: an experimental and theoretical approach, Sci. Rep. (2018).
      Available from: https://doi.org/10.1038/s41598-018-24143-2.

[16] M. Mobin, M. Basik, M. Shoeb, A novel organic-inorganic hybrid complex based on Cissus quadrangu-
      laris plant extract and zirconium acetate as a green inhibitor for mild steel in 1 M HCl solution, Appl.
      Surf. Sci. 469 (2019) 387−403. Available from: https://doi.org/10.1016/j.apsusc.2018.11.008.

[17] S. Mo, L.J. Li, H.Q. Luo, N.B. Li, An example of green copper corrosion inhibitors derived from flavor
      and medicine: vanillin and isoniazid, J. Mol. Liq. (2017). Available from: https://doi.org/10.1016/j.
      molliq.2017.07.081.

[18] M.V. Diamanti, U.V. Velardi, A. Brenna, A. Mele, M.P. Pedeferri, M. Ormellese, Compatibility of
      imidazolium-based ionic liquids for CO_2 capture with steel alloys: a corrosion perspective, Electrochim.
      Acta (2016). Available from: https://doi.org/10.1016/j.electacta.2016.02.003.

[19] M.M. Solomon, H. Gerengi, S.A. Umoren, N.B. Essien, U.B. Essien, E. Kaya, Gum Arabic-silver nanopar-
      ticles composite as a green anticorrosive formulation for steel corrosion in strong acid media,
      Carbohydr. Polym. (2018). Available from: https://doi.org/10.1016/j.carbpol.2017.10.051.

[20] M. Finšgar, J. Jackson, Application of corrosion inhibitors for steels in acidic media for the oil and gas indus-
      try: a review, Corros. Sci. 86 (2014) 17−41. Available from: https://doi.org/10.1016/j.corsci.2014.04.044.

[21] S. Ghareba, S. Omanovic, Interaction of 12-aminododecanoic acid with a carbon steel surface: towards the development of "green" corrosion inhibitors, Corros. Sci. (2010). Available from: https://doi.org/10.1016/j.corsci.2010.02.019.

[22] R. Heidersbach, Corrosion in oil and gas production, Microbiologically Influenced Corrosion in the Upstream Oil and Gas Industry, 2017, pp. 3–34. Available from: http://doi.org/10.1201/9781315157818.

[23] A.J. Szyprowski, Methods of investigation on hydrogen sulfide corrosion of steel and its inhibitors, Corrosion (2003). Available from: https://doi.org/10.5006/1.3277538.

[24] E. Olvera, J. Mendoza-Flores, J. Genesca, CO2 corrosion control in steel pipelines. Influence of turbulent flow on the performance of corrosion inhibitors, J. Loss Prev. Process. Ind. 35 (2015) 19–28. Available from: https://doi.org/10.1016/j.jlp.2015.03.006.

[25] X. Dong, H. Liu, Z. Chen, K. Wu, N. Lu, Q. Zhang, Enhanced oil recovery techniques for heavy oil and oilsands reservoirs after steam injection, Appl. Energy (2019). Available from: https://doi.org/10.1016/j.apenergy.2019.01.244.

[26] K. Nalli, Appendix VI: corrosion and its mitigation in the oil and gas industries, Process Plant Equip. (2012) 673–679. Available from: https://doi.org/10.1002/9781118162569.app6.

[27] N.A. Negm, N.G. Kandile, I.A. Aiad, M.A. Mohammad, New eco-friendly cationic surfactants: synthesis, characterization and applicability as corrosion inhibitors for carbon steel in 1N HCl, Colloids Surf. A Physicochem. Eng. Asp. (2011). Available from: https://doi.org/10.1016/j.colsurfa.2011.09.032.

[28] M.H. Hussin, A. Abdul, M. Nasir, M. Ibrahim, N. Brosse, The capability of ultrafiltrated alkaline and organosolv oil palm (*Elaeis guineensis*) fronds lignin as green corrosion inhibitor for mild steel in 0.5 M HCl solution, Measurement 78 (2016) 90–103. Available from: https://doi.org/10.1016/j.measurement.2015.10.007.

[29] M. Bethencourt, F.J. Botana, J.J. Calvino, M. Marcos, M.A. Rodríguez-Chacón, Lanthanide compounds as environmentally-friendly corrosion inhibitors of aluminium alloys: a review, Corros. Sci. (1998). Available from: https://doi.org/10.1016/S0010-938X(98)00077-8.

[30] P. Karungamye, H.C. Ananda Murthy, Methanolic extracts of *Adansonia digitata* (Baobab) fruit pulp and seeds as potential green inhibitors for mild steel corrosion in 0.5 M H₂SO₄ solution, J. Adv. Chem. Sci. (2017). Available from: https://doi.org/10.22607/IJACS.2017.5040010.

[31] A.S. Yaro, A.A. Khadom, R.K. Wael, Apricot juice as green corrosion inhibitor of mild steel in phosphoric acid, Alex. Eng. J. (2013). Available from: https://doi.org/10.1016/j.aej.2012.11.001.

[32] M. Mobin, M. Basik, J. Aslam, Pineapple stem extract (Bromelain) as an environmental friendly novel corrosion inhibitor for low carbon steel in 1 M HCl, Measurement. 134 (2019) 595–605. Available from: https://doi.org/10.1016/j.measurement.2018.11.003.

[33] P. Muthukrishnan, P. Prakash, B. Jeyaprabha, K. Shankar, Stigmasterol extracted from *Ficus hispida* leaves as a green inhibitor for the mild steel corrosion in 1M HCl solution, Arab. J. Chem. (2015). Available from: https://doi.org/10.1016/j.arabjc.2015.09.005.

[34] M. Mobin, M. Basik, Y. El Aoufir, Corrosion mitigation of mild steel in acidic medium using *Lagerstroemia speciosa* leaf extract: a combined experimental and theoretical approach, J. Mol. Liq. 286 (2019). Available from: https://doi.org/10.1016/j.molliq.2019.110890.

[35] K.K. Anupama, K. Ramya, A. Joseph, Electrochemical measurements and theoretical calculations on the inhibitive interaction of *Plectranthus amboinicus* leaf extract with mild steel in hydrochloric acid, Measurement. (2017). Available from: https://doi.org/10.1016/j.measurement.2016.10.030.

[36] M. Mobin, M. Rizvi, Inhibitory effect of xanthan gum and synergistic surfactant additives for mild steel corrosion in 1M HCl, Carbohydr. Polym. 136 (2015) 384–393. Available from: https://doi.org/10.1016/j.carbpol.2015.09.027.

[37] M. Basik, M. Mobin, Chondroitin sulfate as potent green corrosion inhibitor for mild steel in 1 M HCl, J. Mol. Struct. (2020). Available from: https://doi.org/10.1016/j.molstruc.2020.128231.

[38] R.S. Varma, Greener and sustainable trends in synthesis of organics and nanomaterials, ACS Sustain. Chem. Eng. (2016). Available from: https://doi.org/10.1021/acssuschemeng.6b01623.

[39] N.A.N. Mohamad, N.A. Arham, J. Jai, A. Hadi, Plant extract as reducing agent in synthesis of metallic nanoparticles: a review, Adv. Mater. Res. (2014). Available from: https://doi.org/10.4028/www.scientific.net/AMR.832.350.

[40] M. Nasrollahzadeh, S.M. Sajadi, M. Khalaj, Green synthesis of copper nanoparticles using aqueous extract of the leaves of *Euphorbia esula* L and their catalytic activity for ligand-free Ullmann-coupling reaction and reduction of 4-nitrophenol, RSC Adv. (2014). Available from: https://doi.org/10.1039/c4ra08863h.

[41] S. Bose, L. Fatima, H. Mereyala, Green chemistry approaches to the synthesis of 5-alkoxycarbonyl-4-aryl-3,4-dihydropyrimidin-2(1H)-ones by a three-component coupling of one-pot condensation reaction: comparison of ethanol, water, and solvent-free conditions, J. Org. Chem. 68 (2003) 587−590. Available from: https://doi.org/10.1021/jo0205199.

[42] H. Duan, D. Wang, Y. Li, Green chemistry for nanoparticle synthesis, Chem. Soc. Rev. (2015). Available from: https://doi.org/10.1039/c4cs00363b.

[43] P.O. Ameh, Corrosion inhibition and adsorption behaviour for mild steel by *Ficus glumosa* gum in H_2SO_4 solution, Afr. J. Pure Appl. Chem. (2012). Available from: https://doi.org/10.5897/ajpac12.001.

[44] M. Messali, H. Lgaz, R. Dassanayake, R. Salghi, S. Jodeh, N. Abidi, et al., Guar gum as efficient non-toxic inhibitor of carbon steel corrosion in phosphoric acid medium: electrochemical, surface, DFT and MD simulations studies, J. Mol. Struct. 1145 (2017) 43−54. Available from: https://doi.org/10.1016/j.molstruc.2017.05.081.

[45] M. Solomon, H. Gerengi, S. Umoren, Carboxymethyl cellulose/silver nanoparticles composite: synthesis, characterization and application as a benign corrosion inhibitor for St37 steel in 15% H_2SO_4 medium, ACS Appl. Mater. Interfaces 9 (2017). Available from: https://doi.org/10.1021/acsami.6b14153.

[46] E. Salehi, R. Naderi, B. Ramezanzadeh, Synthesis and characterization of an effective organic/inorganic hybrid green corrosion inhibitive complex based on zinc acetate/*Urtica dioica*, Appl. Surf. Sci. 396 (2017) 1499−1514. Available from: https://doi.org/10.1016/j.apsusc.2016.11.198.

[47] M. Mobin, M.A. Khan, Investigation on the adsorption and corrosion inhibition behavior of gum acacia and synergistic surfactants additives on mild steel in 0.1 M H_2SO_4, J. Dispers. Sci. Technol. 34 (2013) 1496−1506. Available from: https://doi.org/10.1080/01932691.2012.751031.

[48] S.A. Umoren, U.F. Ekanem, Inhibition of mild steel corrosion in H_2SO_4 using exudate gum from pachylobus edulis and synergistic potassium halide additives, Chem. Eng. Commun. 197 (2010) 1339−1356. Available from: https://doi.org/10.1080/00986441003626086.

[49] H. Kumar, V. Yadav, *Musa acuminata* (green corrosion inhibitor) as anti-pit and anti-cracking agent for mild steel in 5M hydrochloric acid solution, Chem. Data Collect. 29 (2020) 100500. Available from: https://doi.org/10.1016/j.cdc.2020.100500.

[50] A. Dehghani, G. Bahlakeh, B. Ramezanzadeh, M. Ramezanzadeh, Potential role of a novel green eco-friendly inhibitor in corrosion inhibition of mild steel in HCl solution: detailed macro/micro-scale experimental and computational explorations, Constr. Build. Mater. 245 (2020) 118464. Available from: https://doi.org/10.1016/j.conbuildmat.2020.118464.

[51] A. Berrissoul, E. Loukili, N. Mechbal, F. Benhiba, A. Guenbour, B. Dikici, et al., Anticorrosion effect of a green sustainable inhibitor on mild steel in hydrochloric acid, J. Colloid Interface Sci. 580 (2020) 740−752. Available from: https://doi.org/10.1016/j.jcis.2020.07.073.

[52] A.H. Tantawy, K.A. Soliman, H.M. Abd El-Lateef, Novel synthesized cationic surfactants based on natural *Piper nigrum* as sustainable-green inhibitors for steel pipeline corrosion in CO2-3.5%NaCl: DFT, Monte Carlo simulations and experimental approaches, J. Clean. Prod. 250 (2020) 119510. Available from: https://doi.org/10.1016/j.jclepro.2019.119510.

[53] K. Muthukumarasamy, S. Pitchai, K. Devarayan, L. Nallathambi, Adsorption and corrosion inhibition performance of *Tunbergia fragrans* extract on mild steel in acid medium, Mater. Today Proc. (2020) 1−5. Available from: https://doi.org/10.1016/j.matpr.2020.06.533.

[54] A. Dehghani, G. Bahlakeh, B. Ramezanzadeh, M. Ramezanzadeh, Integrated modeling and electrochemical study of Myrobalan extract for mild steel corrosion retardation in acidizing media, J. Mol. Liq. (2020). Available from: https://doi.org/10.1016/j.molliq.2019.112046.

[55] S. Chen, S. Chen, B. Zhu, C. Huang, W. Li, *Magnolia grandiflora* leaves extract as a novel environmentally friendly inhibitor for Q235 steel corrosion in 1 M HCl: combining experimental and theoretical researches, J. Mol. Liq. 311 (2020) 113312. Available from: https://doi.org/10.1016/j.molliq.2020.113312.

[56] M.P. Asfia, M. Rezaei, G. Bahlakeh, Corrosion prevention of AISI 304 stainless steel in hydrochloric acid medium using garlic extract as a green corrosion inhibitor: electrochemical and theoretical studies, J. Mol. Liq. 315 (2020) 113679. Available from: https://doi.org/10.1016/j.molliq.2020.113679.

[57] A.A. Fadhil, A.A. Khadom, S.K. Ahmed, H. Liu, C. Fu, H.B. Mahood, *Portulaca grandiflora* as new green corrosion inhibitor for mild steel protection in hydrochloric acid: quantitative, electrochemical, surface and spectroscopic investigations, Surf. Interfaces 20 (2020) 100595. Available from: https://doi.org/10.1016/j.surfin.2020.100595.

[58] A. Berrissoul, A. Ouarhach, F. Benhiba, A. Romane, A. Zarrouk, A. Guenbour, et al., Evaluation of *Lavandula mairei* extract as green inhibitor for mild steel corrosion in 1 M HCl solution. Experimental and theoretical approach, J. Mol. Liq. 313 (2020) 113493. Available from: https://doi.org/10.1016/j.molliq.2020.113493.

[59] M.A. Bidi, M. Azadi, M. Rassouli, A new green inhibitor for lowering the corrosion rate of carbon steel in 1 M HCl solution: Hyalomma tick extract, Mater. Today Commun. 24 (2020) 100996. Available from: https://doi.org/10.1016/j.mtcomm.2020.100996.

[60] A. Dehghani, G. Bahlakeh, B. Ramezanzadeh, M. Ramezanzadeh, *Aloysia citrodora* leaves extract corrosion retardation effect on mild-steel in acidic solution: molecular/atomic scales and electrochemical explorations, J. Mol. Liq. 310 (2020) 113221. Available from: https://doi.org/10.1016/j.molliq.2020.113221.

[61] O.O. Ogunleye, A.O. Arinkoola, O.A. Eletta, O.O. Agbede, Y.A. Osho, A.F. Morakinyo, et al., Green corrosion inhibition and adsorption characteristics of *Luffa cylindrica* leaf extract on mild steel in hydrochloric acid environment, Heliyon 6 (2020) e03205. Available from: https://doi.org/10.1016/j.heliyon.2020.e03205.

[62] A. Thomas, M. Prajila, K.M. Shainy, A. Joseph, A green approach to corrosion inhibition of mild steel in hydrochloric acid using fruit rind extract of *Garcinia indica* (Binda), J. Mol. Liq. 312 (2020) 113369. Available from: https://doi.org/10.1016/j.molliq.2020.113369.

[63] N. Asadi, M. Ramezanzadeh, G. Bahlakeh, B. Ramezanzadeh, Utilizing Lemon Balm extract as an effective green corrosion inhibitor for mild steel in 1M HCl solution: a detailed experimental, molecular dynamics, Monte Carlo and quantum mechanics study, J. Taiwan Inst. Chem. Eng. 95 (2019) 252−272. Available from: https://doi.org/10.1016/j.jtice.2018.07.011.

[64] G. Bahlakeh, B. Ramezanzadeh, A. Dehghani, M. Ramezanzadeh, Novel cost-effective and high-performance green inhibitor based on aqueous *Peganum harmala* seed extract for mild steel corrosion in HCl solution: detailed experimental and electronic/atomic level computational explorations, J. Mol. Liq. (2019). Available from: https://doi.org/10.1016/j.molliq.2019.03.086.

[65] G. Bahlakeh, A. Dehghani, B. Ramezanzadeh, M. Ramezanzadeh, Highly effective mild steel corrosion inhibition in 1 M HCl solution by novel green aqueous Mustard seed extract: experimental, electronic-scale DFT and atomic-scale MC/MD explorations, J. Mol. Liq. 293 (2019) 111559. Available from: https://doi.org/10.1016/j.molliq.2019.111559.

[66] M. Benarioua, A. Mihi, N. Bouzeghaia, M. Naoun, Mild steel corrosion inhibition by Parsley (Petroselium Sativum) extract in acidic media, Egypt. J. Pet. 28 (2019) 155−159. Available from: https://doi.org/10.1016/j.ejpe.2019.01.001.

[67] A. Dehghani, G. Bahlakeh, B. Ramezanzadeh, Green eucalyptus leaf extract: a potent source of bio-active corrosion inhibitors for mild steel, Bioelectrochemistry 130 (2019) 107339. Available from: https://doi.org/10.1016/j.bioelechem.2019.107339.

[68] C.M. Fernandes, T. da, S. Ferreira Fagundes, N. Escarpini dos Santos, T. Shewry de, M. Rocha, et al., *Ircinia strobilina* crude extract as corrosion inhibitor for mild steel in acid medium, Electrochim. Acta. 312 (2019) 137−148. Available from: https://doi.org/10.1016/j.electacta.2019.04.148.

[69] S. Nikpour, M. Ramezanzadeh, G. Bahlakeh, B. Ramezanzadeh, M. Mahdavian, Eriobotrya japonica Lindl leaves extract application for effective corrosion mitigation of mild steel in HCl solution: experimental and computational studies, Constr. Build. Mater. (2019). Available from: https://doi.org/10.1016/j.conbuildmat.2019.06.005.

[70] E. Alibakhshi, M. Ramezanzadeh, G. Bahlakeh, B. Ramezanzadeh, M. Mahdavian, M. Motamedi, *Glycyrrhiza glabra* leaves extract as a green corrosion inhibitor for mild steel in 1 M hydrochloric acid solution: experimental, molecular dynamics, Monte Carlo and quantum mechanics study, J. Mol. Liq. (2018). Available from: https://doi.org/10.1016/j.molliq.2018.01.144.

[71] H. Hassannejad, A. Nouri, Sunflower seed hull extract as a novel green corrosion inhibitor for mild steel in HCl solution, J. Mol. Liq. (2018). Available from: https://doi.org/10.1016/j.molliq.2018.01.142.

[72] A.A. Khadom, A.N. Abd, N.A. Ahmed, *Xanthium strumarium* leaves extracts as a friendly corrosion inhibitor of low carbon steel in hydrochloric acid: kinetics and mathematical studies, South. Afr. J. Chem. Eng. (2018). Available from: https://doi.org/10.1016/j.sajce.2017.11.002.

[73] P. Muthukrishnan, B. Jeyaprabha, P. Prakash, Adsorption and corrosion inhibiting behavior of *Lannea coromandelica* leaf extract on mild steel corrosion, Arab. J. Chem. (2017). Available from: https://doi.org/10.1016/j.arabjc.2013.08.011.

[74] E. Ituen, O. Akaranta, A. James, S. Sun, Green and sustainable local biomaterials for oilfield chemicals: *Griffonia simplicifolia* extract as steel corrosion inhibitor in hydrochloric acid, Sustain. Mater. Technol. 11 (2017) 12−18. Available from: https://doi.org/10.1016/j.susmat.2016.12.001.

[75] L.L. Liao, S. Mo, H.Q. Luo, N.B. Li, Longan seed and peel as environmentally friendly corrosion inhibitor for mild steel in acid solution: experimental and theoretical studies, J. Colloid Interface Sci. 499 (2017) 110−119. Available from: https://doi.org/10.1016/j.jcis.2017.03.091.

[76] K.K. Alaneme, S.J. Olusegun, A.W. Alo, Co rrosion inhibitory properties of elephant grass (*Pennisetum purpureum*) extract: effect on mild steel corrosion in 1 M HCl solution, Alex. Eng. J. 55 (2016) 1069−1076. Available from: https://doi.org/10.1016/j.aej.2016.03.012.

[77] M. Jokar, T.S. Farahani, B. Ramezanzadeh, Electrochemical and surface characterizations of morus alba pendula leaves extract (MAPLE) as a green corrosion inhibitor for steel in 1M HCl, J. Taiwan Inst. Chem. Eng. 63 (2016) 436−452. Available from: https://doi.org/10.1016/j.jtice.2016.02.027.

[78] M. Prabakaran, S.H. Kim, K. Kalaiselvi, V. Hemapriya, I.M. Chung, Highly efficient *Ligularia fischeri* green extract for the protection against corrosion of mild steel in acidic medium: electrochemical and spectroscopic investigations, J. Taiwan Inst. Chem. Eng. 59 (2016) 553−562. Available from: https://doi.org/10.1016/j.jtice.2015.08.023.

[79] N. El Hamdani, R. Fdil, M. Tourabi, C. Jama, F. Bentiss, Alkaloids extract of *Retama monosperma* (L.) Boiss. seeds used as novel eco-friendly inhibitor for carbon steel corrosion in 1 M HCl solution: electrochemical and surface studies, Appl. Surf. Sci. (2015). Available from: https://doi.org/10.1016/j.apsusc.2015.09.159.

[80] M. Faustin, A. Maciuk, P. Salvin, C. Roos, M. Lebrini, Corrosion inhibition of C38 steel by alkaloids extract of *Geissospermum laeve* in 1M hydrochloric acid: electrochemical and phytochemical studies, Corros. Sci. (2015). Available from: https://doi.org/10.1016/j.corsci.2014.12.005.

[81] N.A. Odewunmi, S.A. Umoren, Z.M. Gasem, Utilization of watermelon rind extract as a green corrosion inhibitor for mild steel in acidic media, J. Ind. Eng. Chem. 21 (2015) 239−247. Available from: https://doi.org/10.1016/j.jiec.2014.02.030.

[82] A. Dehghani, G. Bahlakeh, B. Ramezanzadeh, M. Ramezanzadeh, Potential of Borage flower aqueous extract as an environmentally sustainable corrosion inhibitor for acid corrosion of mild steel: electrochemical and theoretical studies, J. Mol. Liq. (2019). Available from: https://doi.org/10.1016/j.molliq.2019.01.008.

[83] R. Haldhar, D. Prasad, A. Saxena, R. Kumar, Experimental and theoretical studies of Ficus religiosa as green corrosion inhibitor for mild steel in 0.5 M H_2SO_4 solution, Sustain. Chem. Pharm. (2018). Available from: https://doi.org/10.1016/j.scp.2018.07.002.

[84] R. Haldhar, D. Prasad, A. Saxena, *Myristica fragrans* extract as an eco-friendly corrosion inhibitor for mild steel in 0.5 M H2SO4 solution, J. Environ. Chem. Eng. (2018). Available from: https://doi.org/10.1016/j.jece.2018.03.023.

[85] A.I. Ikeuba, P.C. Okafor, Green corrosion protection for mild steel in acidic media: saponins and crude extracts of *Gongronema latifolium*, Pigment. Resin Technol. (2019). Available from: https://doi.org/10.1108/PRT-03-2018-0020.

[86] A. Saxena, D. Prasad, R. Haldhar, Investigation of corrosion inhibition effect and adsorption activities of *Cuscuta reflexa* extract for mild steel in 0.5 M H_2SO_4, Bioelectrochemistry (2018). Available from: https://doi.org/10.1016/j.bioelechem.2018.07.006.

[87] Z. Sanaei, M. Ramezanzadeh, G. Bahlakeh, B. Ramezanzadeh, Use of *Rosa canina* fruit extract as a green corrosion inhibitor for mild steel in 1 M HCl solution: a complementary experimental, molecular dynamics and quantum mechanics investigation, J. Ind. Eng. Chem. (2019). Available from: https://doi.org/10.1016/j.jiec.2018.09.013.

[88] C.M. Reddy, B.D. Sanketi, S. Narendra Kumar, Corrosion inhibition of mild steel by *Capsicum annuum* fruit paste, Perspect. Sci. (2016). Available from: https://doi.org/10.1016/j.pisc.2016.06.033.

[89] K.R. Ansari, D.S. Chauhan, M.A. Quraishi, M.A.J. Mazumder, A. Singh, Chitosan Schiff base: an environmentally benign biological macromolecule as a new corrosion inhibitor for oil & gas industries, Int. J. Biol. Macromol. 144 (2020) 305–315. Available from: https://doi.org/10.1016/j.ijbiomac.2019.12.106.

[90] K. Haruna, T.A. Saleh, M.A. Quraishi, Expired metformin drug as green corrosion inhibitor for simulated oil/gas well acidizing environment, J. Mol. Liq. 315 (2020) 113716. Available from: https://doi.org/10.1016/j.molliq.2020.113716.

[91] D.S. Chauhan, M.A.J. Mazumder, M.A. Quraishi, K.R. Ansari, Chitosan-cinnamaldehyde Schiff base: a bioinspired macromolecule as corrosion inhibitor for oil and gas industry, Int. J. Biol. Macromol. 158 (2020) 127–138. Available from: https://doi.org/10.1016/j.ijbiomac.2020.04.200.

21

Corrosion inhibitors for high temperature corrosion in oil and gas industry

Jeenat Aslam[1], Ruby Aslam[2]

[1]DEPARTMENT OF CHEMISTRY, COLLEGE OF SCIENCE, YANBU, TAIBAH UNIVERSITY, AL-MADINA, SAUDI ARABIA [2]CORROSION RESEARCH LABORATORY, DEPARTMENT OF APPLIED CHEMISTRY, FACULTY OF ENGINEERING AND TECHNOLOGY, ALIGARH MUSLIM UNIVERSITY, ALIGARH, INDIA

Chapter outline

21.1 Introduction

Corrosion is a stochastic process that involves interdisciplinary principles including surface science, thermodynamics, electrochemistry, and kinetics, hydrodynamics, mechanics, and chemistry, including metallurgy/materials science. It costs the oil and gas industry tens of billions of dollars in lost income and the cost of treatment per year [1]. In addition, high-temperature corrosion is among the most critical problems in the handling of materials, structure design, and life span assessment of engineering parts in the oil and gas industry that are exposed to high-temperature environments. Corrosion at high temperatures is the degradation of materials due to salt or ash deposition. The formation of corrosion products such as oxides, nitrides, carbides, sulfides, or their mixtures in general results in a loss of cross-sectional load bearing, decreases reliability and stability, and consequently shortens the service life of engineering components [2]. The inward diffusion of oxygen, nitrogen, sulfur, or carbon can also result in unwanted deposition of compounds that are typically brittle in nature. High-temperature corrosion thus deteriorates materials and therefore has an effect on the efficiency of engineering components and on environmental factors, such as global warming and air pollution, directly or indirectly.

The selection of inhibitors for high-temperature corrosions are about the most difficult inhibitors to formulate owing to the complex field environment fixed with high reactions rate at high temperature. Additionally, the selection of suitable inhibitors and the laboratory simulations for measuring the essential performance of inhibitors at high temperature are also extremely demanding as the processes of evaluation appears complex [3−5]. Therefore attempts made near corrosion inhibitors evaluation at high temperatures have been restricted despite the high requirements of inhibitors in oil and gas industries. Furthermore, problems in understanding the high-temperature corrosion mechanisms in order to match its mitigation with suitable inhibitors did not assist the condition either [6,7].

Despite the inherent challenges, several developments have been made in the earlier period in order to consider the mechanisms at which the metals corrode at high temperature in addition to formulating numerous inhibitors for mitigation [8]. Rather some effective inhibitors for the high-temperature corrosion have been patented [9−13]. It is common to use inhibitors to combat such corrosion. Different acetylenic alcohols, fluorinated surfactants, quaternary derivatives of heterocyclic nitrogen bases, halo methylated aromatic compounds, formamides, and surface active agents, alone or in combination with other materials, are the materials that have been proposed for the purpose in the past [14].

In the present chapter, the mechanism, diverse operating environments and the development in the formulation of corrosion inhibitors for high temperature corrosion in oil and gas industry has been discussed.

21.2 Various forms of high-temperature corrosion

Corrosion occurs when different metals are exposed to the oil and gas environment at raised temperature, also called high-temperature corrosion. It might also be identified as high-temperature tarnishing, or high-temperature scaling, high-temperature oxidation. High-temperature corrosion arises in the absence of liquid electrolyte. Though it might begin at around $\sim 250°C$, it usually happens at temperatures above $\sim 450°C$. The reaction products of gas and metal usually deposit onto the surface of metal as a scale. The scale is protective when: it forms as a solid that fully wraps the surface of metal; it adheres onto the surface of metal; it has a high melting point so that it withstands higher-temperature; and species diffusion (e.g., metal cations or other anions) across.

Once a scale with the above features is produced, the rate of corrosion gradually reduces with time. Conversely, if the scale is noncompact, metal anions and cations can diffuse across it. The rate at which ions diffuse across the scale depends on its porosity as well as the species diffusion rate, electrical potential difference, and concentration gradient across it. The protectiveness of the scale often relies on temperature. The scale is protective below its melting point and therefore the rate of corrosion will gradually decrease over time under this condition. Above its melting point, the scale becomes nonprotective and under this condition the rate of corrosion linearly increases with time.

Depending on the type of scale, it is possible to categorized high-temperature corrosion as (1) sulfidation, (2) oxidation, (3) chlorination, (4) nitridation, or (5) carburization. This categorization is based on the behavior of stable surface product produced on the metal at the end of the reaction in the presence of the hot gases.

21.2.1 Sulfidation

The sulfidation process is commonly used to transform oxides into sulfides, but it is also susceptible to corrosion and surface modification. Sulfidation is based on the formation of metal-sulfur scales in the presence of H_2S or another sulfur-containing gaseous atmosphere. Most sulfur-containing atmospheres also include oxygen; hence both the sulfide and oxide scales form together. The kinetics of the formation of oxide scales is slower than those of the formation of sulfide scales. The melting point of the oxide scale is greater than that of the sulfide scale, so the formation of the sulfide scale contributes to increased corrosion.

21.2.2 Oxidation

Oxidation is one of the most general forms of high-temperature corrosion. In the presence of oxygen, air, and steam, metals undergo oxidation to produce oxide scales. The protectiveness of the oxide scale relies on the temperature, climate, metal, and other impurities. For instance, Fig. 21−1 indicates the effect of the chromium content of steel on the category of oxides formed at high temperatures [15].

21.2.3 Chlorination

Chlorination is the metallic chlorides formation process. The metallic chlorides are unstable due to their significant vapor pressures at high temperatures; thus their formation extinguishes otherwise in constant oxide scale. Additionally, in the gaseous phase, the oxygen reacts with metallic chlorides to form chlorine gas, which increases the localized corrosion via extinguishing the metal oxide scale:

$$2FeCl_2 + O_2 \rightarrow Cl_2 + 2FeO \tag{21.1}$$

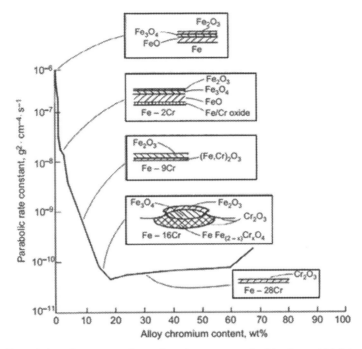

FIGURE 21–1 The effect of chromium content in steel on the category of oxides form at high-temperature. Source: *S. Papavinasam, Chapter 5—Mechanisms, in: S. Papavinasam (Ed.), Corrosion Control in the Oil and Gas Industry, Gulf Professional Publishing, 2014, pp. 249–300.*

21.2.4 Nitridation

The development of metal nitrides is called nitridation. Nitridation arises in the presence of ammonia or nitrogen. The kinetics of nitridation in a nitrogen atmosphere is higher than that of an ammonia atmosphere. The nitridation increases corrosion owing to the metallic nitride's precipitation, for example, nitrides of titanium, aluminum, chromium, and niobium. Metal dusting in stainless steels is caused by nitridation.

21.2.5 Carburization

Carburization is the formation of metallic carbides via the reaction between carbon and metal at temperatures of $\sim700°C–800°C$. Carburization generally happens in hydrocarbon-comprising environments. The high-temperature hydrocarbon (methane) decomposition forms carbon. Under most situations, the surface of metal is protected via oxide scale; though in hydrocarbon environments carburization happens at temperatures of $\sim700°C–850°C$.

$$CH_4 \rightarrow 2H_2 + C \tag{21.2}$$

In recent decades, a rising demand for high temperature corrosion inhibitors can be seen [4,5,16,17]. The causes are the rising drilling depth with increasingly hostile atmospheres, and moreover, the systems of corrosion inhibitor have to become more efficient and more complex. A few corrosion inhibitor formulations for high-temperature applications exist in the market to serve this demand, but most of the products start to fail at temperatures higher than 250°C [18]. The most generally utilized organic inhibitors are still fatty imidazolines prepared from polyamines and fatty acids. Their derivatives (salts) are also utilized. These products start to decompose at 250°C due to the imidazoline ring structure being thermally unstable. The decomposition causes performance loss and can have negative effects such as an increase in viscosity and the formation of insoluble deposits.

High temperatures are able to increase the inhibitors desorption and rate of corrosion, hence ensuing the decrease the inhibition efficiency [19,20]. In order to tackle high-temperature corrosion in the oil and gas industry, it is therefore primarily important to select and improve inhibitor properties to match the aggressive environment.

Herein, we discuss some nonpatented and patented corrosion inhibitors formulations used in different areas of the oil and gas industries where various mechanisms of high-temperature corrosion take place under the following heads. Table 21−1 represents the high-temperature corrosion inhibitors studied.

21.3 Well acidization inhibitors

The acidizing of wellbore is an inspiring exercise in the oil and gas industry. It is a procedure that liquefies rocks and scale blockages with hot acids in order to permit the oil and gas to reach near the oil well. For this process, hydrochloric acid (HCl) or a combination of HCl and hydrofluoric acid (HF) are commonly used to remove blockages and dissolve rocks, but at the same time they quite easily corrode the casing, tubing, and downhole equipment. This condition is bad in the high-temperature oil wells, and thus acidization sometimes is not successful or is not used. The deep wells can operate at temperatures up to ∼200°C or higher due to the primary well's exhaustion. Thus, to combat the high-temperature corrosion, some corrosion inhibitors and their formulations can be applied.

Chen [21] investigated imidazoline and its precursor amide as a highly effective inhibitor for use in the oil and gas industries. The efficiency of their high-temperature corrosion inhibition was assessed and compared. The results show that at 65.5°C, imidazoline and amide give an inhibition efficiency greater than 95%. The minimum concentration of inhibitors required to provide maximum protection for imidazoline and amide depends on the temperature. The concentration needed at a high temperature of 65°C for imidazoline and amide is approximately 40 times that at 65.5°C. At 65.5°C, to give 95% maximum protection, a minimum of 25 ppm inhibitor concentration is required; at 148.8°C/3000 psig, it requires 1000 ppm to give a maximum of 90% protection. However, amide tends to have a somewhat better performance at high temperatures, suggesting better long-term thermal stability.

Table 21–1   Some reported high-temperature corrosion inhibitors.

S. no.	Inhibitor	Experimental conditions	Maximum protection	Reference
1.	Imidazoline derivative	$T = 148.8°C$, Pressure = 3000 psig, oil and gas steel, 25 ppm concentration	I.E. 90%	[21]
3.	3-Undecane-4-aryl-5-mercapto-1,2,4-triazole (triazole 1)	Oil well steel (N80) and cold rolled mild steel, 15% HCl, $T = 105°C$, 5000 ppm concentration	I.E. 60.2%	[22]
4.	3(heptadeca-8-ene)-4-aryl-5-mercapto-1,2,4-triazole (triazole 2)	Oil well steel (N80) and cold rolled mild steel, 15% HCl, T = 105°C, 5000ppm concentration	I.E. 90.5%	[22]
5.	3(deca-9-ene)-4-aryl-5-mercapto-1,2—4-triazole (triazole 3)	Oil well steel (N80) and cold rolled mild steel, 15% HCl, $T = 105°C$, 5000 ppm concentration	I.E. 96.2%	[22]
6.	2-undecane-5-mercapto-1-oxa-3,4-diazole (UMOD)	N80 Mild steel, 15% HCl, $T = 105°C$, 500 ppm concentration	I.E. 98.9%	[23]
7.	2-heptadecene-5-mercapto-1-oxa-3,4-diazole (HMOD)	N80 Mild steel, 15% HCl, $T = 105°C$, 500 ppm concentration	I.E. 69.1%	[23]
8.	2-decene-5-mercapto-1-oxa-3,4-diazole (DMOD)	N80 Mild steel, 15% HCl, $T = 105°C$, 500 ppm concentration	I.E. 97.7%	[23]
9.	Lauric acid hydrazide (LAH)	N80 oil well steel, mild steel, 15% HCl, $T = 105°C$, 5000 ppm concentration	I.E. 71.3%	[24]
10.	Oleic acid hydrazide (OAH)	N80 oil well steel, mild steel, 15% HCl, $T = 105°C$, 5000 ppm concentration	I.E. 84.4%	[24]
11.	Undecenoic acid hydrazide (UAH)	N80 oil well steel, mild steel, 15% HCl, $T = 105°C$, 5000 ppm concentration	I.E. 90.4%	[24]
12.	1-Undecane-4-phenyl-thiosemicarbazide (UPTS)	N80 oil well steel, mild steel, 15% HCl, $T = 105°C$, 5000 ppm concentration	I.E. 77.8%	[24]
13.	1-Heptadecene-4-phenyl-thiosemicarbazide (HPTS)	N80 oil well steel, mild steel, 15% HCl, $T = 105°C$, 5000 ppm concentration	I.E. 86.1%	[24]
14.	1-Decene-4-phenyl-thiosemicarbazide (DPTS)	N80 oil well steel, mild steel, 15% HCl, $T = 105°C$, 5000 ppm concentration	I.E. 96.0%	[24]

Abayarathna et al. [25] reported a study of some imidazoline-based corrosion inhibitor formulation applied to C1018 steel in a 3.3% NaCl solution saturated with 150 psi of partial CO_2 pressure 135°C at. The system was made more aggressive by purposely polluting with 10 g L^{-1} elemental sulfur or 30 psi H_2S. On the basis of weight loss studies at 2000 rpm in a revolving cylinder autoclave, a corrosion inhibition efficiency of 500 ppm concentration was calculated. In the presence of CO_2 and H_2S, although the formulation offered adequate inhibition effectiveness, the corrosion rate decreased from 256 mpy in the absence of an inhibitor to 27 mpy in the presence of a 500 ppm inhibitor. In the presence of elemental sulfur, however, the formulation performed very poorly, obtaining a corrosion rate of 153 mpy with severe localized corrosion.

3-Phenyl-2-propyl-1-ol has been identified as an effective inhibitor of API J55 oilfield pipes for the corrosion control of HCl acidification by Finšgar and Jackson [26]. This inhibitor has a peculiar property of dehydration to form β-hydroxypropiophenone on the steel surface and can be adsorbed to it by removing further water molecules to further convert phenyl vinyl ketone. Phenyl vinyl ketone can also oligomerize to create a stronger layer. Others consist of quinolinium and pyridinium composites that can inhibit the oilfield corrosion at temperatures as high as ~204°C. For the better performance, this inhibitor can possibly be united with aliphatic acids and aromatic ketones.

Schauhoff and Kissel [7] synthesized four compounds. The compounds were 1:1 reaction product of four different tall oil fatty acid amino amides and imidazolines and tested their corrosion protection as high-temperature corrosion inhibitors for the oil industry using the rotating wheel test at 340°C for mild steel. After a 24 hours pretreatment at 315°C corrosion protection rates of almost 80% can be reached with the new amino amides at 100 ppm corrosion inhibitor concentration, which is comparable to the results without thermal pretreatment.

Quraishi and Jamal [22] also studied the action of corrosion inhibition of 500 ppm of three fatty acid triazoles, namely, 3-undecane-4-aryl-5-mercapto-1,2,4-triazole (triazole 1), 3(heptadeca-8-ene)-4-aryl-5-mercapto-1,2,4-triazole (triazole 2), and 3(deca-9-ene)-4-aryl-5-mercapto-1,2-4 triazole (triazole 3) as inhibitors for oil well steel (N80) in 15% HCl at temperatures above 105°C. These inhibitors exhibited the drastic corrosion current reduction. At 5000 ppm all three fatty acid triazoles exhibited strong corrosion inhibition, in the following order: Triazole 3 (96.2% inhibition efficiency) > Triazoles 2 (90.5%) > Triazoles 1 (60.2%). The highest inhibition performance exhibited by Triazole 3, may be due to the double bond at the terminal position. The results have shown that these triazoles do not cause any major shift in corrosion potential values, suggesting that they are mixed-type inhibitors.

The anticorrosive properties of three long-chain fatty acid oxadiazoles, namely, 2-undecan-5-mercapto-1-oxa-3,4-diazole (UMOD), 2-heptadecene-5-mercapto-1-oxa-3,4-diazole (HMOD), and 2-decene-5-mercapto-1-oxa-3,4-diazole (DMOD) (Fig. 21–2), were reported by Quraishi and Jamal [23] on mild steel in 15% HCl at $105 \pm 2°C$ using the weight loss method. Inhibition tests were also conducted on N80 steel in 15% HCl containing 5000 ppm

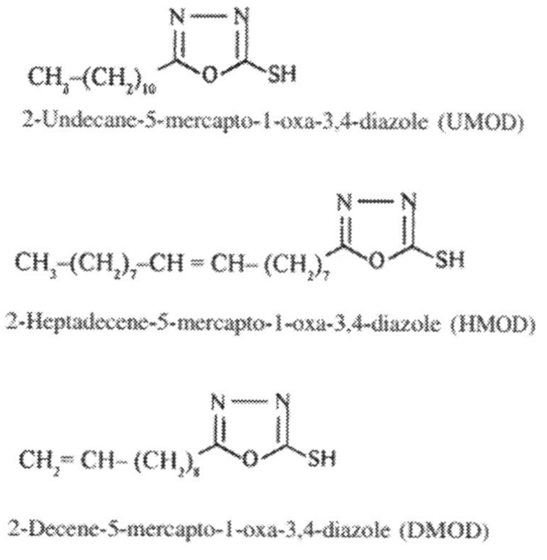

2-Undecane-5-mercapto-1-oxa-3,4-diazole (UMOD)

2-Heptadecene-5-mercapto-1-oxa-3,4-diazole (HMOD)

2-Decene-5-mercapto-1-oxa-3,4-diazole (DMOD)

FIGURE 21–2 Structure and name of used inhibitors. Source: *M.A. Quraishi, D. Jamal, Mater. Chem. Phys., 71 (2001) 202–205.*

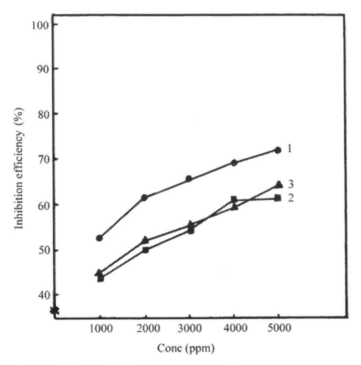

FIGURE 21–3 Inhibition efficiency variation on mild steel with different inhibitor concentrations in 15% boiling HCl at 0.5 h immersion time from weight loss measurements: (1) UMOD; (2) HMOD; (3) DMOD. Source: *M.A. Quraishi, D. Jamal, Mater. Chem. Phys., 71 (2001) 202–205.*

of UMOD under the same circumstances (Fig. 21−3). These results showed that oxadiazole derivatives are good corrosion inhibitors, and that UMOD was the strongest corrosion inhibitor among them.

In another study, Quraishi et al. [24] tested some hydrazides, namely, undecenoic acid hydrazide (UAH), oleic acid hydrazide (OAH), lauric acid hydrazide (LAH), and semithiocarbazides, namely, 1-decene-4-phenyl-thiosemicarbazide (DPTS), 1-heptadecene-4-phenyl-thiosemicarbazide (HPTS), and 1-undecane-4-phenyl-thiosemicarbazide (UPTS), as corrosion inhibitors on mild steel and oil well steel (N-80) in boiling 15% HCl solution at 105°C. Among the compounds tested as inhibitors for the corrosion of mild steel in boiling 15% HCl, the order of percent inhibition efficiency at 5000 ppm follows the order: for hydrazides, UAH (90.4) > OAH (84.4%) > LAH (71.3%); for semithiocarbazides, DPTS (96.0%) > HPTS (86.1%) > UPTS (77.8%).

Amosa et al. [27] reported the corrosion inhibition efficacy of synthetic magnetite and ferrous gluconate for oil well steel (N-80) in a range of hydrogen sulfide (HF) concentrations at temperatures 66, 135, and 177°C. Fig. 21−4 suggests that the results of corrosion inhibition depend on the concentration of the scavenger and the pH of medium. From the findings, it is evident that the ferrous complex exhibiting 99.2% inhibition efficiency was found to be a good corrosion inhibitor at 177°C compared to synthetic magnetite showing 75.1% inhibition efficiency.

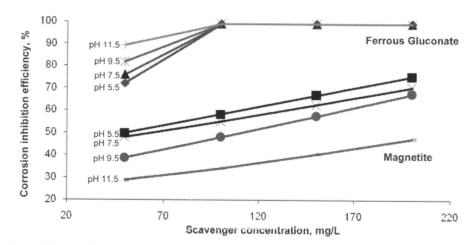

FIGURE 21–4 Inhibition efficiency of the two scavengers in 50 mg L^{-1} sulfide at different concentrations of scavenger at 177°C and 6000 psi. Source: *M.K. Amosa, I.A. Mohammed, S.A. Yaro, A.O. Arinkoola, O.O. Ogunleye, Nafta 61(5) (2010) 239–246.*

Abdel Ghany et al. [28] tested coal tar distillation products (CTDP) and aqueous extract of ginger for the inhibition of mild steel in 15% and 20% HCl solutions at 65°C and 90°C. The results showed a reduction in the rate of corrosion and an improvement in the efficiency of inhibition as the concentration of the two inhibitors increased. And the maximum inhibition efficiency values are 98.6% and 90.3% for CTDP and 99.3% and 96% for Ginger extract at 65°C and 95°C, respectively. With an increase in HCl concentration to 20%, no marked increase in the efficiency of the two formulations was achieved.

21.4 Ash deposit and calcium carbonate scale inhibitors

The Fe-, Ni-, and Co-based super alloys are generally used for the making of apparatuses which are applied in steam boilers and the hostile environments of gas turbines. These alloys are susceptible to the degradation of hot corrosion. There are a lot of corrosion inhibitors which are commercially available and able to decrease the harshness of oil ash corrosion such as Mn- and Mg-based additives. MnO_2, CaO, ZnO, Al_2O_3, PbO, BaO, BaO, SiO_2, Ba $(OH)_2$, $CaCO_3$, $Ca(OH)_2$, and oil-soluble Al, Ni, Fe, and other compounds [29] are also suitable for the various units of oil and gas industries hot corrosion.

Goya et al. [30] reported the effect of superficially applied ZrO_2 inhibitor on the high-temperature corrosion performance of some Co-, Fe-, and Ni-based super alloys as a corrosion inhibitor for ash deposits in the presence of Na_2SO_4 and V_2O_5. The weight loss experiment explained that the corrosion rate decreased and followed the trend Ni < Fe < Co, verifying that the inhibitor has the most efficient performance on the Ni-based super alloy. This study was ascribed to the additional chromium-rich layer formation via Ni-based alloy as well as the for-mation of refractory $Ni(VO_3)_2$, which are all shielding in nature. Thus the application of ZrO_2 in

FIGURE 21–5 SEM images representing the surface morphology of alloy A (Superfer 800H), alloy B (Superco 605), and alloy C (Superni 75) subjected to cyclic oxidation at 900°C for 50 cycles in (A) $Na_2SO_4-60\%V_2O_5$ environment and (B) $Na_2SO_4-60\%V_2O_5 + ZrO_2$ environment. Source: *G. Goya, H. Singh, S. Prakash, Appl. Surf. Sci. 254 (2008) 6653–6661.*

the oil and gas possibly will be useful in the Fe- and Ni-based arrangements. The SEM micrographs for the alloys A, B, and C exposed to $Na_2SO_4-60\%V_2O_5$ induced corrosion without and with additive are shown in Fig. 21–5A and B. In the absence of an additive, Fig. 21–5A shows that the scale seems to have fallen down, leaving the surface cavities. The SEM micrograph for alloy A, on the other hand, shows a mixture of crystalline and globular structures, while for alloy B, some transgranular cracks and granular structures display some porosity in the grains. With big nodules, the alloy C exhibits a dense scale (Fig. 21–5B).

Singh et al. [31] studied the high-temperature corrosion performance of some Fe-, Co-, and Ni-based super alloys in the presence of Y_2O_3 as an inhibitor for salt or ash deposits in the presence of Na_2SO_4 and V_2O_5 environment. The rate of corrosion for the Co-based alloy was found to be maximum and for the Ni-based alloy was minimum. Superficially applied Y_2O_3 was found to be helpful in decreasing the high temperature alloys corrosion. Jones [32] patented SnO_2 a as high-temperature corrosion inhibitor via the liquefied vanadate—sulfate deposits and gaseous SO_3 on engines and other high-temperature equipment, where materials including sulfur, vanadium, and sodium are burnt, mainly in the petroleum industry. In this application, SnO_2 was formulated as a stabilizer for zirconia as a thermal barrier coating to generate a zirconia-stabilized SnO_2 that is greatly resilient to corrosion via molten vanadate—sulfate hot corrosion in the presence of SO_3 gas.

Du et al. [19] tested the effectiveness of HZG inhibitor. Imidazoline derivative, organic amine and fluorocarbon imidazoline, which has strong corrosion-resistant strength, are the key constituents of HZG, with an average corrosion-resistant percentage of 90% at 130°C. The field application demonstrates that the HZG can solve the corrosion problems of most oil wells and has a strong corrosion resistant effect.

21.5 Naphthenic acid inhibitors

The high-temperature corrosion gets accelerated in the presence of naphthenic acid. Naphthenic acids corrosion commonly arises in the temperature ranging between 200°C and 400°C. Besides the temperature, a few significant features that influence the strictness of naphthenic acid corrosion include the concentration of naphthenic acid, phase behavior, metallurgy, velocity, flow characteristics, molecular structure, and the type and concentration of sulfur species present. The extent of corrosion in the distillation unit due to naphthenic acid is lower than the temperature of 204°C. The collaborative corrosivity of high temperature and naphthenic acid become more important above 204°C. However, the traditional nitrogen-based corrosion inhibitors at this high temperature are therefore ineffective. Naphthenic acid is present in its vapor form at elevated temperatures. Therefore with marginal elevation in the corrosion rate, the process becomes different from that of the liquid phase. This possibly will also be aggravated in the presence of sulfur-containing compounds, for example, H_2S. Hence, the corrosion inhibitors formulation for sulfur and naphthenic acid high-temperature corrosion for oil and gas is extremely essential and could be attained by the following combinations: alkaline earth metal phenatesulfide—phosphonate and trialkyl-phosphatein—a ratio of 1:1 up to 5:1 as reported by Elizabeth et al. [11]. The examples includes (1) combination of sulfur dichloride (3 mol) + monoalkyl-substituted phenol (4 mol); (2) sulfur dichloride (1 mol) + alkyl phenol substituted (2 mol) with one or more alkyl groups; and (3) sulfur dichloride and alkyl phenol ratio (1:1), etc. This group also patented another method [11] for the reduction of high-temperature naphthenic acid corrosion via the use of a mercaptotriazine compound. This method is used to manage the corrosion of internal metallic surfaces of oil and gas processing crude tools or high temperature petroleum distillates at a temperature range between 177°C and 482°C. In the use of mercaptotriazine, attention is paid to the concentration, which could be in the range of 1—5000 ppm and the mercaptotriazine derivative could be 2,4,6-trimercapto-1,3,5 triazine [11]. Some mercaptan and sulfur-containing compounds, such as simple alkyl mercaptans, dialkyl polysulfide, and di-t-nonylpentasulfide (TNPS), have previously been known for the purpose of high-temperature naphthenic acid corrosion. Among these only TNPS has a performance closer to mercaptotriazine compounds that show better high-temperature corrosion inhibition in the presence of elemental sulfur, H_2S, and mercaptans that formerly present in distillates and crude oil. Efficient corrosion inhibitors are also estimated to be sulfur-substituted mercapto-triazine compounds in which one or two mercaptohydrogens are changed by aryl, alkyl, and cycloalkyl.

Some other inhibitors contain the use of phosphite-based compounds including at least one aryl group. The inhibition efficiency of all these compounds depends mainly on the operating circumstances of flow regime, temperature range, and the inhibitor concentration. The inhibitor is therefore recommended to be placed in the range of 2000−5000 ppm dose at a reasonably high concentration at the beginning. When it is found that the inhibitor has formed a corrosion protective layer on the metal surface, the subsequent dose will be reduced to a range of 100−1500 ppm. Isooctyl diphenyl phosphite, diphenyl phosphite, and triphenylphosphite are examples of phosphite compounds for high-temperature corrosion inhibitors [33].

Via the use of sulfidizing agents, naphthenic acid corrosion may also be controlled to increase the performance of phosphorus in the management of high-temperature corrosion in the oil and gas industry. In this case, a stronger and more successful match was found to be the mixture of organic polysulfide and phosphate ester. In other combinations, a phosphate group for the phosphate ester can be selected from the following: diphosphate, phosphate, thiophosphate, triphosphate, trithiophosphate, and dithiophosphate. Others consist of ethyl phosphate, methyl phosphate, isopropyl phosphate, butyl phosphate, n-propyl phosphate, etc., whereas the sulfide group can be chosen from terpene polysulfides or olefin polysulfides as the preferred groups. In combination with the different stoichiometric ratio of sulfide, the polysulfide contains cycloalkyl (aromatics contain 6−30 carbon atoms) or alkyl. The percentage of sulfur content in the polysulfide can vary from 10% to 60%, but the selected range is 25%−50% for better results. Up to 30%−80% or 25−2000 ppm petroleum streams can easily be incorporated into the oil-soluble polysulfide [34].

21.6 Some nonspecific inhibitors

Some nonspecific high-temperature corrosion inhibitors were used in oil and gas corrosion management of various units. These inhibitors consist of various oxotungstates by Popoola et al. [1], which were studied on aluminum and its alloys at temperature among 90°C and 300°C in a water environment.

Zagidullin et al. [35] patented a method based on terephthalic acid for corrosion inhibition. By the interaction of polyethylene-polyamine with terephthalic acid, an acid inhibitor is formulated in a 2:1 molar ratio at a temperature range of 150°C−190°C for 4−8 hours, accompanied by a reaction with benzyl chloride at 80°C for 5 hours. Along with urotropin and neonol in water, the product was used as a component. The inhibitor can resist corrosion at high temperatures, particularly at higher concentrations.

Reznik et al. [36] have patented 2-thioxo-4-thiazolidinone (rhodanine) and its 3- or 5-derivatives as an important steel structure Fe corrosion inhibitor for the high-temperature environment of carbonic acid derivative oil refining equipment.

1,8-Diamino-p-methane, isophorone diamine, diethyleneetriamine, hydroxyethyl-ethylene diamine, and bis-aminomethyl-norbomylene are some amino amides and imidazoline-based corrosion inhibitors for application in oil and gas production, especially in deep hot wells or in refining processes [37].

21.7 Conclusions

This chapter is based on the high-temperature corrosion inhibitors, which can manage and control corrosion in the oil and gas industry. The significance was on the understanding of the mechanism, vulnerable units, and selection of suitable corrosion inhibitors based on the various contributing factors accountable for the corrosion of a specific unit at high temperature. Based on this understanding, inhibitors formulation can be made by preparing a novel compound or combination of compounds with essential properties to withstand such atmosphere for efficient corrosion inhibition. The beginning experiments for selection of high-temperature corrosion inhibitors emphasized in this chapter have revealed that electron density and structure are essential conditions for selection of high-temperature corrosion inhibitors mitigation. In order to resist the high-temperature degradation effect, the inhibitors should have high molecular weight and should have large electron cloud at the same time to enable better adsorption. For formulations involving the combination of two or more inhibitors to improve temperature tolerance, the combination ratio is most critical for the high efficiency to be achieved. In the choice of inhibitor, the units or area of activity is also very critical since the metal components of each structural unit and the presence of other corrosive agents, such as CO_2, H_2S, acids, and other hot gases, appear to exacerbate the rate of high-temperature corrosion. The corrosion inhibitors such as ZrO_2 were therefore found to be very strong in the inhibition of oil ash corrosion related to alloys based on Fe, Co, and Ni. Formulations such as trimer acid salts of primary amines may be excellent for acid intensified corrosion units, such as acidification units, while for naphthenic acids, CO_2, and H_2S accelerated corrosion the 1:1 to 5:1 ratio of alkaline earth metal phosphonate-phenatesulfide and trialkyl phosphate combination may be used for high efficiency.

References

[1] L.T. Popoola, A.S. Grema, G.K. Latinwo, B. Gutti, A.S. Balogun, Intl. J. Indus. Chem. 4 (2013) 1−35.

[2] W. Gao, Z. Li, Developments in High Temperature Corrosion and Protection of Materials, in: Woodhead publishing series in metals and surface engineering, 2008, pp. 1−5.

[3] S.P. Sharp, L. Yarborough: German Patent DE-O 3 029 790, 1981 Standard Oil. Jim Jr. Maddox, W. Schoen: US Patent US 3 687 847, Texaco, 1972.

[4] D.S. Treybig, T.W. Glass: US Patent US 594 518, Dow Chemical, 1984.

[5] D. Redmore, B.T. Outlaw: US Patent US 4 315 087, Petrolite, 198.

[6] R.C. John, A.D. Pelton, A.L. Young, W.T. Thompson, I.G. Wright, T.M. Besmann, Mater. Res. 7 (2004) 163–173.

[7] S. Schauhoff, C.L. Kissel, Mater. Corros. 51 (2002) 141–146.

[8] N. Hackerman, Corrosion 4 (1948) 45–60.

[9] J.G. Edmondson, High temperature corrosion inhibitor. US Patent 5,500,107, filed 15 March 1994 and issued 19 March 1996, 1996.

[10] T. Hong, W.P. Jepson, Corros. Sci. 43 (2001) 1839–1849.

[11] B.K. Elizabeth, S. Joe, G.H. John, F. Sam, High temperature corrosion inhibitors simulator. US Patent 5,503,006, filed 23 October 1980 and issued 04 February 1986, 1996.

[12] V.S. Saji, Rec. Pat. Corros. Sci. 2 (2010) 6–12.

[13] J.Z. Michael, T. Benjamin, Corrosion inhibitors for use in hot hydrocarbons, US Patent 4,941,994, filed 18 July 1989 and issued 17 July 1990, 1990.

[14] D.S. Sullivan, C.E. Strubelt, K.W. Becker, High temperature corrosion inhibitor, US Patent US4028268A, 1975.

[15] S. Papavinasam, Chapter 5 - mechanisms, in: S. Papavinasam (Ed.), Corrosion Control in the Oil and Gas Industry, Gulf Professional Publishing, 2014, pp. 249–300.

[16] J. Jr. Maddox, W. Schoen: US Patent US 3 687 847, Texaco, 1972.

[17] S.P. Sharp, L. Yarborough: German Patent DE-O 3 029 790, Standard Oil, 1981.

[18] J.R. Lindsay Smith, A.U. Smart, M.V. Twigg, J. Chem. Soc. Perkin Trans. 2 (1992) 939.

[19] Q. Du, G. Xie, X. Yan, L. Zhang, Y. Pei, F. Shang, et al., J. Chem. Pharm. Res. 7 (2015) 1062–1068.

[20] C. Fiaud, A. Harch, D. Mallouh, M. Tzinmann, Corros. Sci. 35 (1993) 1437–1444.

[21] H.J. Chen, High temperature corrosion inhibition performance of imidazoline and amide, Corrosion, Paper no. 00035, 2000.

[22] M.A. Quraishi, D. Jamal, J. Am. Oil Chem. Soc. 77 (2000) 1107–1111.

[23] M.A. Quraishi, D. Jamal, Mater. Chem. Phys. 71 (2001) 202–205.

[24] M.A. Quraishi, D. Jamal, M.T. Saeed, JAOCS 77 (2000) 3.

[25] D. Abayarathna, A. Naraghi, N. Obeyesekere, Inhibition of corrosion of carbon steel in the presence of CO_2, H_2S and S, in: NACE International Conference, 16–20 March. National Association of Corrosion Engineers, San Diego, California, USA (2003).

[26] M. Finšgar, J. Jackson, Corros. Sci. 86 (2014) 17–41.

[27] M.K. Amosa, I.A. Mohammed, S.A. Yaro, A.O. Arinkoola, O.O. Ogunleye, Nafta 61 (5) (2010) 239–246. Available from: https://hrcak.srce.hr/index.php?show=clanak&id_clanak_jezik=82178.

[28] N.A. Abdel Ghany, M.F. Shehata, R.M. Saleh, A.A. El Hosary, Mater. Corros. 68 (2016) 355–360. Available from: https://doi.org/10.1002/maco.201609146.

[29] L.D. Paul, R.R. Seeley, Corrosion 47 (2) (1991) 152.

[30] G. Goya, H. Singh, S. Prakash, Appl. Surf. Sci. 254 (2008) 6653–6661.

[31] H. Singh, Gitanjaly, S. Singh, S. Prakash, Appl. Surf. Sci. 255 (2009) 7062–7069.

[32] R.L. Jones, Corrosion inhibition in high temperature environment, NC 75, 053, 22217-5660, 1993. < https://apps.dtic.mil/dtic/tr/fulltext/u2/d015910.pdf >.

[33] I.D. Robertson, L.M. Dean, G.E. Rudebusch, N.R. Sottos, S.R. White, J.S. Moore, ACS Macro Lett. 6 (2017) 609–612.

[34] E. Greyson, J. Manna, S.C. Mehta, Scale and corrosion inhibitors for high temperature and pressure conditions. US Patent 8,158,561 B2, filed 13 August 2010 and issued 17 April 2012, 2012.

[35] R.N. Zagidullin, U.S. Rysaev, M.Y. Abdrashitov, D.U. Rysaev, Y.P. Kozyreva, I.S. Mazitova, et al., Method of preparing corrosion inhibitor by reaction of polyethylene-polyamine with terephathalic acid, RU Patent 2357007, 2009.

[36] V.S. Reznik, Y.P. Khodyrev, V.D. Akamsin, R.M. Galiakberov, R.K. Giniyatullin, V.E. Semneov, et al., Method of iron corrosion inhibition of oil-refining equipment in carbonic acid derivatives. RU Patent 2351690, 2009.

[37] D.S. Sullivan, E.C. Strubelt, K.W. Becker, High temperature corrosion inhibitor. US Patent 4,028,268, filed 03 December 1975 and issued 07 June 1977, 1977.

Index

Note: Page numbers followed by "*f*" and "*t*" refer to figures and tables, respectively.

Printed in the United States
by Baker & Taylor Publisher Services